Novel Techniques for Food Quality Analysis and Safety

Novel Techniques for Food Quality Analysis and Safety: Fundamentals, Scope, Feasibility, and Challenges provides a timely response to the growing demand for a shift away from conventional analytical approaches to emerging and rapid techniques for ensuring food quality and safety. Traditional food analysis suffers from limitations to speed, sensitivity, portability, and real-time decision-making. This book aims to bridge this critical gap by bringing together innovative analytical tools and technologies that are reshaping the future of food analysis.

Features:

- A systematic overview of the emerging techniques of food analysis.
- Fundamental principles discussed in a manner that will allow readers to grasp the working mechanisms behind novel analytical techniques.
- Balanced treatment of theory and application.
- Recent research results and some pointers to future advancements in food analytical techniques.

This book is intended as a reference for academics, as well as for researchers, scientists, and professionals working in the fields of food safety and quality control; it may also be used as a textbook for postgraduate students of food engineering and technology.

Novel Techniques for Food Quality Analysis and Safety

Fundamentals, Scope, Feasibility, and Challenges

Edited by
Arun Kumar Gupta, Yashi Srivastava, and Poonam Mishra

CRC Press
Taylor & Francis Group
Boca Raton London New York

CRC Press is an imprint of the
Taylor & Francis Group, an informa business

Designed cover image: Book's authors

First edition published 2027
by CRC Press
2385 NW Executive Center Drive, Suite 320, Boca Raton FL 33431

and by CRC Press
4 Park Square, Milton Park, Abingdon, Oxon, OX14 4RN

CRC Press is an imprint of Taylor & Francis Group, LLC

ISBN: 978-1-032-77352-0 (hbk)
ISBN: 978-1-032-77353-7 (pbk)
ISBN: 978-1-003-48261-1 (ebk)

DOI: 10.1201/9781003482611

Typeset in Times
by Apex CoVantage, LLC

Dedicated to My Mother Mrs. Bindu Mishra

— Dr Poonam Mishra

Contents

Preface ix
Editor Biographies xi
List of Contributors xiii

Chapter 1 Introduction 1
Arun Kumar Gupta and Poonam Mishra

Chapter 2 Surface-Mediated C Dots/Quantum Dots: Synthesis, Characterization, and Application in Food Quality and Safety Analysis 13
Anupama Bora, Himali Horo, Susmita Rabha, Apurba Das, and Ankur Pandey

Chapter 3 Application of Biosensors in Food Analysis 34
Radhika Theagarajan, Tanya Kunwar, Saurabh Kharayat, Vivian Singh, Arun Kumar Gupta, and Poonam Mishra

Chapter 4 Laser and Laser Focusing Optics: Principal, Preparation, Analysis, Spectral Acquisition, and Application in Food Quality and Safety 57
Manpreet Kaur and Yashi Srivastava

Chapter 5 Nanocomposites: Recent Trends of Nanocomposites in Food Analysis and Safety 87
Rajamanickam R., Mahesh Jaiwar, Anbarasan R., Sajeevkumar V.A., and Kumar R.

Chapter 6 Smartphone-Based Techniques for Food Analysis: Application of Computer Vision and Artificial Intelligence as Rapid Detection Techniques in the Food Sector 104
Monica Yumnam, Keshram Dulait, and Poonam Mishra

Chapter 7 Aerogel: An Emerging Functional Material for Food Quality and Safety Analysis 126
Subhamoy Dhua, Keshram Dulait, Arnab Bayan, and Poonam Mishra

Chapter 8 Rapid Microbial Techniques: Methods of Detection of Microbial in Food Processing 152

Neetu Gupta and Yashi Srivastava

Chapter 9 NMR Techniques in Food Product Analysis 175

Avik Mazumder, Manpreet Kaur, and Yashi Srivastava

Chapter 10 Molecular-Based Techniques: Implication of Genomic and Proteomic Techniques in Food Quality and Safety Analysis 209

Priya Kalyan, Pratiksha, and Jyoti Rustagi

Chapter 11 Principle and Application of Surface-Based Analysis in Food Quality and Safety Analysis 237

Milad Tavassoli, Behnam Bahramian, Reza Abedi-Firoozjah, Ali Ehsani, and Ehsan Sadeghi

Chapter 12 Principle, Fundamentals, and Application of Hyphenated Techniques in Food Analysis 265

Rhuthuparna Malayil, Amitha Thankachan, Sandeep Singh, and Anjana Munshi

Chapter 13 Assessing the Feasibility of Emerging Techniques in Food Industries 287

Priyanka Singh

Index **303**

Preface

Novel Techniques for Food Quality Analysis and Safety: Fundamentals, Scope, Feasibility and Challenges provides a timely response to the growing demand for a shift away from conventional analytical approaches to emerging and rapid techniques for ensuring food quality and safety. Traditional food analysis suffers from limitations to speed, sensitivity, portability, and real-time decision-making. This book aims to bridge this critical gap by bringing together innovative analytical tools and technologies that are reshaping the future of food analysis.

Currently, most textbooks on food analysis primarily focus on classical chemical, microbiological, or instrumental methods, while the rapidly evolving domains of nanotechnology, biosensing, spectroscopy, molecular biology, and digital tools are often discussed in isolation. More importantly, the practical feasibility, industrial relevance, and translational challenges of these emerging techniques are rarely addressed in a unified manner. This book is structured to establish a clear connection between fundamental principles, recent technological advancements, and their real-world applicability in food quality and safety assessment. Accordingly, this book is intended for academicians, as well as for researchers, scientists, and professionals working in food safety and quality control.

The chapters are framed to guide readers progressively from an introduction to innovative food analysis techniques to advanced tools such as quantum dot- and carbon dot-based functional materials, biosensors, laser-induced breakdown spectroscopy, nanocomposites, aerogel-based sensors, and smartphone-enabled analytical platforms integrating computer vision and artificial intelligence. The book also addresses hyphenated techniques and critically examines the feasibility, scalability, and challenges associated with adopting emerging technologies in the food industry.

A distinguishing feature of this book is its balanced treatment of theory and application. Fundamental principles are discussed in a manner that allows readers to grasp the working mechanisms behind each technique, followed by practical considerations related to sensitivity, selectivity, sample preparation, data interpretation, and industrial deployment.

This book is the result of collective efforts from contributors with diverse expertise across food science, analytical chemistry, nanotechnology, biosensing, and molecular biology. Their contributions have enriched the content and ensured that the book reflects both academic rigor and practical relevance. We sincerely acknowledge the support and cooperation of all contributors and the editorial team whose efforts have been instrumental in bringing this work to completion.

We hope that this book will serve as a valuable resource for students, educators, researchers, and industry professionals, fostering a deeper understanding of novel analytical techniques and encouraging their thoughtful adoption to enhance food quality and safety in an increasingly complex global food system.

Editors

Editor Biographies

Dr. Arun Kumar Gupta earned his MSc (food science and technology) degree from Pondicherry University, Puducherry, and his PhD in food engineering and technology from the Tezpur University, Assam, India. He is currently serving as Assistant Professor in the Department of Food Process Technology at Sardar Vallabhbhai Patel University of Agriculture and Technology, Meerut, India. He has made significant contributions to the field of food science and technology, as reflected in his extensive body of work, which includes 75 original research and review articles published in peer-reviewed international journals, 20 Scopus-indexed book chapters, and seven patents filed. Dr. Gupta has also been actively involved in academic publishing, serving as editor of five books with esteemed publishers such as Elsevier, CRC Press, and Springer. He is a life member of the Association of Food Scientists & Technologists (India), the International Society of Food Engineering, and the Institute of Scholars. Dr. Gupta has been recognized with the prestigious SRISTI-GYTI Award 2021 and has also received accolades such as the National Summer Research Fellowship, Best Poster Award, Best Oral Paper Award, and Young Investigator Award at various national conferences. Additionally, he is a regular reviewer for several reputable journals in his field.

Dr. Yashi Srivastava completed a PhD from Mysore University, MSc and BSc from Allahabad University, and currently she is working as Associate Professor in the Centre of Food Technology, University of Allahabad, Prayagraj. She worked as Assistant Professor, Department of Applied Agriculture (food science and technology), Central University of Punjab, Bathinda. She has more than 9.5 years of work experience that includes academic and administration experience. Dr. Srivastava's research areas include fruit and vegetable processing, millet processing, food packaging, edible oil processing, food equipment design, food regulation, and quality characterization. Her scientific achievements include h-index 13, i-10 index 15, and citation 644. She has published many journal articles in international publications. She has presented her research work in many international symposiums. She has also authored many book chapters, short communications, and review articles. She has received fellowships from renowned government scientific bodies such as APEDA and ICAR. She is a lifetime member of the Food Science and Technology Institution (AFSTi), India.

Prof. Poonam Mishra received her MTech degree from HBTI Kanpur and PhD in food engineering and technology from Tezpur University, India. In 2004, Prof. Mishra joined the Center of Food Technology, University of Allahabad as a lecturer, where she has been engaged in teaching and R&D work related to food engineering and technology. Prof. Mishra was also actively involved in establishing the Food Quality Control Laboratory at the Center of Food Technology. In 2009, she joined Tezpur University as Assistant Professor, and presently she is working as Professor in the Department of Food Engineering and Technology, Tezpur Central University, India. Her fields of interest include fabrication of functional nanomaterials, functional food, food analysis, and biosensors. She has handled a number of external funded projects successfully. She has published more than 65 research papers in peer-reviewed national and international journals, several papers in national and international conferences, book chapters, and popular articles. Prof. Mishra edited three books for publishers CRC Press, Elsevier, and Springer. Prof. Mishra supervised a number of research scholars, postgraduates, and undergraduate students. She has three patents to her name and has received several national and international awards. She is a life member of the Association of Food Scientists & Technologists (India) and the Nutrition Society of India.

Contributors

Reza Abedi-Firoozjah
Kermanshah University of Medical Sciences
Kermanshah, Iran

Behnam Bahramian
Tabriz University of Medical Sciences
Tabriz, Iran

Arnab Bayan
Department of Food Engineering and Technology
Tezpur University
Tezpur, Assam, India

Anupama Bora
Department of Vocational Studies and Skill Development
School of Agricultural Sciences, Nagaland University
Nagaland, India

Apurba Das
Department of Physics
Handique Girls' College
Guwahati, Assam, India

Subhamoy Dhua
Department of Food Engineering and Technology
Tezpur University
Tezpur, Assam, India

Keshram Dulait
Department of Food Engineering and Technology
Tezpur University
Tezpur, Assam, India

Ali Ehsani
Department of Food Science and Technology
Faculty of Nutrition and Food Science
Tabriz University of Medical Sciences
Tabriz, Iran

Arun Kumar Gupta
Department of Food Process Technology
College of Post-Harvest Technology & Food Processing
Sardar Vallabhbhai Patel University of Agriculture and Technology
Meerut, Uttar Pradesh, India

Neetu Gupta
Central University of Punjab
Bhatinda, Punjab, India

Himali Horo
Department of Chemical Engineering
Thapar Institute of Engineering and Technology
Patiala, Punjab, India

Mahesh Jaiwar
Department of Applied Agriculture
Central University of Punjab
Bathinda, Punjab, India

Priya Kalyan
Department of Food Science Technology & Processing
Amity University Punjab
Mohali, Punjab, India

Manpreet Kaur
Department of Applied Agriculture
Central University of Punjab
Bathinda, Punjab, India

Saurabh Kharayat
Department of Food Science & Technology
Graphic Era (Deemed to be University)
Dehradun, Uttarakhand, India

Tanya Kunwar
Department of Food Science & Technology
Graphic Era (Deemed to be University)
Dehradun, Uttarakhand, India

Rhuthuparna Malayil
Department of Human Genetics and Molecular Medicine
Central University of Punjab
Bathinda, Punjab, India

Avik Mazumder
Defence Research and Development Establishment
Gwalior, Madhya Pradesh, India

Poonam Mishra
Department of Food Engineering and Technology
Tezpur University
Tezpur, Assam, India

Anjana Munshi
Department of Human Genetics and Molecular Medicine
Central University of Punjab
Bathinda, Punjab, India

Ankur Pandey
Department of Physics
Handique Girls' College
Guwahati, Assam, India

Pratiksha
Department of Food Science and Technology
Amity University Punjab
Mohali, Punjab, India

Anbarasan R.
Food Engineering and Packaging Technology Division
Defence Institute of Bio-Defence Technologies
Mysore, Karnataka, India

Kumar R.
Food Engineering and Packaging Technology Division
Defence Institute of Bio-Defence Technologies
Mysore, Karnataka, India

Rajamanickam R.
Food Engineering and Packaging Technology Division
Defence Institute of Bio-Defence Technologies
Mysore, Karnataka, India

Susmita Rabha
Department of Physics
National Institute of Technology
Silchar, Assam, India

Jyoti Rustagi
Animal Biotechnology Division
National Dairy Research Institute
Karnal, Haryana, India

Ehsan Sadeghi
Department of Food Science and Technology
Faculty of Nutrition and Food Science
Tabriz University of Medical Sciences
Tabriz, Iran

Priyanka Singh
ICAR-Indian Institute of Sugarcane Research
Lucknow, Uttar Pradesh, India

Sandeep Singh
MD Anderson Cancer Centre
Houston, Texas

Vivian Singh
Department of Food Science & Technology
Graphic Era (Deemed to be University)
Dehradun, Uttarakhand, India

Yashi Srivastava
Associate Professor in Centre of Food Technology
University of Allahabad
Prayagraj, Uttar Pradesh, India

Milad Tavassoli
Department of Food Science and Technology
Faculty of Nutrition and Food Science
Tabriz University of Medical Sciences
Tabriz, Iran

Amitha Thankachan
Department of Orthopedics
All India Institute of Medical Sciences
Bathinda, Punjab, India

Radhika Theagarajan
Department of Food Technology
Manakula Vinayagar Institute of Technology
Puducherry. India

Sajeevkumar V.A.
Freeze Drying and Animal Products Technology Division
Defence Institute of Bio-Defence Technologies
Mysore, Karnataka, India

Monica Yumnam
Department of Food Engineering and Technology
Tezpur University
Tezpur, Assam, India

1 Introduction

Arun Kumar Gupta and Poonam Mishra

1.1 INTRODUCTION

Ensuring food quality and safety is a major priority in the global food industry, requiring continuous advancements in analytical techniques. Traditional food analysis methods are often labor-intensive and time-consuming, necessitating the development of novel approaches that enhance accuracy, efficiency, and sensitivity (Gupta et al., 2023). Emerging technologies such as spectral techniques, molecular-based techniques, surface-based analysis, and hyphenated techniques are revolutionizing food analysis by providing rapid and precise assessments of food composition, contaminants, and overall quality. The integration of these novel techniques into food quality assessment and safety monitoring systems represents a significant leap forward in ensuring consumer health and regulatory compliance (Rakovský et al., 2014). As technology continues to evolve, further advancements in miniaturization, automation, and data processing will enhance the feasibility and accessibility.

1.2 QUANTUM/C DOTS-BASED FUNCTIONAL MATERIALS

Quantum dots (QDs) and carbon dots (C-dots) are nanoscale semiconductor particles exhibiting unique optical and electronic properties. Their application in food analysis includes fluorescence-based sensing for detecting contaminants and quality assessment (Wang et al., 2020). The synthesis of QDs often involves methods such as hydrothermal, microwave, and solvothermal approaches, ensuring high quantum yield and stability (Zhao et al., 2021b, 2021c). Characterization techniques include UV-Vis spectroscopy, fluorescence spectroscopy, and electron microscopy (Liu et al., 2009). These functional materials enhance food safety and traceability by enabling rapid and sensitive detection methods. The synthesis of QDs and C-dots involves various methods to achieve high quantum yield and stability. Hydrothermal synthesis treats precursor materials in a sealed vessel at elevated temperatures and pressures, promoting the formation of nanocrystals with controlled size and morphology. Microwave-assisted synthesis utilizes microwave radiation to rapidly heat the precursors, resulting in uniform nucleation and growth of nanoparticles within minutes. Solvothermal synthesis, similar to hydrothermal methods but using organic solvents, allows for the creation of QDs with diverse compositions and properties. For C-dots, green synthesis approaches using natural precursors have gained attention due to their eco-friendliness and cost-effectiveness. For instance, hydrothermal treatment of plant extracts can produce water-soluble, highly fluorescent C-dots suitable for sensing applications (Kalita et al., 2021).

DOI: 10.1201/9781003482611-1

Accurate characterization of QDs and C-dots is essential to understand their properties and ensure their suitability for specific applications. Common techniques include UV-Vis spectroscopy to assess optical absorption properties; fluorescence spectroscopy to evaluate emission characteristics; and electron microscopy (TEM/SEM) to provide detailed images of particle size, shape, and morphology at the nanoscale. These characterization methods ensure that the synthesized nanoparticles meet the desired specifications for their intended applications (Gupta et al., 2023). While the application of QDs and C-dots in food analysis presents numerous advantages, it is imperative to consider potential safety and environmental impacts. Ensuring that these nanomaterials do not pose risks to consumers or the environment is crucial. Ongoing research focuses on understanding their toxicity profiles and developing guidelines for their safe use in food-related applications. The development and application of QDs and C-dots in food analysis represent a significant advancement in ensuring food safety and quality. Their unique properties, combined with advanced synthesis and characterization techniques, make them indispensable tools in modern analytical methodologies.

1.3 BIOSENSORS

Biosensors integrate biological elements with a transducer to detect foodborne pathogens, toxins, and contaminants (Gupta et al., 2022). They function through electrochemical, optical, and piezoelectric principles, offering rapid and sensitive detection (Singh et al., 2022). Over the past three decades, biosensor development has expanded into various fields, including clinical research, medical diagnostics, environmental monitoring, and, more recently, food quality and safety. Optical biosensors, which integrate a biorecognition element with an optical transducer, offer real-time, label-free detection with high sensitivity and specificity (Damborsky et al., 2016). Techniques such as surface plasmon resonance (SPR), fiber optic sensors, and fluorescence microscopy enhance their applicability in food safety and agriculture. The growing interest in biosensors is evident from a significant rise in research publications, with a 4,000% increase (Turner, 2013). Their adoption in food and agriculture is driven by the need for early detection of contaminants and toxins, as well as the perishability of food products (Sharma & Mutharasan, 2013; Biji et al., 2015). Electrochemical biosensors, which convert biochemical information into electrical signals, are widely used due to their high sensitivity, low power requirements, and cost-effectiveness (Zhang et al., 2012). These sensors operate through amperometric, potentiometric, voltammetric, and impedimetric measurements (Madufor, 2019). Fluorescence-based biosensors leverage fluorescent conjugated polymers to detect biological and chemical substances, allowing for signal amplification (Esser & Swager, 2010). Similarly, colorimetric sensor arrays provide a low-cost, disposable, and robust method for detecting gases such as carbon monoxide and nitrogen dioxide (Lonsdale et al., 2017; Schmitt et al., 2016). Infrared gas sensors are also employed in fruit ripening and storage monitoring, offering high selectivity and a broad detection range (Skolik et al., 2019). With advancements in biosensor technology, their role in food safety, pathogen detection, and environmental monitoring continues to expand, reinforcing their importance in sustainable food systems.

1.4 LASER-INDUCED BREAKDOWN SPECTROSCOPY (LIBS)

LIBS is an atomic emission spectroscopic technique that utilizes a focused pulsed laser beam to generate plasma from a material. The plasma consists of atoms, ions, and free electrons that emit electromagnetic radiation as they cool. This emitted light is then analyzed using a spectrometer to generate a characteristic spectrum of the elements present in the sample. LIBS operates on the principles of optical emission spectroscopy (OES), where the laser pulse is absorbed by the analyte surface, causing a small portion of the material to be ablated, atomized, and ionized, forming a plasma plume. As the plasma cools, it emits radiation at specific wavelengths corresponding to the elements in the target material (Rakovský et al., 2014). Recent advancements in LIBS have focused on improving high-resolution spectrometers, charge-coupled device (CCD) detectors, and miniaturization, leading to the development of compact, low-power, and portable systems (Mohaidat et al., 2011). These innovations have expanded LIBS applications across various sectors, particularly in food analysis. LIBS has emerged as a promising analytical tool due to its multi-elemental detection capability, rapid response, minimal sample preparation, low operational cost, and ease of use. These attributes make it highly suitable for food quality assessment, safety monitoring, and authenticity verification.

The food industry has explored LIBS for a range of at-, on-, and in-line applications, primarily for mineral composition analysis but also for qualitative assessments such as authenticity control, sample discrimination, bacterial contamination detection, depth profiling, and elemental surface mapping (Kaiser et al., 2012). Due to its noncontact nature and ability to analyze samples with minimal destruction, LIBS has gained interest as a rapid and reliable tool for food characterization. Researchers have increasingly investigated its potential for micro-destructive analysis, recognizing its advantages as a process analytical technology (PAT) or a portable food analysis tool (Markiewicz-Keszycka et al., 2017). The continued development of LIBS instrumentation and data processing techniques is expected to further enhance its precision, sensitivity, and applicability in the food sector. As a result, LIBS is positioned as a versatile technique for ensuring food quality, safety, and traceability in modern food production systems.

1.5 NANOCOMPOSITES

Nanocomposite materials consist of multiple phases, where at least one dimension falls within the nanometer range. Reducing material dimensions to the nanoscale creates phase interfaces that significantly enhance material properties. The ratio between the surface area and volume of the reinforcing material used in nanocomposite preparation plays a critical role in understanding the structure-property relationship. Nanocomposites present innovative opportunities across various sectors, including medicine, pharmaceuticals, food packaging, electronics, and energy industries (Omanović-Mikličanin et al., 2020). In food packaging, nanomaterials—particularly nanoparticles—are gaining prominence due to their mechanical, optical, catalytic, and antimicrobial properties. Among the most widely used nanomaterials in food packaging are silver nanoparticles (AgNPs) and nanoclay, which are incorporated into packaging materials to enhance durability and functionality. Additionally, nano-zinc oxide (ZnO) and titanium-based nanomaterials are available commercially, though to a lesser extent. These nanomaterials exhibit antimicrobial properties, helping to control microbial growth, while also improving barrier properties to extend the shelf life and freshness of packaged food products. Despite the advantages of nanomaterials in food packaging, concerns have arisen regarding potential safety risks associated with the migration of nanoparticles from food-contact materials into food products, particularly in fish and seafood (Adeyeye et al., 2023). Consequently, research efforts have intensified to develop biodegradable, sustainable alternatives to petroleum-based polymers. Polymers derived from natural sources are considered among the most promising materials for this purpose. However, natural polymer-based materials, such as polymeric films, often exhibit poor mechanical strength and high water sensitivity. A promising solution to overcome these limitations is the reinforcement of natural polymeric films with nanoparticles, leading to the development of nano-biocomposite polymers. These materials offer improved physical properties, making them ideal for food packaging applications. This chapter provides a comprehensive review of the latest advancements in biopolymer-based nanocomposites, emphasizing their potential as sustainable packaging materials (Valencia & do Amaral Sobral, 2018).

1.6 AEROGELS

Aerogels have gained prominence across multiple industries due to their unique properties, including low bulk density, high porosity (primarily mesoporous), and a high specific surface area ($\geq$150 m^2/g). These characteristics make them ideal for applications in food fortification and functional packaging (Dhua et al., 2022). The food industry is increasingly adopting aerogels for advanced packaging solutions aimed at enhancing food preservation and sustainability. One promising application of aerogels in food packaging is their use as absorbent pads, designed to minimize excess liquid released by fresh products during storage, thereby extending shelf life (Benito-González et al., 2021). These pads typically feature a non-permeable synthetic polymer outer layer (e.g., polyethylene) and a hydrophilic interlayer capable of absorbing exudates. Additionally, aerogel-based active packaging systems can incorporate antimicrobial compounds, further enhancing food safety (Budtova et al., 2020).

The synthesis of aerogels involves carefully selecting precursor materials, cross-linkers, and processing techniques to optimize their structural and functional properties. Bio-based polymers such as cellulose, starch, and hemicelluloses, along with biodegradable synthetic polymers like polyvinyl alcohol (PVA), serve as sustainable alternatives to petroleum-derived plastics (Budtova et al., 2020). Cross-linking agents, including glutaraldehyde, calcium chloride, and citric acid, help tailor aerogel properties for specific applications (Wang et al., 2022). Aerogel drying methods significantly impact their final properties. Freeze drying, a two-step process involving freezing and sublimation, helps create a porous structure, while supercritical CO_2 drying, widely used due to its nontoxic nature and efficiency, preserves aerogel integrity without surface tension effects (Dhua et al., 2022). In food packaging, key performance parameters of aerogels include high porosity (>95%), low density, and superior water absorption capacity, all of which contribute to effective exudate absorption and controlled release of active compounds (He et al., 2026). Balancing water absorption and retention ensures optimal performance in fresh food packaging applications, reinforcing aerogels as a promising solution for sustainable and functional food packaging systems.

1.7 SMARTPHONE-BASED TECHNOLOGIES

Smartphone-based technologies have emerged as innovative tools in food safety and quality control, offering portable, cost-effective, and user-friendly solutions for on-site analysis. These technologies leverage the advanced capabilities of smartphones, such as high-resolution cameras, powerful processors, and wireless connectivity, to conduct rapid food analysis and contamination detection (Zhao et al., 2021a, 2021b). One of the most common methods involves optical assays, where smartphone cameras capture images of colorimetric or fluorescent reactions indicative of contaminants such as pesticides, heavy metals, and microbial pathogens. For instance, smartphone-integrated biosensors have been developed to detect *Salmonella* in chicken meat through magnetic particle immunoseparation coupled with image analysis, significantly improving food safety monitoring (Hossain et al., 2022).

Another widely explored approach is electrochemical sensing, where smartphones are connected to portable electrochemical detectors that measure food freshness by detecting volatile organic compounds (VOCs) associated with spoilage. This has been particularly effective in assessing meat and seafood quality in real time (Luo et al., 2021). Additionally, microfluidic paper-based analytical devices (μPADs) have been incorporated into smartphone systems for the rapid detection of chemical contaminants such as nitrates, antibiotics, and mycotoxins in food samples (Yetisen et al., 2020).

Recent innovations also include machine learning-based smartphone applications that analyze food images to determine adulteration, nutritional composition, and spoilage indicators (Sun et al., 2022a, 2022b, 2022c). Furthermore, smartphone spectroscopy has been explored for nondestructive quality assessment of fruits, vegetables, and dairy products, allowing producers and consumers to quickly assess food freshness and nutritional quality without laboratory intervention (Zhu et al., 2023).

The integration of smartphone technology with artificial intelligence (AI), cloud computing, and the Internet of Things (IoT) has further expanded its applications in food safety. Real-time data collection and remote monitoring through smartphone-connected sensors enhance traceability and food quality assurance in supply chains (Xu et al., 2021a, 2021b, 2021c). Despite these advancements, challenges such as sensor sensitivity, data standardization, and regulatory approvals remain crucial for widespread adoption. Nonetheless, smartphone-based food safety and quality control technologies continue to evolve, providing innovative, accessible, and efficient solutions for ensuring food integrity from farm to fork.

1.8 RAPID MICROBIAL TECHNIQUES

Rapid microbial techniques have revolutionized food safety and quality assurance by enabling the swift detection and quantification of microbial contaminants in food and beverages. Traditional culture-based methods are time-consuming, requiring 24–72 hours for results, whereas rapid techniques leverage molecular biology, biosensors, and advanced imaging systems to provide real-time or near-instantaneous microbial detection (Fusco et al., 2022). These methods not only enhance food safety but also support regulatory compliance, reduce foodborne illness risks, and improve supply chain efficiency.

One of the most widely adopted rapid microbial techniques is polymerase chain reaction (PCR) and its variations, such as quantitative PCR (qPCR) and loop-mediated isothermal amplification (LAMP). These techniques amplify target DNA sequences of pathogens like *Salmonella*, *Listeria monocytogenes*, and *Escherichia coli* with high sensitivity and specificity (Zhao et al., 2021a, 2021b, 2021c, 2021d). LAMP, in particular, has gained traction due to its rapid turnaround time, simplicity, and ability to be used in field settings without sophisticated equipment (Wang et al., 2023a, 2023b, 2023c).

Another significant advancement is biosensor technology, which includes electrochemical, optical, and nanomaterial-based biosensors for microbial detection. Smartphone-integrated biosensors, SPR sensors, and enzyme-linked immunosorbent assays (ELISA) have been developed to detect microbial contaminants in milk, meat, and fresh produce within minutes (Kumar et al., 2021). Additionally, microfluidic devices and lab-on-a-chip technologies allow simultaneous detection of multiple pathogens using minimal sample volumes, further reducing testing time and resource requirements (Sun et al., 2022b, 2022c).

Flow cytometry and impedance-based microbial detection techniques have also emerged as valuable tools in food microbiology. These methods analyze bacterial cell morphology, viability, and metabolic activity in real time, making them highly effective for spoilage detection and shelf-life estimation (Chen et al., 2020a, 2020b). Moreover, next-generation sequencing (NGS) and metagenomics offer a comprehensive approach to microbial analysis, providing insights into the microbial composition of food products and detecting emerging pathogens (Singh et al., 2022). Recent innovations include AI-powered image-based microbial detection systems, which utilize machine learning to analyze colony morphology and growth patterns, significantly reducing the need for manual intervention (Liu et al., 2023b, 2023c). Furthermore, real-time monitoring using IoT-connected microbial sensors has enhanced food safety tracking in processing plants, storage facilities, and retail environments (Ahmed et al., 2021a). While these rapid microbial techniques offer significant advantages, challenges such as standardization, regulatory approvals, and cost-effectiveness must be addressed for widespread adoption. Nonetheless, the continued development of rapid microbial detection methods is crucial for ensuring global food safety and minimizing foodborne disease outbreaks.

1.9 SPECTRAL TECHNIQUES IN FOOD SAFETY AND QUALITY CONTROL

Spectral techniques have gained significant attention in food analysis due to their rapid, nondestructive, and high-throughput capabilities for detecting contaminants, assessing quality, and monitoring food composition. These techniques include near-infrared (NIR), mid-infrared (MIR), Raman, and ultraviolet-visible (UV-Vis) spectroscopy, which rely on the interaction of electromagnetic radiation with food components to provide detailed chemical and structural information (Baeten et al., 2022).

NIR spectroscopy is widely used for assessing moisture, protein, fat, and carbohydrate content in cereals, dairy, and meat products. It offers real-time, on-site analysis without requiring sample preparation (Xu et al., 2021a, 2021c). Similarly, MIR spectroscopy has been employed to detect food adulteration, such as identifying melamine in milk and fraud in edible oils, by analyzing molecular vibrations (Dixit et al., 2023a, 2023b). Raman spectroscopy, known for its high specificity, has been applied in food authentication and pathogen detection, particularly in combination with surface-enhanced Raman scattering (SERS) (Guo et al., 2022a, 2022b). Recent innovations in spectral techniques include hyperspectral imaging (HSI) and Fourier-transform infrared (FTIR) spectroscopy, which integrate imaging with spectroscopy for spatial and chemical analysis. These methods enable rapid detection of contaminants, pesticide residues, and food fraud at the molecular level (Gowen et al., 2021). Future advancements include AI-driven spectral analysis, improving the accuracy and automation of food safety assessments.

1.10 MOLECULAR-BASED TECHNIQUES FOR FOOD SAFETY

Molecular-based techniques are crucial in detecting foodborne pathogens, allergens, and genetically modified organisms (GMOs) with high specificity and sensitivity. These techniques include PCR, qPCR, LAMP, and NGS (Chen et al., 2022b). PCR-based methods are extensively used for pathogen identification, including *Salmonella*, *Listeria monocytogenes*, and *E. coli*. qPCR enhances detection sensitivity by quantifying microbial DNA in real time, reducing analysis time significantly compared to culture-based methods (Wang et al., 2023a, 2023b). LAMP has emerged as a cost-effective and field-deployable alternative, providing rapid and robust detection of foodborne pathogens without requiring complex instrumentation (Zhao et al., 2021a, 2021c). NGS and metagenomics have revolutionized microbial analysis by enabling whole genome sequencing and microbiome profiling of food samples, allowing comprehensive identification of pathogens and spoilage organisms (Singh et al., 2022). The integration of AI and cloud computing in molecular-based techniques is enhancing data analysis, improving food traceability, and ensuring regulatory compliance.

1.11 SURFACE-BASED ANALYSIS FOR FOOD CONTAMINANT DETECTION

Surface-based analysis techniques are instrumental in detecting trace contaminants, food fraud, and microbial pathogens by analyzing the surface properties of food samples. These techniques include atomic force microscopy (AFM), SPR, and scanning electron microscopy (SEM), which provide high-resolution imaging and molecular interaction analysis (Li et al., 2022). SPR-based biosensors have been widely utilized for detecting food allergens, toxins, and antibiotic residues due to their label-free, real-time analysis capability (Ahmed et al., 2021a). AFM allows nanoscale imaging of food matrices, enabling the identification of microstructural changes due to contamination or degradation (Chen et al., 2020b, 2020c). SEM, often combined with energy-dispersive X-ray spectroscopy (EDS), has been used to analyze nanoparticle contamination and characterize food microstructures (Liu et al., 2023a). Advancements in nanotechnology have further improved surface-based analysis, with nanostructured sensors enhancing detection sensitivity for heavy metals, pesticides, and bacterial toxins in food (Sun et al., 2022b). The future of surface-based techniques lies in the development of portable, AI-integrated analytical tools for on-site food safety assessment.

1.12 HYPHENATED TECHNIQUES IN FOOD ANALYSIS

Hyphenated techniques combine multiple analytical methods to enhance detection accuracy and provide comprehensive food characterization. Examples include gas chromatography-mass spectrometry (GC–MS), liquid chromatography-mass spectrometry (LC–MS), and Fourier-transform infrared spectroscopy-mass spectrometry (FTIR–MS) (Wang et al., 2023b, 2023c). GC–MS is widely employed for detecting pesticide residues, mycotoxins, and VOCs in food due to its high sensitivity and specificity (Dixit et al., 2023a). LC–MS is a powerful tool for analyzing food adulterants, bioactive compounds, and contaminants such as antibiotics and heavy metals (Xu et al., 2021b, 2021c). FTIR-MS, by integrating vibrational spectroscopy with MS, provides detailed molecular insights into food composition and adulteration (Guo et al., 2022b). Recent advancements include AI-powered data analysis and miniaturized portable systems for real-time food analysis. The growing application of hyphenated techniques in food authentication, quality control, and regulatory enforcement highlights their importance in modern food safety.

1.13 CHALLENGES AND FUTURE SCOPE OF THE INNOVATIVE TECHNIQUES

Despite significant advancements in food safety analysis, several challenges hinder the widespread adoption of novel techniques: high costs, infrastructure requirements, and the need for trained personnel limit accessibility, especially for small-scale industries (Wang et al., 2023c). Standardization and regulatory approvals remain critical bottlenecks, as many emerging technologies lack universal validation protocols (Zhao et al., 2021c, 2021d). Sensitivity and specificity issues in biosensors and molecular-based assays can lead to false positives or negatives, impacting reliability (Chen et al., 2022a, 2022b). Additionally, scalability remains a challenge, with lab-based methods such as PCR and MS requiring adaptation for field applications (Singh et al., 2022). The increasing integration of AI and IoT raises concerns about data security and privacy, while consumer skepticism regarding nanotechnology and genetic testing presents hurdles for acceptance (Xu et al., 2021a, 2021c).

Future innovations in food analysis will focus on miniaturization, AI-driven automation, and real-time monitoring. Portable biosensors, smartphone-based detection systems, and lab-on-a-chip devices will enhance field deployability and accessibility (Liu et al., 2023c). AI and big data analytics will improve predictive food safety models, while blockchain-enabled traceability systems will enhance supply chain transparency (Sun et al., 2022a, 2022d). Nanotechnology applications, including nanocomposites for food packaging and nano-biosensors, will provide ultrasensitive detection capabilities (Guo et al., 2022b). Green analytical chemistry and sustainable packaging solutions will gain traction to minimize environmental impact (Dixit et al., 2023a, 2023b). As technology advances, interdisciplinary collaboration and regulatory harmonization will be key to ensuring the seamless integration of these innovations into the global food industry.

REFERENCES

Adeyeye, S. A. O., Babu, A. S., & Ganesh, P. S. (2023). Starch nanocrystal and its food packaging applications. *Current Research in Nutrition and Food Science Journal*, *11*(1), 01–21.

Ahmed, M. U., Saaem, I., Wu, P. C., & Brown, A. S. (2021a). IoT-enabled microbial sensors for real-time food safety monitoring. *Trends in Food Science & Technology*, *118*, 95–108. https://doi.org/10.1016/j.tifs.2021.09.012

Ahmed, M. U., Saaem, I., Wu, P. C., & Brown, A. S. (2021b). SPR-based biosensors for real-time food allergen detection. *Trends in Food Science & Technology*, *118*, 95–108. https://doi.org/10.1016/j.tifs.2021.09.012

Baeten, V., Pierna, J. A. F., Dardenne, P., & Dufour, J. P. (2022). Advances in spectral techniques for food safety. *Food Control*, *132*, 108731. https://doi.org/10.1016/j.foodcont.2022.108731.

Benito-Gonzalez, I., López-Rubio, A., Galarza-Jiménez, P., & Martinez-Sanz, M. (2021). Multifunctional cellulosic aerogels from Posidonia oceanica waste biomass with antioxidant properties for meat preservation. *International Journal of Biological Macromolecules*, *185*, 654–663.

Biji, K. B., Ravishankar, C. N., Mohan, C. O., & Srinivasa Gopal, T. K. (2015). Smart packaging systems for food applications: A review. *Journal of food science and technology*, *52*(10), 6125–6135.

Budtova, T., Aguilera, D. A., Beluns, S., Berglund, L., Chartier, C., Espinosa, E., . . . & Buwalda, S. J. (2020). Biorefinery approach for aerogels. *Polymers*, *12*(12), 2779.

Chen, H., Wu, H., Wang, X., & Zhang, Y. (2020a). Advances in AFM-based microbial detection in food safety. *Biosensors and Bioelectronics*, *152*, 112015. https://doi.org/10.1016/j.bios.2020.112015

Chen, H., Wu, H., Wang, X., & Zhang, Y. (2020b). Advances in impedance-based microbial detection in food safety. *Biosensors and Bioelectronics*, *152*, 112015. https://doi.org/10.1016/j.bios.2020.112015

Chen, H., Wu, H., Wang, X., & Zhang, Y. (2022). Advances in biosensor-based microbial detection for food safety. *Biosensors and Bioelectronics*, *195*, 113015. https://doi.org/10.1016/j.bios.2022.113015

Chen, Y., Liu, Z., & Wang, Q. (2022). Molecular detection of foodborne pathogens using PCR-based approaches. *Food Microbiology*, *105*, 103978. https://doi.org/10.1016/j.fm.2022.103978

Damborský, P., Švitel, J., & Katrlík, J. (2016). Optical biosensors. *Essays in biochemistry*, *60*(1), 91–100.

Dhua, S., Gupta, A. K., & Mishra, P. (2022). Aerogel: Functional emerging material for potential application in food: A review. *Food and Bioprocess Technology*, *15*(11), 2396–2421.

Dixit, Y., Casado-Gavalda, M. P., & Cama-Moncunill, R. (2023a). Applications of infrared spectroscopy in food adulteration detection. *Analytical Chemistry*, *95*(6), 3431–3445. https://doi.org/10.1021/acs.analchem.2c00978

Dixit, Y., Casado-Gavalda, M. P., & Cama-Moncunill, R. (2023b). Green analytical chemistry in food quality assessment. *Analytical Chemistry*, *95*(6), 3431–3445. https://doi.org/10.1021/acs.analchem.2c00978

Esser, B., & Swager, T. M. (2010). Detection of ethylene gas by fluorescence turn-on of a conjugated polymer. *Angewandte Chemie International Edition*, *49*(47), 8872–8875.

Fusco, V., Chieffi, D., Fanelli, F., & Quero, G. M. (2022). Rapid methods for microbial detection and identification in food safety and quality. *Food Microbiology*, *103*, 103941. https://doi.org/10.1016/j.fm.2022.103941

Gowen, A. A., Tsuta, M., & O'Donnell, C. P. (2021). Hyperspectral imaging for food safety and quality monitoring. *Critical Reviews in Food Science and Nutrition*, *61*(12), 1897–1912. https://doi.org/10.1080/10408398.2020.1746724

Guo, L., Chen, H., & Zhang, J. (2022a). Nanotechnology applications in food packaging and quality assessment. *Food Chemistry*, *381*, 132168. https://doi.org/10.1016/j.foodchem.2022.132168

Guo, L., Chen, H., & Zhang, J. (2022b). Raman spectroscopy and SERS for food authentication and quality control. *Food Chemistry*, *381*, 132168. https://doi.org/10.1016/j.foodchem.2022.132168

Gupta, A. K., Das, T., Kumar, H., Rastogi, S., Espinosa, E., Rincón, E., . . . & Mishra, S. (2023). Novel food materials: Fundamentals and applications in sustainable food systems for food processing and safety. *Food Bioscience*, *55*, 103013.

Gupta, A. K., Koch, P., Yumnam, M., Medhi, M., Madufor, N. J., Opara, U. L., & Mishra, P. (2022). Biosensors involved in fruit and vegetable processing industries. In *Biosensors in Food Safety and Quality* (pp. 111–134). CRC Press.

He, Y., Liu, H., & Ying, W. (2026). Construction of lignin-based aerogel with superelasticity and high reusability for air filtration and water pollution treatment. *Industrial Crops and Products*, *247*, 123525.

Hossain, M. M., Rana, M. M., Kim, H., & Kim, Y. (2022). Smartphone-based biosensors for food safety and quality monitoring: A comprehensive review. *Biosensors and Bioelectronics*, *203*, 114034. https://doi.org/10.1016/j.bios.2022.114034

Kaiser, J., Novotný, K., Martin, M. Z., Hrdlička, A., Malina, R., Hartl, M., . . . & Kizek, R. (2012). Trace elemental analysis by laser-induced breakdown spectroscopy—Biological applications. *Surface Science Reports*, *67*(11–12), 233–243.

Kalita, M., Medhi, M., Yumnam, M., Gupta, A. K., & Mishra, P. (2021). Synthesis of carbon dots from *Citrus limon* peel by microwave-assisted process and its application for detection of ferric ion (Fe^{3+}) and development of pH paper. In *Food Loss and Waste Reduction* (pp. 83–103). Apple Academic Press.

Kumar, S., Thakur, M. S., & Reddy, K. R. (2021). Recent developments in biosensors for foodborne pathogen detection. *TrAC Trends in Analytical Chemistry*, *144*, 116432. https://doi.org/10.1016/j.trac.2021.116432

Li, Q., Sun, X., & Zhao, X. (2022). Application of SEM and AFM in food contamination analysis. *Micron*, *152*, 103967. https://doi.org/10.1016/j.micron.2022.103967

Liu, J., Zhang, X., Li, Q., & Chen, Y. (2023a). AI-driven food safety monitoring systems. *Journal of Food Science and Technology*, *60*(5), 1347–1361. https://doi.org/10.1007/s13197-023-05479-8

Liu, J., Zhang, X., Li, Q., & Chen, Y. (2023b). AI-driven image analysis for microbial detection in food safety. *Journal of Food Science and Technology*, *60*(5), 1347–1361. https://doi.org/10.1007/s13197-023-05479-8

Liu, J., Zhang, X., Li, Q., & Chen, Y. (2023c). AI-driven imaging techniques for microbial detection in food safety. *Journal of Food Science and Technology*, *60*(5), 1347–1361. https://doi.org/10.1007/s13197-023-05479-8

Liu, Y. J., He, L. L., Mustapha, A., Li, H., Hu, Z. Q., & Lin, M. S. (2009). Antibacterial activities of zinc oxide nanoparticles against Escherichia coli O_{157}: H_7. *Journal of applied microbiology, 107*(4), 1193–1201.

Lonsdale, C. R., Hegarty, J. D., Cady-Pereira, K. E., Alvarado, M. J., Henze, D. K., Turner, M. D., . . . & Scarino, A. J. (2017). Modeling the diurnal variability of agricultural ammonia in Bakersfield, California, during the CalNex campaign. *Atmospheric Chemistry and Physics, 17*(4), 2721–2739.

Luo, H., Liu, Z., Yang, H., & Wang, Y. (2021). Smartphone-based electrochemical sensors for food safety analysis. *Sensors and Actuators B: Chemical, 334*, 129665. https://doi.org/10.1016/j.snb.2021.129665

Madufor, K. J. N. (2019). *Design, fabrication and testing of an electrochemically based DNA biosensor for detection of Coniella Granati fungal disease of pomegranate fruit and aster yellows bacterial grapevine disease,* Stellenbosch University, (Doctoral dissertation)

Markiewicz-Keszycka, M., Cama-Moncunill, X., Casado-Gavalda, M. P., Dixit, Y., Cama-Moncunill, R., Cullen, P. J., & Sullivan, C. (2017). Laser-induced breakdown spectroscopy (LIBS) for food analysis: A review. *Trends in Food Science & Technology, 65*, 80–93.

Mohaidat, Q., Palchaudhuri, S., & Rehse, S. J. (2011). The effect of bacterial environmental and metabolic stresses on a laser-induced breakdown spectroscopy (LIBS) based identification of Escherichia coli and Streptococcus viridans. *Applied Spectroscopy, 65*(4), 386–392.

Omanović-Mikličanin, E., Badnjević, A., Kazlagić, A., & Hajlovac, M. (2020). Nanocomposites: A brief review. *Health and Technology, 10*(1), 51–59.

Rakovský, J., Čermák, P., Musset, O., & Veis, P. (2014). A review of the development of portable laser-induced breakdown spectroscopy and its applications. *Spectrochimica Acta Part B: Atomic Spectroscopy, 101*, 269–287. https://doi.org/10.1016/j.sab.2014.09.015

Schmitt, K., Tarantik, K., Pannek, C., Sulz, G., & Wöllenstein, J. (2016). Colorimetric gas sensing with enhanced sensitivity. *Procedia Engineering, 168*, 1237–1240.

Sharma, H., & Mutharasan, R. (2013). Half antibody fragments improve biosensor sensitivity without loss of selectivity. *Analytical chemistry, 85*(4), 2472–2477.

Singh, R., Sharma, P., Kumar, M., & Singh, N. (2022a). Metagenomic approaches for foodborne pathogen detection and monitoring. *Food Control, 135*, 108767. https://doi.org/10.1016/j.foodcont.2022.108767

Singh, R., Sharma, P., Kumar, M., & Singh, N. (2022b). Metagenomics and proteomics in food safety monitoring. *Food Control, 135*, 108767. https://doi.org/10.1016/j.foodcont.2022.108767

Skolik, P., Morais, C. L., Martin, F. L., & McAinsh, M. R. (2019). Determination of developmental and ripening stages of whole tomato fruit using portable infrared spectroscopy and chemometrics. *BMC Plant Biology, 19*(1), 236.

Sun, X., Wang, Y., Zhang, C., & Yang, X. (2022). Machine learning-assisted smartphone applications for food quality assessment. *Trends in Food Science & Technology, 120*, 132–145. https://doi.org/10.1016/j.tifs.2022.01.012

Sun, X., Yang, Y., & Zhang, C. (2022a). Advances in nanostructured sensors for food safety applications. *Analytical Chemistry, 94*(18), 6892–6904. https://doi.org/10.1021/acs.analchem.2c01278

Sun, X., Yang, Y., & Zhang, C. (2022b). Machine learning-based predictive analytics for food safety. *Analytical Chemistry, 94*(18), 6892–6904. https://doi.org/10.1021/acs.analchem.2c01278

Sun, X., Yang, Y., & Zhang, C. (2022c). Microfluidic lab-on-a-chip devices for rapid pathogen detection. *Analytical Chemistry, 94*(18), 6892–6904. https://doi.org/10.1021/acs.analchem.2c01278

Turner, A. P. (2013). Biosensors: Sense and sensibility. *Chemical Society Reviews, 42*(8), 3184–3196

Valencia, G. A., & do Amaral Sobral, P. J. (2018). Recent trends on nano-biocomposite polymers for food packaging. In *Polymers for Food Applications* (pp. 101–130). Springer International Publishing. https://doi.org/10.1007/978-3-319-94625-2_5

Wang, R., Li, C., & Zhao, X. (2023a). Advances in LAMP-based pathogen detection for food safety. *International Journal of Food Microbiology, 383*, 109975. https://doi.org/10.1016/j.ijfoodmicro.2023.109975

Wang, R., Li, C., & Zhao, X. (2023b). Advances in LAMP-based rapid microbial detection in food safety. *International Journal of Food Microbiology, 383*, 109975. https://doi.org/10.1016/j.ijfoodmicro.2023.109975

Wang, R., Li, C., & Zhao, X. (2023c). Advances in quantum dot biosensors for food safety. *International Journal of Food Microbiology, 383*, 109975. https://doi.org/10.1016/j.ijfoodmicro.2023.109975

Xu, J., Wang, S., & Li, H. (2021a). IoT and AI-enabled food traceability systems. *Food Analytical Methods, 14*(7), 1785–1802. https://doi.org/10.1007/s12161-021-02013-x

Xu, J., Wang, S., & Li, H. (2021b). Near-infrared and mid-infrared spectroscopy for food quality analysis. *Food Analytical Methods, 14*(7), 1785–1802. https://doi.org/10.1007/s12161-021-02013-x

Xu, W., Zhu, C., Li, X., & Chen, X. (2021). Integration of smartphone sensors with cloud computing for real-time food quality monitoring. *Food Control, 125*, 107973. https://doi.org/10.1016/j.foodcont.2021.107973

Yetisen, A. K., Akram, M. S., & Lowe, C. R. (2020). Paper-based microfluidic devices in food safety applications. *Analytical Chemistry, 92*(5), 3252–3262. https://doi.org/10.1021/acs.analchem.9b04743

Zhang, D., Yan, Y., Li, Q., Yu, T., Cheng, W., Wang, L., . . . & Ding, S. (2012). Label-free and high-sensitive detection of Salmonella using a surface plasmon resonance DNA-based biosensor. *Journal of Biotechnology, 160*(3–4), 123–128.

Zhao, X., Lin, C. W., Wang, J., & Oh, D. H. (2021a). Advances in rapid detection methods for foodborne pathogens using smartphone-based biosensors. *Food Microbiology, 97*, 103738. https://doi.org/10.1016/j.fm.2021.103738

Zhao, X., Lin, C. W., Wang, J., & Oh, D. H. (2021b). Future prospects of rapid detection methods for foodborne pathogens. *Food Microbiology, 97*, 103738. https://doi.org/10.1016/j.fm.2021.103738

Zhao, X., Lin, C. W., Wang, J., & Oh, D. H. (2021c). Recent advances in qPCR-based rapid detection methods for foodborne pathogens. *Food Microbiology, 97*, 103738. https://doi.org/10.1016/j.fm.2021.103738

Zhao, X., Lin, C. W., Wang, J., & Oh, D. H. (2021d). Recent advances in PCR-based rapid detection methods for foodborne pathogens. *Food Microbiology, 97*, 103738. https://doi.org/10.1016/j.fm.2021.103738

Zhu, Y., Wang, J., Fan, Z., & Li, H. (2023). Non-destructive food quality assessment using smartphone-based spectroscopy. *Food Analytical Methods, 16*(3), 675–690. https://doi.org/10.1007/s12161-023-02345-1

2 Surface-Mediated C Dots/Quantum Dots

Synthesis, Characterization, and Application in Food Quality and Safety Analysis

Anupama Bora, Himali Horo, Susmita Rabha, Apurba Das, and Ankur Pandey

2.1 INTRODUCTION

Carbon quantum dots (CQDs) or quantum dots (QDs) are spherical particles with a size of less than 10 nm; they are nontoxic and environment-friendly, so they have good potential in the food industry. QDs are categorized on the basis of materials used in carbon quantum dots and semiconductor QDs (Gidwani et al., 2021). QDs have specific features like they are highly stable against photo bleaching and size-dependent and excitation-independent photoluminescence (PL). QDs also have good photostability and longer life because of which they can be considered instead of fluorescent index dyes and proteins (Sistani and Shekarchizadeh, 2023). Although semiconductor QDs are used for various applications in detection, tracking, and labeling, one of their drawbacks is that they are toxic due to their heavy metal for environmental and health concerns (Gidwani et al., 2021). Graphene QDs have low toxicity and are environmentally friendly due to their small size and higher oxygen content, so they are used in various fields, especially in the food sector. Graphene QDs have potential characteristics such as good biocompatibility, low cytotoxicity, and photobleach resistance, so they are used as fluorescent probes for many analytes. One of the most environmentally safe alternate nanomaterials that offer good biocompatibility, nontoxicity, and good optical properties are carbon quantum dots (CQDs). CQDs are less than 10 nm in size and have high optical stability over organic dyes (Sobiech et al., 2021).

Food analysis is vital for food safety control and quality assessment, and the common techniques for analysis are spectroscopy, culture, immunoassay, colony counting, chromatography, electrochemical methods, etc. These methods are advantageous due to their reproducibility, high selectivity, and sensitivity. QDs-based fluorescence sensors are vital in ensuring food safety and quality (Luo et al., 2020).

DOI: 10.1201/9781003482611-2

2.2 MECHANISM OF CARBON DOTS (CDs)/ QDs IN FOOD ANALYSIS

Surface-mediated CDs and QDs have garnered significant attention in food quality and safety analysis due to their exquisite properties that can be exploited in various detection and sensing applications (Kumar and Rhim, 2024). The mechanism typically involves the optical, electronic, and chemical properties of the CDs and QDs related to fluorescence quenching and recovery mechanisms. Figure 2.1 shows the overall mechanism and challenges associated to use of QDs in the food industry. A brief overview is presented in Figure 2.1.

2.2.1 Fluorescence Quenching/Enhancement

QDs and CDs both have strong fluorescence properties (Khan et al., 2023). On interaction with various analytes found in food samples, such as pesticides, heavy metals, or pathogens, the fluorescence of the dots can either amplify or quench (Khan et al., 2023). This change in fluorescence intensity offers a sensitive and quick way to analyze food safety by detecting and quantifying the target analytes (Tian et al., 2023).

2.2.2 Selective Binding and Sensing

Tailoring the surface functionalization of CDs or QDs such that they bind to particular pollutants or analytes in food samples has been shown to be possible by various research groups (Sun et al., 2024). To facilitate specific binding to target molecules, functional groups such as carboxyl, amino, or thiol groups are added to the surface of the dots (Sun et al., 2024). The presence and concentration of the target analytes can then be determined by monitoring changes in the optical or electrical characteristics of the dots caused by this selective binding.

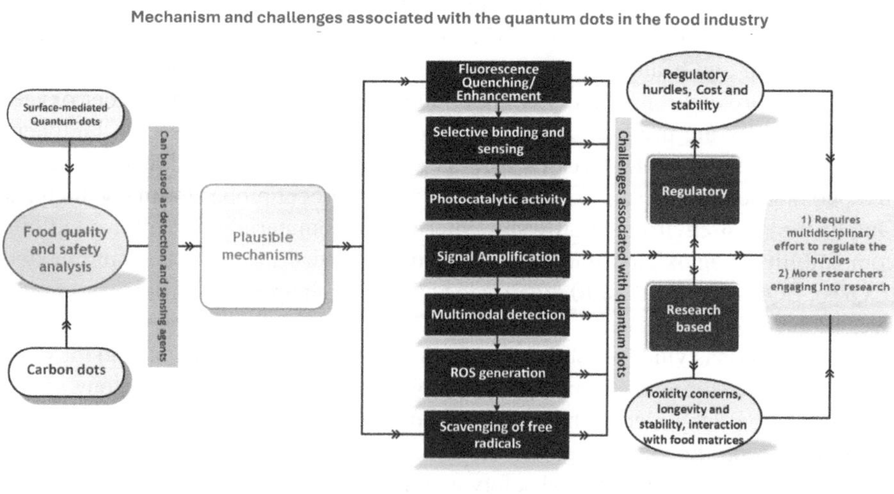

FIGURE 2.1 Mechanism and challenges associated to use of quantum dots in the food industry

2.2.3 Photocatalytic Activity

Under specific conditions, some CDs show photocatalytic activity. These characteristics have been used to break down impurities or pollutants found in food samples. This mechanism enables improving the safety and caliber of food products by irradiating CDs with light, which can produce reactive oxygen species (ROS) or other radicals that can break down organic pollutants or microbiological contaminants (Sun et al., 2024).

2.2.4 Signal Amplification

In several detection tests, surface-mediated CDs and QDs can also be used to enhance signals (Tian et al., 2023). Their fluorescent properties, for example, can be utilized as labels in immunoassays or DNA assays, where the binding events between the target analyte and the recognition element (e.g., antibody or DNA probe) increase the signal. The sensitivity and accuracy of food quality and safety analysis are improved by such signal amplification (Tian et al., 2023).

2.2.5 Multimodal Detection

In order to provide supplementary information about the food samples, surface-modified CDs and QDs can be incorporated into multimodal detection platforms along with other sensing components or methodologies (Zou et al., 2016). To obtain improved detection performance and reliability, for instance, they can be employed in conjunction with mass spectrometry (MS), chromatography techniques, or electrochemical sensors (Zou et al., 2016).

Additionally, it has been demonstrated that CDs functions as an active addition in the field of food packaging because of their exceptional biocompatibility, antibacterial, and antioxidant activity. The mechanism behind such antibacterial and antioxidant properties includes the generation of ROS under specific conditions such as exposure to light or in presence of oxygen (Khan et al., 2023). The surface functional groups and defects in the CDs usually act as catalytic sites for the generation of ROS that contain singlet oxygen ($^1O_2$), superoxide radicals ($O_2{\bullet}^-$), and hydroxyl radicals (•OH), which are highly reactive molecules responsible for oxidative damage to microbial cells and biomolecules (Khan et al., 2023). The CDs also undergo photodynamic reactions, which enable them to generate ROS that induce oxidative stress and cellular death in microbial pathogens.

In contrast to ROS responsible for the antibacterial activity of CDs, the antioxidant property is due to scavenging of the free radicals thereby inhibiting the oxidative stress (Tian et al., 2023). The π-conjugated structure of the CDs neutralizes the free radicals by acting as sites for electron donors and acceptors. The neutralization process safeguards the oxidative stress to biomolecules like lipids, proteins, and DNA. Some CDs reportedly exhibit metal ion chelation by sequestering the transition metals involved in the generation of the ROS through Fenton-like reactions (Zou et al., 2024; Zhang et al., 2022). Thus, the ROS generation is inhibited, thereby mitigating the oxidative-stress induced damage.

2.3 SURFACE MODIFICATIONS OF C DOTS/QDs

QDs are nanostructured particles of typical size 10 nm to 50 nm. QDs possess extraordinary optoelectronic properties due to the restricted movement of charge carriers within a small dimension, that is, the size of the QD (Cheng et al., 2014). The size of the QDs is decided by the bulk exciton Bohr radius. They are generally composed of II-VI or III-V group semiconducting materials. Their optical and electronic properties are dependent on both the size and composition of the QDs. These inorganic QDs possess peculiar chemical properties as compared to bulk owing to their high surface-to-volume ratio. These QDs are widely exploited for various optical, optoelectronic, and energy-harvesting applications. Recently, fluorescent inorganic QDs have found applications in biomedical and food science as well. In food science, QDs are widely used for quality and safety assessment of food. As the QDs are composed of inorganic materials, direct application of these in biological samples is not a good idea.

It is known that surface atoms are less tightly bonded and produce defect states or impurities in the structure. It is always important to reduce such defects in structure by using some coating or shielding to the outside environment to achieve both physical and chemical stabilities of the structure (i.e., doping). As QDs possess a higher surface-to-volume ratio, they are more prone to defects leading to compromised electrical and optical responses. The surface chemistry of QD highly depends on the synthesis method, and different approaches are effective for each. In addition, for biomedical and food science applications, surface modification of QD is done to achieve the desired biocompatibility and functionality for intended applications (Xing and Yang, 2014). As the need for surface modification in QD is understood, it is also important to understand the methods and techniques for the same. There are a few techniques that are widely utilized for surface modifications of QD. Some of the commonly used techniques are as follows:

2.3.1 LIGAND EXCHANGE TECHNIQUE

Ligand exchange technique is one of the most widely used methods for surface modification of QDs. It is the methodology of replacement of the original ligands or functional groups with one or more types of appropriate ligands in the surface bonded to the central structure of QDs (Dubois et al., 2007). This is done to achieve the required properties in QD for the intended applications. For example, QDs synthesized for biomedical applications must be stable in terms of biological physiochemistry, biocompatibility, and water-solubility. In general, in such cases, monodentate thiols such as mercaptoacetic acid, bidentatethiols (e.g., dihydrolipoic acid, carbodithioic acids), mono- and bidentate amines, phosphonic acids, and carboxylic acids and multidentate ligands (e.g., oligomers and polymers) are used. All these ligands make the QDs hydrophilic in nature, which is necessary for the effective application of QDs in biological environments. Among various hydrophilic ligands, mercapto-groups-based ligands are preferred due to the high binding affinity to the QDs' surface (Xing and Yang, 2014). Mercaptoalkyloligo (ethylene glycol) carboxylic acid, mercaptoundecanoic acid (MUA), cysteine, a sugar-functionalized thiol, a biotin-functionalized oligo (ethylene glycol) thiol, hydrophilic dendron functionalized thiols, and glutathione are some examples of ligands that are used in biological application.

2.3.1.1 In-Situ Functionalization

As the name suggests, the surface modification of the QD dot is done during the synthesis process itself. This method eliminates the complexity associated with other methods where multiple steps are involved—for example, synthesis of water-soluble CdSe/ZnS core/shell coated with (MPA) in a single step where the preparation of ZnS shell is done in oleyl amine-capped CdSe QD environment in presence of mercaptopropionic-acid-stabilizing ligand.

2.3.1.2 Silica Coating

It is a handy technique where a coating of silica layer is done on the QD to make it water soluble and biocompatible. In this process, the outer silica layer provides a superficial bioconjugation ability with biomolecules. Silica coating on QD can be performed in several ways such as sol-gel, surface silanization, and micellization of siloxane surfactants.

2.3.1.3 Polymer Coating

Surface modification of as-prepared QDs by polymer coating has a great importance in achieving certain properties in a single element. Polyethylene glycol and maleic anhydride-based amphiphilic polymers were used in such applications. As discussed in the previous sections, one of the approaches of surface modification is ligand exchange, where the original ligand associated with QD is replaced by desired type of ligand to achieve a particular characteristic. In some cases, the properties of QD gets affected as the original ligand was removed, e.g.: reduction in fluorescence intensity when thiol groups directly bonded to fluorescent core CdSe in CdSe/ZnS core/shell QD (Speranskaya et al., 2014). Encapsulation of QDs with amphiphilic polymers makes the QD more efficient in two ways: first, it does not remove the original ligand that helps in retaining the original characteristics of QD and, second, the introduction of multiple hydrophobic units associated with a single polymer chain, which results in strong interaction with the initial organic coatings. Overall, polymer coating provides better stability over a large range of pH of solutions (Hezinger et al., 2008).

2.3.1.4 Click Chemistry

Click chemistry is a universal technique by which any reaction is facilitated to occur with highest efficiency without using highly reactive or cross-reactive intermediates and chemicals at moderate temperature without damaging the original composition or structure (Chem, 2011). In addition, a precise control over the number of reacting molecules is possible by click chemistry. The use of click chemistry in surface modification of QDs had provided insight of the reaction in individual molecules at the cellular level because the response of QDs highly depends on interfacial properties rather than bulk property. Click chemistry is applicable in various types of surface modification techniques, e.g., self-assembled monolayers via azide/alkyne-functionalized surfaces, polymeric surfaces, block copolymeric surfaces (Bernardin et al., 2010; Gassensmith et al., 2013; Schieber et al., 2012).

2.3.1.5 Biological Conjugation

Surface modification of inorganic QDs is must via bioconjugation for biological diagnostics and imaging applications. In this method, bonding of QDs with biomolecules is done. Typically, all sorts of biological compounds starting from protein (e.g., antibodies, enzymes), structural protein (e.g., collagen), receptor, carrier (e.g., myoglobin); peptides (e.g., glutathione), cell-penetrating peptides, enzyme inhibitors; nucleic acids such as DNA, RNA, ribozymes, DNAzymes; along with fatty acids, lipids, carbohydrates, and all related members of each of these super-families of biomolecules are being used (Blanco-canosa et al., 2014). Bio-conjugated QDs are utilized in two ways either passively or actively. In the passive case, the main response depends on QDs' intrinsic property, e.g., QD-antibody conjugates in immunoassays, flow cytometry, and immunohistochemistry. In immunohistochemistry, the biorecognition is done by the antibody. Overall, in all forms of in vitro immunoassays, hybridization probes, and in vivo imaging, passive utility of bioconjugate QDs is found (Jennings et al., 2011). In active use, the QDs play as the center for conjugation with one or more types of molecules and as a transducer to monitor any changes in inherent property. For instance, multiple conjugations of peptides or nucleic acids with QDs locally boost the avidity in association with improved sensitivity (Medintz et al., 2009; Sapsford et al., 2011).

2.3.1.6 Doping

Like other methods, doping is also one method by using which the limitations and challenges associated with QDs practical application are possible to tackle. For instance, it is well known that carbon quantum dots are excellent candidates for less toxic biological imaging owing to better biocompatibility and photostability. The degree of toxicity is much less when processed from carbon-rich food wastes. However, as synthesized carbon quantum dots associate with low quantum yield (QY) and poor stability, these shortcomings were successfully overcome by doping with sulfur and nitrogen (Chen et al., 2022).

2.3.1.7 Layer-by-Layer Assembly

Those QDs intended for certain applications (e.g., optoelectronic devices) require higher temperature sustainability. For testing, thin film samples are used. In this regard, instead of mixing the QDs in polymer thin film, consequent QD and polymers can be assembled in layer-by-layer structure to obtain a homogeneous structure. The optimized sample may be used for devising (Kim et al., 2009).

2.3.1.8 Hybrid Nanostructure

In energy harvesting, QDs are utilized for photocatalysis and photo electrolysis. As QDs exhibit their size-dependent optical bandgap, their range of application can be tuned according to requirements (Xiao et al., 2015).

2.4 CHEMISTRY OF SURFACE MODIFICATION C DOTS/QDS

The protocols available for the synthesis of CDs are in plenty and often use external energies in the forms of hydrothermal, microwave, ultrasound, or laser ablation (Pundi and Chang, 2022). Moreover, all these protocols use organic precursors, and oxidizing agents engendering the abundant presence of the carboxyl groups are the CDs' surface. The presence of carboxyl groups is undesirable because it can lead to high water solubility, aggregation with an increase in particle size, poor fluorescence properties, and low quantum yield (QY), thereby rendering CDs not suitable for many applications (Yan et al., 2018). Therefore, it is required to tailor the surface properties of the CDs by attaching the specific moieties to the carboxyl groups present at the surface of the CDs. The modified CDs have a carbon core attached to a functional group. Both core and attached functional groups exhibit their own set of properties. These properties can be tailored on demand based as per the applicability (Wang et al., 2023).

The functionalization methods can be broadly classified into covalent and non-covalent methods based on the interactions between the carbon core and functionalization molecules. The covalent interactions, as the name suggests, involve linkage of groups to the core via covalent bonds. On the contrary, non-covalent interactions entail the bonding of functional groups through mechanisms such as π interactions, electrostatic attractions, complexations, hydrogen bonding, and similar processes.

2.4.1 Functionalization of CQDs Using Covalent Bonds

A variety of functional groups can be attached to the CDs' surface using covalent bonds. These bonds follow any one of the following reaction mechanisms including, but not limited to, amide coupling reactions, esterification, sulfonylation, or copolymerization reactions. These are discussed in subsequent sections.

2.4.1.1 Amide Coupling Reactions

The amide coupling reactions are well known in chemistry. These set of reactions involved the linkage between amide groups and carboxyl groups using 1–(3-dimethylaminopropyl)-3-ethyl carbodiimide (EDC) and N-hydroxysuccinimide (NHS) as coupling agents (V et al., 2019). The surface modification via EDC-NHS coupling mechanism significantly improves the fluorescence properties and QY (Dong et al., 2015). Moreover, EDC-NHS opens up the playground and enables the attachments of a variety of functional groups like dyes (Deng et al., 2016; Jin et al., 2018; Kim et al., 2017, 2015), supramolecules (Sun et al., 2017), and dopamine (Arreguin et al., 2009; Palanisamy et al., 2013) on the surface of CDs.

2.4.1.2 Silylation Reaction

The active hydrogen present on the surface of the CDs acts like a linkage point where silane groups can be attached via covalent bonds (Wang et al., 2023). The silane groups are biocompatible and nontoxic material that can act like silicone shell which enables the synthesis of designer CDs with multicolored fluorescence probes and nanoparticles. Also, the silane groups like tetraethyl orthosilicate (TEOS) and

3-aminopropyl triethoxysilane (APTES) prevent agglomeration by increasing the surface area, enhancing the water solubility and reducing the cytotoxicity (Rao et al., 2016).

2.4.1.3 Esterification

The esterification reactions can be utilized to host the functional groups over the CDs' surface containing plentiful carboxyl or hydroxyl groups. The wettability of CDs can also be altered from hydrophilic to lipophilic behavior. Esterification can be used to fabricate hydroxyl-functionalized CDs using lactose, then modified them with mercapto succinic acid (MSA) to enhance QY and detect Ag^+ ions through static quenching (Algarra et al., 2014). Moreover, CDs modified with carboxymethyl cellulose can be used for fluorescence quenching to detect pollutants like 2,4,6-trinitrophenol (TNP) (Gao et al., 2018).

2.4.1.4 Copolymerization Reaction

In certain biological applications, larger molecular weight CDs are required. To fabricate such CDs, surface is modified by copolymerization route that involves reaction between active hydrogen groups and cyclic or double-bond compounds. For example, polymerization reaction used to fabricate multi-hydroxy hyper branched polyglycerol groups attached to the CDs. Here, cyclodextrin is used to generate large amounts of hydroxyl groups on the surface of the CDs; later, these prepared CDs are used as initiator for polymerization reaction of glycidol. Later, these bulky fluorescent CDs are utilized for cell imaging applications (Li et al., 2017).

2.4.1.5 Schiff's Base Reaction

The pristine CDs can also be decorated via Schiff's base reaction between aldehydic and amine groups to form strong imine (-C=N-) bonds. The elastomeric silicone-based composites with embedded CDs prepared via Schiff's base reaction exhibited self-healing and fluorescent properties. The Schiff's base reaction provided necessary bonding between the $-NH_2-$ group in the elastomer (bis(amine)-terminated poly(dimethysiloxane) (H_2N-PDMS-NH_2)) and the aldehyde group present on the CDs (Lu et al., 2023), as shown in Figure 2.2.

2.4.2 Functionalization of CDs Using Non-Covalent Bonds

The CDs can also be decorated via non-covalent interactions such as π-π interactions, complexation, or electrostatic interactions. This type of interactions keeps the structural integrity intact and provides enhanced photostability, biocompatibility, and new interfacial properties suitable for various applications.

2.4.2.1 π-π Interactions

The CDs have extended π system that can be modified through π interactions with aromatic molecules, resulting in enhanced fluorescence properties. For example, Si-doped CDs, modified with dopamine via π interactions, exhibit exceptional

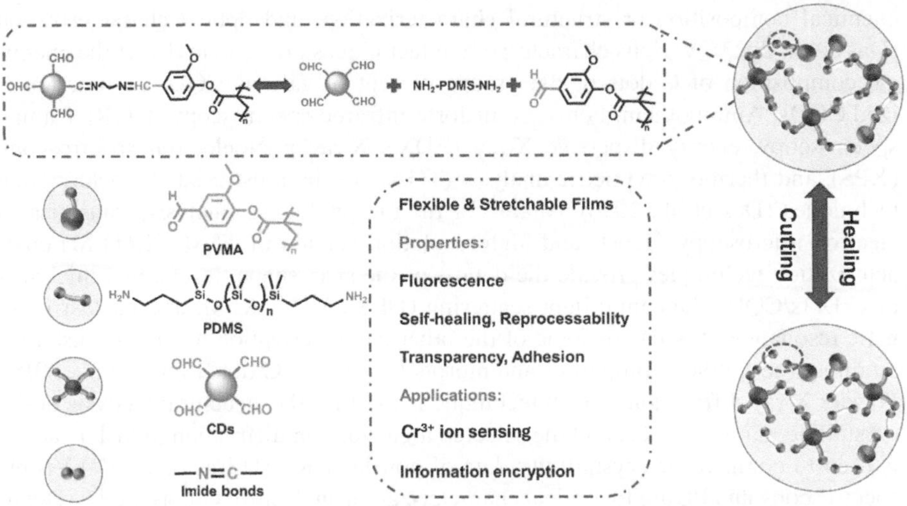

FIGURE 2.2 Scheme depicting Schiff Base Reaction for surface modification of CDs. Reprinted (adapted) with permission from Lu et al. (2023). Copyright 2023 American Chemical Society

fluorescence that can detect Ag^+ ions with superior sensitivity, making them useful for intracellular imaging and bioimaging of HeLa cells (Jiang et al., 2016).

2.4.2.2 Complexation Interactions

The complexation involves the coordination of metal ions with functional groups on the CDs' surface, such as amino, hydroxyl, or carboxyl groups. The CDs prepared via complexation route have enhanced stability against various solvents. These prepared CDs find various applications such as CDs-COOH complexed with Na^+ ions and have been used for seawater desalination (Guo et al., 2014). Likewise, lanthanide-modified CDs are used in dual-modal fluorescence/magnetic resonance imaging for bioimaging (Shi et al., 2015).

2.4.2.3 Electrostatic Interactions

The oppositely charged molecules can be utilized to modify CDs surfaces through electrostatic interactions, introducing other fluorophores and targeting molecules. For example, CDs modified with positively charged polyethyleneimine (PEI) and Nile Blue exhibit two-photon fluorescence and can detect sulfide ions in biological fluids (Jin et al., 2017). Another example involves CDs modified with rhodamine B for the detection of Hg^{2+} ions, useful for cell imaging (Lan et al., 2014).

2.5 CHARACTERIZATION OF THE C DOTS/QDS

CDs or carbon quantum dots can be characterized using numerous techniques. Three types of characterization used for C-dots are morphological characterization, chemical composition or structural characterization, and optical characterization (Das et al., 2023). Various characterization techniques are used to detect the chemical composition of C-dots or the groups present in various CQDs or functionalized CQDs. Among them, Fourier-transform infrared spectroscopy (FTIR), Raman spectroscopy, energy dispersive X-ray (EDX), X-ray photoelectron spectroscopy (XPS), and thermo gravimetric analysis (TGA) are the most used characterization techniques (Das et al., 2023). Again, for the morphological analysis, transmission electron microscopy (TEM) and high-resolution version of TEM (HRTEM) characterization techniques provide the details of the size, shape, layers, and thickness of C-Dots/CQDs. Dynamic light scattering (DLS), SEM, AFM, and nuclear magnetic resonance (NMR) are some of the other characterization techniques used for attaining the synthetic properties and morphology of the C-dots (Singh et al., 2018). Powder X-ray diffraction (XRD) technique is used for the structural analysis of the crystalline nanomaterials, and the selected area electron diffraction (SAED) pattern is used to confirm the crystallinity data of powder XRD. Ultraviolet (UV)-Visible spectroscopy and PL are for optical characterization analysis of C-dots/CQDs. Detail of the same can be studied in other books.

2.5.1 PL Spectroscopy

PL spectroscopy is used to observe phosphorescence and fluorescence abilities and the QY of the QDs/C-dots. PL is one of the most enthralling characterizations of C-dots concerning properties and applications. C-dots have optical properties that impact the sample from various particles at maximum excitation wavelength; these abilities indicate the conversion of a carbon source into QDs. As the size of CQDs increases, the color of fluorescence varies accordingly, demonstrating a red shift in emissive radiation caused by a decrease in the HOMO-LUMO gap (Wang and Hu, 2014). Other techniques are not in purview of the chapter.

2.6 APPLICATION OF THE ENGINEERED C DOTS/QDS FOR FOOD QUALITY AND SAFETY ANALYSIS

C-dots/CQDs with or without any surface functionalization have gained attention due to their wide applications in various fields. Due to their small size and shape, they are widely used in areas like bioimaging, drug delivery, chemical sensing, environmental applications, electrical devices, food analysis, food packaging (Das et al., 2023). Here, we will discuss about the application of C-dots for food quality and safety analysis.

2.6.1 QDs in Food Analysis

QDs and CQDs are used in the analysis of food samples via nanomaterial-based method. An effective way to enhance the selectivity of these nanomaterials is by combining them with molecularly imprinted polymers (MIPs). This integration influences the unique optical properties of QD/CQD and the high selectivity of MIPs, offering a significant advantage. Magnetic nanoparticles (MNPs) with QDs have potential for food analysis. MNPs exhibit a strong magnetic response and can be rapidly recovered under an external magnetic field, making them highly promising for efficiently controlling the adsorption and release of target analytes at low concentrations in food samples that are complex in nature (Xiong et al., 2022). Magnetic fluorescent QD nanocomposites have fluorescence properties by which they quickly detect target analytes and separate them from the complex food matrices using magnetic field. These nanocomposites simplify the pretreatment process and inhibit analyte wastage by decreasing time of detection and increasing detection efficiency (Xiong et al., 2022).

2.6.1.1 Detection of Food Additives and Pathogens

Food additives are commonly used in food processing to enrich flavor, extend shelf life, and boost nutritional value. Developing a rapid and reliable diagnostic method is essential for improving the monitoring of food additives (Ye et al., 2017). Some of the QDs/C-dots used for the detection of food additives are listed in Table 2.1. Specific species culturing and examining their biochemical and immunological properties are common methods for pathogen detection. So, QD-based fluorescence nanosensors are found to be effective for pathogen detection in food (Huang et al., 2019). Table 2.1 represents some of the QDs/CQDs used for food-pathogen detection. The detection strategy leverages a fluorescence-quenching mechanism utilizing carbon quantum dots (CQDs) and MNPs. CQDs, synthesized from orange peel via a green approach, exhibit a strong green fluorescence under 420 nm excitation. These CQDs are functionalized with an *E. coli*-specific aptamer (aptamer-CQDs), while MNPs are conjugated with complementary DNA (cDNA-MNPs). In the absence of *E. coli*, aptamer-CQDs hybridize with cDNA-MNPs, preserving fluorescence intensity. However, upon the introduction of *E. coli*, the aptamer preferentially binds to the bacterial surface, displacing CQDs from MNPs. This displacement reduces the

TABLE 2.1
Application of QDs in the Detection of Food Additives and Pathogens

Quantum Dots (QDs)	Analytes	Limit of Detection (LOD)	Linear Range	Samples	Reference
CQDs	Nitrite	0.15 µmol/L	0.50–50 µmol/L	Corn sausage, ham sausage, pickle, and hot dog	(Zhang et al., 2022b)
CQDs	Nitrite	0.47 µM	5–80 µM	-	(Deng et al., 2021)
CQDs	Borax	8.0 µmol/L	0.025–10 mmol/L	Flour, bread, instant noodle	(Zhou et al., 2022)
Au@CQDs	Melamine	3.6 nM	1M µto 10 µM	Milk	(Hu et al., 2019)
CQDs@MNPs	*E. coli*	487 CFU/mL	500–10^6 CFU/mL	Milk	(Hu et al., 2021)
CQDs@gold nanoparticles	*S. aureus*	10 CFU/mL	10^8–10^1CFU/mL	Milk, orange juice	(Pebdeni et al., 2020)

concentration of fluorescent CQDs-MNPs, leading to a measurable decrease in fluorescence intensity. The fluorescence reduction exhibits a strong correlation with *E. coli* concentration within the range of 500–10^6 CFU/mL, with a detection limit of 487 CFU/mL. The method was successfully validated in milk samples, demonstrating high accuracy and agreement with conventional plate-counting techniques, highlighting its potential for reliable food safety monitoring (Hu et al., 2021).

2.6.1.2 Heavy Metal Detection in Water and Beverages

The consumption of heavy metals such as mercury, arsenic, cadmium, and lead in food and drinking water poses severe threats to human health. Heavy metal toxicity impairs vital organs, including the brain, lungs, kidneys, and liver, leading to fatigue and degenerative diseases such as Parkinson's and Alzheimer's. Prolonged exposure can also heighten cancer risk (Jaishankar et al., 2014). Detecting and quantifying these contaminants in food can prevent these potential health problems and ensure food safety. Fluorescent sensors have gained considerable attention in this field over other techniques such as atomic absorption spectroscopy, inductively coupled plasma atomic emission spectrometry, and inductively coupled plasma mass spectrum in terms of their higher sensitivity, faster response, and ease of use. QDs and CDs (CDs) have been used as fluorescent tools to detect heavy metals in foods. CDs derived from various precursors of natural and synthetic origin have been explored, which showed promising results with very low detection limits. Nitrogen-doped CDs synthesized using citric acid and ethylene diamine have been used to detect mercury ions (Hg^{2+}) based on the mechanism of static quenching of the CDs by Hg^{2+} ions. The CDs were synthesized using one-step hydrothermal method and had an average size of 16.4 nm and a very high QY of 67.4%. The detection test was conducted at room temperature, where different concentrations of Hg^{2+} were added to the CDs-PBS solution (0.1 M, pH 7) and examined using fluorescence spectroscopy. The PL intensity at 445 nm gradually reduced upon increasing the concentration of the ion from 0 to 10 μM. The selectivity of the nanoparticles toward the Hg^{2+} ion was also examined using other metal ions, amino acids, and some additives used in beverages. Hg^{2+} caused a tremendous decrease in the PL intensity as compared to the other materials. Under optimal conditions, the CDs showed a good linearity between quenched PL intensity and concentration of Hg^{2+} in a range of 0.3 to 2.0 μM with a detection limit of 0.24 μM. The method was effectively applied to check Hg^{2+} in beverage samples, which indicated its high potency in detection (Fu et al., 2022). Similarly, a CD-magnetic nanocomposite (Fe_3O_4/CDs) was used for the detection of Hg^{2+} with the lowest detectable limit of 0.3 nM. The blue emission of Fe_3O_4/CDs could be quenched by Hg^{2+} and can be restored using saturated EDTA solution. The recovery of Hg^{2+} in Chagan Lake water, tap water, and beverages was measured at 96.5% to 108.8%, proving the Fe_3O_4/CDs sensing system's effectiveness in natural samples. Additionally, Fe_3O_4/CDs facilitate the efficient removal of Hg^{2+} from these samples, making them a highly promising tool for environmental monitoring (Xie et al., 2021).

CDs synthesized from cranberry beans using hydrothermal method have been studied for the detection of Ferric ions (Fe^{3+}). The CDs were spherical in shape, with an average size of 3.96 nm and a QY of approximately 10. 85%. The selectivity of the CDs toward metal ions was examined by adding 500 μM of various ions such as Ca^{2+}, Zn^{2+}, Na^{+}, K^{+}, Ni^{+}, Co^{2+}, Cd^{2+}, Pb^{2+}, Hg^{2+}, Sn^{2+}, Fe^{2+}, Cu^{2+}, and Fe^{3+} to 100 μL of CDs solution in ultrapure water (1:9). The fluorescence emission spectra of the samples were checked at 380 nm for quenching. Significant quenching can be observed in the case of Fe^{3+} compared to other metal ions, which showed negligible quenching. With increasing concentration of Fe^{3+}, there was an increase in the quenching of the emission peak with a limit of detection value of 9.55 μM. The sensor was used to detect the presence of Fe^{3+} in ground and tap water, which indicated its potency in measuring real-world samples (Zulfajri et al., 2019). Similarly, graphene quantum dots (GQDs) synthesized using watermelon rind waste were also used for the detection of Fe^{3+} ions. The nanoparticles were 6.46 nm sized and having a high QY of 37%. For Fe^{3+} quenching studies, 1 mL aliquot of GQD solution (1 ppm) was mixed with 1 mL of varying concentrations of Fe^{3+} in 0.1 M Tris-HCl buffer solution and analyzed using fluorescence spectroscopy. There was 43% decrease in the PL intensity with a concentration of 100 μM while negligible quenching was observed in the case of other metal ions. The limit of detection of this sensor was 0.28 μM (Rodwihok et al., 2022). A signal shut-off probe for the detection of Cr(IV) was developed from Si, N-codoped carbon quantum dots (Si, N-CQDs) without any biomolecules or labeling materials. The Si, N-CQDs exhibited a very high QY of 75% and were synthesized from diacetone acrylamide by hydrothermal method. The main quenching mechanism of Si, N-CQDs is attributed to Förster Resonance Energy Transfer (FRET). The Si, N-CQDs showed high sensitivity toward Cr(VI) within the linear range of 0–200 μM and the detection limit of 0.995 μM (Zhang et al., 2022b). In a work reported by Danil et al., $AgInS_2$/ZnS ternary QDs were integrated with super paramagnetic Fe_3O_4 nanoparticles embedded in porous calcium carbonate microspheres. The material was employed to detect toxic cobalt (Co^{2+}), nickel (Ni^{2+}), and lead (Pb^{2+}) ions in water. For testing, $CaCO_3$-Fe_3O_4-$AgInS_2$/ZnS microspheres were placed in aqueous solutions with varying concentrations of Co^{2+}, Ni^{2+}, and Pb^{2+} ions. Through PL quenching, the possibility of metal ions detection using the microsphere was obtained up to a concentration of Co^{2+}, Ni^{2+}, and Pb^{2+} as low as ≈0.01 ppm, ≈0.1 ppm, and ≈0.01 ppm, respectively (Kurshanov et al., 2020).

2.6.1.3 Detection of Insecticide and Antibiotic Residues in Food

Insecticides can cause neurotoxicity, endocrine disorders, mutagenicity, acute toxicity, and chronic health problems and increase the risk of cancer. Antibiotics can harm human health through the food chain and the environmental cycle and can lead to the development of antibiotic-resistant bacteria, making infections more challenging to treat and leading to serious health consequences (Ahmad et al., 2024; Sistani and Shekarchizadeh, 2023).

MIPs integrated with CDs have been widely explored for the detection of various types of insecticides in food industry. A combination of MIP and CDs polymerized on an MNP surface (Fe_3O_4-SiO_2-CQDs) was synthesized for the detection of 4-nitrophenol (4-NP) in food samples. 4-NPs are used in insecticides and herbicides and are highly carcinogenic and genotoxic even at lower concentrations. Silane-functionalized CDs were developed by hydrothermal method using trisodium citrate as the precursor and APTES as the functionalizing agent. The schematic representation of the synthesis and working is shown in Figure 2.3.

The nitro group of 4-NP can quench the fluorescence intensity of CDs, and this mechanism is used to detect 4-NP using the sensor. The sensor exhibited a decent linear range of 0.08–10 μmol/L and a low detection limit ($3\sigma/k$) of 23.45 nmol/L. Real-world samples of drinking water and fish were analyzed, showing 4-NP detection capacity, satisfactory detection recoveries, and low relative standard deviations (RSDs) (Zhu et al., 2021). Another study using a combination of MIP and CDs has been reported for rapid detection of lambda-cyhalothrin (LC), a pyrethroid insecticide used frequently in vegetables and tea plants. S-doped CDs were synthesized using a hydrothermal technique using sodium thiosulfate and sodium citrate solutions. The CDs were modified using room-temperature ionic liquid, 1-vinyl-3-butylimidazolium tetrafluoroborate [VBIm][BF_4], and imprinted with MIP. Upon addition of LC to the sensor, significant quenching was observed. The sensor was tested in green tea and spinach samples. The presence of LC was below the maximum residue limit in the samples. The evaluation results presented recoveries within a range of 98.90 to 116.93% from standard addition, signifying high accuracy and selectivity of the probe. The results were also validated by comparing with HPLC results (Zhang et al., 2019). QDs-doped covalent organic frameworks in MIP (QDs-doped COFs-MIP)

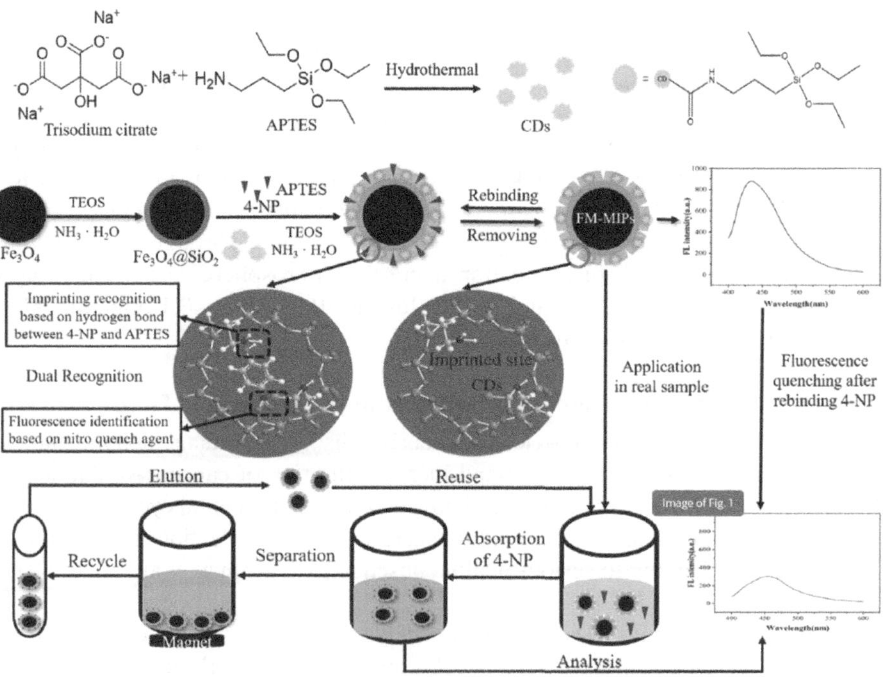

FIGURE 2.3 Schematic representation of synthetic and working of Fe_3O_4-SiO_2-CQDs for the detection of 4-nitrophenol [Adopted with permission from Zhang et al. (2020)] Copyright (2021) Elsevier

were synthesized for the detection of nereistoxin (NRT)-related insecticide in tap water. The NRT-related insecticides consist of cartap, thiocyclam, bensultap, monosultap, and thiosultap and are mostly used for pest management. The QDs surface was coated with silica, and an amino group was anchored on the surface as an antenna. The fluorescence of the probe is observed to be quenched after selectively interacting with thiosulfate due to photoinduced charge transfer. Upon testing with tap water, a detection limit of 1.60 μg/L and 86.5–106.5% recoveries of spiked tap water were achieved (Zhang et al., 2020).

Cerium and nitrogen co-doped carbon quantum dots (Ce–N–CDs) with high QY of 93.14% were developed for the detection of tetracyclines (TCs) residues, an antibiotic used to resist various bacterial infections in humans and animals. Extreme use of TCs may lead to the presence of its residues in meat and milk, which can cause severe threat to human health, leading to liver damage, allergies, and teeth yellowing. The doped CDs were synthesized using citric acid as the carbon source, ethylene diamine as nitrogen source, and cerium nitrate as dopants. Fluorescence quenching efficiency of the Ce–N–CDs exhibited a good linear relationship with tetracyclines member and doxycycline concentration within 0.4–35 μM with a detection limit of 0.25 μM. Upon testing with milk and pork samples, the recovery rate of standard samples ranged from 92 to 107.06%, with a reproducibility of 1.05–2.84% (Zhang and Fan, 2021). Similar studies were conducted for the detection of TCs in milk using nitrogen-doped graphene quantum dots and aptamer-conjugated graphene quantum dots (GQDs) and palladium nanoparticles (Pd NPs). The nitrogen-doped QDs showed a detection limit of 9.735×10^{-13} M, while aptamer conjugated QDs showed 45 ng/mL (Ahmed et al., 2021; Xie et al., 2022).

2.7 CHALLENGES OF CDS AND QDS IN THE FIELD OF FOOD QUALITY AND SAFETY

Both CDs and QDs have drawn a lot of interest in several areas, including food safety and quality. They do, however, provide several difficulties when applied in this domain. *Toxicity concerns*: The possible toxicity of QDs is one of the main issues they present. Heavy metals like cadmium, lead, or selenium are present in many QDs and can be toxic if consumed. It is essential to guarantee the safety of QDs used in applications connected to food. *Regulatory hurdles*: The use of CDs and QDs in food applications is currently governed by limited regulations. This lack of regulation can pose a serious challenge in the widespread adoption of the CDs and QDs for food safety (Zhang et al., 2022c). *Longevity and stability*: Both CDs and QDs must be able to withstand changes in temperature, pH, and light exposure in order to retain their stability and fluorescence properties. It can be difficult to guarantee these nanoparticles' stability in food matrices under such circumstances (Sun et al., 2024). *Standardization of detection methods*: For ensuring accuracy and dependability, standardized procedures for the identification and measurement of CDs and QDs in food samples must be developed. Achieving consistency across various detection platforms and techniques is a significant challenge that needs to be researched thoroughly for the generation of cost-effective technology (Tian et al., 2023). *Interaction with food matrices*: Food matrices may interact with CDs and

QDs, changing their optical characteristics or interfering with other components in the sample. Reliable detection and analysis depend on an understanding of these interactions and effective mitigation of them (Khan et al., 2023).

2.8 FUTURE SCOPE OF C DOTS/QDS IN FOOD QUALITY AND SAFETY ANALYSIS

QDs/C dots show potential in food quality and safety analysis, leveraging their unique optical properties for rapid and subtle detection. These particles can be engineered to fluoresce in response to specific pathogens, toxins, or contaminants, enabling real-time, on-site analysis with exceptional precision. This capability significantly reduces the time needed for food safety assessments, allowing for faster interventions and better-quality public health outcomes. QDs can monitor food quality by indicating spoilage through changes in fluorescence generated by biochemical shifts in the food. This can be incorporated into smart packaging, providing a visual indicator of freshness and reducing food waste by offering more precise information than traditional expiration dates.

QDs also hold ability in contending food deceit by embedding unique spectral fingerprints into products, ensuring their authenticity and traceability throughout the supply chain. They can also be used in sensors to detect environmental conditions during storage and transport, preserving optimal conditions and ensuring product integrity. As research addresses challenges like toxicity and scalability, QDs are poised to become essential tools in modern food safety protocols, providing better precision, dependability, and transparency in food quality assurance.

REFERENCES

Ahmad, M.F., Ahmad, F.A., Alsayegh, A.A., Zeyaullah, M., AlShahrani, A.M., Muzammil, K., Saati, A.A., Wahab, S., Elbendary, E.Y., Kambal, N., Abdelrahman, M.H., Hussain, S., 2024. Pesticides impacts on human health and the environment with their mechanisms of action and possible countermeasures. *Heliyon* 10, e29128. https://doi.org/10.1016/j.heliyon.2024.e29128

Ahmed, S.R., Kumar, S., Ortega, G.A., Srinivasan, S., Rajabzadeh, A.R., 2021. Target specific aptamer-induced self-assembly of fluorescent graphene quantum dots on palladium nanoparticles for sensitive detection of tetracycline in raw milk. *Food Chem.* 346, 128893. https://doi.org/10.1016/j.foodchem.2020.128893

Algarra, M., Campos, B.B., Radotić, K., Mutavdžić, D., Bandosz, T., Jiménez-Jiménez, J., Rodriguez-Castellón, E., Esteves da Silva, J.C.G., 2014. Luminescent carbon nanoparticles: Effects of chemical functionalization, and evaluation of Ag+ sensing properties. *J. Mater. Chem. A* 2, 8342–8351. https://doi.org/10.1039/C4TA00264D

Arreguin, S., Nelson, P., Padway, S., Shirazi, M., Pierpont, C., 2009. Dopamine complexes of iron in the etiology and pathogenesis of Parkinson's disease. *J. Inorg. Biochem.* 103, 87–93. https://doi.org/10.1016/j.jinorgbio.2008.09.007

Bernardin, A., Guyon, L., Delannoy, P., Bonnaffe, D., Texier, I., 2010. Copper-free click chemistry for highly luminescent quantum dot conjugates: Application to in vivo metabolic imaging. *Chem. Eur. J.*, 583–588.

Blanco-Canosa, J.B., Wu, M., Susumu, K., Petryayeva, E., Jennings, T.L., Dawson, P.E., Algar, W.R., Medintz, I.L., 2014. Recent progress in the bioconjugation of quantum dots. *Coord. Chem. Rev.* 263–264, 101–137. https://doi.org/10.1016/j.ccr.2013.08.030

Chem, J.M., 2011. Click-chemistry for nanoparticle-modification. *J. Mater. Chem.*, 16717–16734. https://doi.org/10.1039/c1jm11558h

Chen, S., Ouyang, W., Zhu, Y., He, L., Zou, L., Ao, X., Liu, S., 2022. Facile synthesis of N, S-doped carbon quantum dots from food waste as fluorescent probe for sensitive detection of thiamphenicol and its analogues in real food samples along with an application in bioimaging. *Foods* 11, 2414. https://doi.org/10.3390/foods11162414

Cheng, X., Lowe, S.B., Ciampi, S., Magenau, A., Gaus, K., Reece, P.J., Gooding, J.J., 2014. Versatile "click chemistry" approach to functionalizing silicon quantum dots: Applications toward fluorescent cellular imaging. *Langmuir* 30(18), 5209–5216. https://doi.org/10.1021/la500945f

Das, C., Sillanpää, M., Zaidi, S.A., Khan, M.A., Biswas, G., 2023. Current trends in carbon-based quantum dots development from solid wastes and their applications. *Environ. Sci. Pollut. Res.* 30, 45528–45554. https://doi.org/10.1007/s11356-023-25822-y

Deng, M., Wang, S., Liang, C., Shang, H., Jiang, S., 2016. A FRET fluorescent nanosensor based on carbon dots for ratiometric detection of Fe3+ in aqueous solution. *RSC Adv.* 6, 26936–26940. https://doi.org/10.1039/C6RA02679F

Deng, Y., Zhou, Y., Li, Q., Qian, J., 2021. One-step hydrothermal synthesis of nitrogen-doped carbon dots for high-sensitivity visual detection of nitrite and ascorbic acid. *Anal. Methods* 13, 3685–3692. https://doi.org/10.1039/D1AY00885D

Dong, W., Zhou, S., Dong, Y., Wang, J., Ge, X., Sui, L., 2015. The preparation of ethylenediamine-modified fluorescent carbon dots and their use in imaging of cells. *Luminescence* 30, 867–871. https://doi.org/10.1002/bio.2834

Dubois, F., Doris, E., Mioskowski, C., 2007. A versatile strategy for quantum dot ligand exchange. *Angew. Chem. Int. Ed.*, 482–483.

Fu, W.J., Peng, Z.X., Dai, Y., Yang, Y.F., Song, J.Y., Sun, W., Ding, B., Gu, H.W., Yin, X.L., 2022. Highly fluorescent N doped C-dots as sensor for selective detection of Hg2+ in beverages. *Spectrochim. Acta—Part A Mol. Biomol. Spectrosc.* 265, 120392. https://doi.org/10.1016/j.saa.2021.120392

Gao, K., Guo, Y., Niu, Q., Han, L., Zhou, L., Wang, L., 2018. Quaternary ammonium-functionalized carbon dots for sensitive and selective detection of 2,4,6-trinitrophenol in aqueous medium. *Sens. Actuators B Chem.* 262, 298–305. https://doi.org/10.1016/j.snb.2018.02.008

Gassensmith, J.J., Erne, P.M., Paxton, W.F., Frasconi, M., Donakowski, M.D., Stoddart, J.F., 2013. Patterned assembly of quantum dots onto surfaces modified with click patterned assembly of quantum dots onto surfaces modifi ed with click microcontact printing. https://doi.org/10.1002/adma.201202606

Gidwani, B., Sahu, V., Shukla, S.S., Pandey, R., Joshi, V., Jain, V.K., Vyas, A., 2021. Quantum dots: Prospectives, toxicity, advances and applications. *J. Drug Deliv. Sci. Technol.* 61, 102308. https://doi.org/10.1016/j.jddst.2020.102308

Guo, C.X., Zhao, D., Zhao, Q., Wang, P., Lu, X., 2014. Na+-functionalized carbon quantum dots: A new draw solute in forward osmosis for seawater desalination. *Chem. Commun.* 50, 7318–7321. https://doi.org/10.1039/c4cc01603c

Hezinger, A.F.E., Teßmar, J., Go, A., 2008. Polymer coating of quantum dots—A powerful tool toward diagnostics and sensorics. 68, 138–152. https://doi.org/10.1016/j.ejpb.2007.05.013

Hu, X., Li, Yanxiao, Xu, Y., Gan, Z., Zou, X., Shi, J., Huang, X., Li, Z., Li, Y., 2021. Green one-step synthesis of carbon quantum dots from orange peel for fluorescent detection of Escherichia coli in milk. *Food Chem.* 339, 127775. https://doi.org/10.1016/j.foodchem.2020.127775

Hu, X., Shi, J., Shi, Y., Zou, X., Arslan, M., Zhang, W., Huang, X., Li, Z., Xu, Y., 2019. Use of a smartphone for visual detection of melamine in milk based on Au@Carbon quantum dots nanocomposites. *Food Chem.* 272, 58–65. https://doi.org/10.1016/j.foodchem.2018.08.021

Huang, C.-C., Hung, Y.-S., Weng, Y.-M., Chen, W., Lai, Y.-S., 2019. Sustainable development of carbon nanodots technology: Natural products as a carbon source and applications to food safety. *Trends Food Sci. Technol.* 86, 144–152. https://doi.org/10.1016/j.tifs.2019.02.016

Jaishankar, M., Tseten, T., Anbalagan, N., Mathew, B.B., Beeregowda, K.N., 2014. Toxicity, mechanism and health effects of some heavy metals. *Interdiscip. Toxicol.* 7, 60–72. https://doi.org/10.2478/intox-2014-0009

Jennings, T.L., Becker-Catania, S.G., Triulzi, R.C., Tao, G., Scott, B., Sapsford, K.E., Spindel, S., Oh, E., Jain, V., Delehanty, J.B., Prasuhn, D.E., Boeneman, K., Algar, W.R., Medintz, I.L., 2011. Reactive semiconductor nanocrystals for chemoselective biolabeling and multiplexed analysis. *ACS Nano.* 5, 5579–5593. https://doi.org/10.1021/nn201050g

Jiang, Y., Wang, Z., Dai, Z., 2016. Preparation of silicon-carbon-based dots@dopamine and its application in intracellular Ag(+) detection and cell imaging. *ACS Appl. Mater. Interfaces* 8, 3644–3650. https://doi.org/10.1021/acsami.5b08089

Jin, H., Gui, R., Wang, Y., Sun, J., 2017. Carrot-derived carbon dots modified with polyethyleneimine and nile blue for ratiometric two-photon fluorescence turn-on sensing of sulfide anion in biological fluids. *Talanta* 169, 141–148. https://doi.org/10.1016/j.talanta.2017.03.083

Jin, M., Liu, X., Zhang, X., Wang, L., Bing, T., Zhang, N., Zhang, Y., Shangguan, D., 2018. Thiazole orange-modified carbon dots for ratiometric fluorescence detection of G-quadruplex and double-stranded DNA. *ACS Appl. Mater. Interfaces* 10, 25166–25173. https://doi.org/10.1021/acsami.8b07869

Khan, A., Ezati, P., Kim, J.T., Rhim, J.-W., 2023. Biocompatible carbon quantum dots for intelligent sensing in food safety applications: Opportunities and sustainability. *Mater. Today Sustain.* 21, 100306. https://doi.org/10.1016/j.mtsust.2022.100306

Kim, D., Okahara, S., Shimura, K., Nakayama, M., 2009. Layer-by-layer assembly of colloidal CdS and ZnS—CdS quantum dots and improvement of their photoluminescence properties. *Langmuir.* 7015–7018.

Kim, Y., Jang, G., Lee, T.S., 2015. New fluorescent metal-ion detection using a paper-based sensor strip containing tethered rhodamine carbon nanodots. *ACS Appl. Mater. Interfaces* 7, 15649–15657. https://doi.org/10.1021/acsami.5b04724

Kim, Y., Jang, G., Lee, T.S., 2017. Carbon nanodots functionalized with rhodamine and poly(ethylene glycol) for ratiometric sensing of Al ions in aqueous solution. *Sens. Actuators B Chem.* 249, 59–65. https://doi.org/10.1016/j.snb.2017.04.055

Kumar, J.V., Rhim, J.-W., 2024. Fluorescent carbon quantum dots for food contaminants detection applications. *J. Environ. Chem. Eng.* 12, 111999. https://doi.org/10.1016/j.jece.2024.111999

Kurshanov, D.A., Khavlyuk, P.D., Baranov, M.A., Dubavik, A., Rybin, A.V., Fedorov, A.V., Baranov, A.V., 2020. Magneto-fluorescent hybrid sensor caco3-fe3o4-agins2/zns for the detection of heavy metal ions in aqueous media. *Materials (Basel).* 13, 1–11. https://doi.org/10.3390/ma13194373

Lan, M., Zhang, J., Chui, Y.-S., Wang, P., Chen, X., Lee, C.-S., Kwong, H.-L., Zhang, W., 2014. Carbon nanoparticle-based ratiometric fluorescent sensor for detecting mercury ions in aqueous media and living cells. *ACS Appl. Mater. Interfaces* 6, 21270–21278. https://doi.org/10.1021/am5062568

Li, S., Guo, Z., Feng, R., Zhang, Y., Xue, W., Liu, Z., 2017. Hyperbranched polyglycerol conjugated fluorescent carbon dots with improved in vitro toxicity and red blood cell compatibility for bioimaging. *RSC Adv.* 7, 4975–4982. https://doi.org/10.1039/C6RA27159F

Lu, X., Dong, Y., Yang, K., Song, L., Wu, Y., Deng, J., 2023. Multifunctional fluorescent self-healing elastomers consisting of polydimethylsiloxane and carbon dots linked via a schiff base reaction. *ACS Appl. Polym. Mater.* 5, 10458–10468. https://doi.org/10.1021/acsapm.3c02322

Luo, X., Han, Y., Chen, X., Tang, W., Yue, T., Li, Z., 2020. Carbon dots derived fluorescent nanosensors as versatile tools for food quality and safety assessment: A review. *Trends Food Sci. Technol.* 95, 149–161. https://doi.org/10.1016/j.tifs.2019.11.017

Medintz, I.L., Farrell, D., Susumu, K., Trammell, S.A., Deschamps, J.R., Brunel, F.M., Dawson, P.E., Mattoussi, H., 2009. Multiplex charge-transfer interactions between quantum dots and peptide-bridged ruthenium complexes. *Anal. Chem.* 81, 4831–4839. https://doi.org/10.1021/ac900412j

Palanisamy, S., Ku, S., Chen, S.-M., 2013. Dopamine sensor based on a glassy carbon electrode modified with a reduced graphene oxide and palladium nanoparticles composite. *Microchim. Acta* 180, 1037–1042. https://doi.org/10.1007/s00604-013-1028-1

Pebdeni, A.B., Hosseini, M., Ganjali, M.R., 2020. Fluorescent turn-on aptasensor of *Staphylococcus aureus* based on the FRET between green carbon quantum dot and gold nanoparticle. *Food Anal. Methods.* 13, 2070–2079. https://doi.org/10.1007/s12161-020-01821-4

Pundi, A., Chang, C.J., 2022. Recent advances in synthesis, modification, characterization, and applications of carbon dots. *Polymers (Basel).* 14. https://doi.org/10.3390/polym14112153

Rao, H., Liu, W., Lu, Z., Wang, Yanying, Ge, H., Zou, P., Wang, X., He, H., Zeng, X., Wang, Y., 2016. Silica-coated carbon dots conjugated to CdTe quantum dots: A ratiometric fluorescent probe for copper(II). *Microchim. Acta* 183, 581–588. https://doi.org/10.1007/s00604-015-1682-6

Rodwihok, C., Tam, T. Van, Choi, W.M., Suwannakaew, M., Woo, S.W., Wongratanaphisan, D., Kim, H.S., 2022. Preparation and characterization of photoluminescent graphene quantum dots from watermelon rind waste for the detection of ferric ions and cellular bio-imaging applications. *Nanomaterials.* 12. https://doi.org/10.3390/nano12040702

Sapsford, K.E., Spindel, S., Jennings, T., Tao, G., Triulzi, R.C., Russ Algar, W., Medintz, I.L., 2011. Optimizing two-color semiconductor nanocrystal immunoassays in single well microtiter plate formats. *Sensors* 11, 7879–7891. https://doi.org/10.3390/s110807879

Schieber, C., Bestetti, A., Lim, J.P., Ryan, A.D., Nguyen, T., Eldridge, R., White, A.R., Gleeson, P.A., Donnelly, P.S., Williams, S.J., Mulvaney, P., 2012. Conjugation of transferrin to azide-modified CdSe / ZnS Core—shell quantum dots using cyclooctyne click chemistry. *Angew. Chem. Int. Ed.* 1–6. https://doi.org/10.1002/ange.201202876

Shi, Y., Pan, Y., Zhong, J., Yang, J., Zheng, J., Cheng, J., Song, R., Yi, C., 2015. Facile synthesis of gadolinium (III) chelates functionalized carbon quantum dots for fluorescence and magnetic resonance dual-modal bioimaging. *Carbon N. Y.* 93, 742–750. https://doi.org/10.1016/j.carbon.2015.05.100

Singh, I., Arora, R., Dhiman, H., Pahwa, R., 2018. Carbon quantum dots: Synthesis, characterization and biomedical applications karbon kuantum noktaları: Sentez, karakterizasyon ve biyomedikal. *Turkish J. Pharm. Sci.* 15, 219–230. https://doi.org/10.4274/tjps.63497

Sistani, S., Shekarchizadeh, H., 2023. Applications of quantum dots in the food industry. In: *Quantum Dots—Recent Advances, New Perspectives and Contemporary Applications*, pp. 1–25. https://doi.org/10.1016/j.colsurfa.2011.12.014

Sobiech, M., Luliński, P., Wieczorek, P.P., Marć, M., 2021. Quantum and carbon dots conjugated molecularly imprinted polymers as advanced nanomaterials for selective recognition of analytes in environmental, food and biomedical applications. *TrAC Trends Anal. Chem.* 142, 116306. https://doi.org/10.1016/j.trac.2021.116306

Speranskaya, E.S., Beloglazova, N. V, Lenain, P., Saeger, S. De, Wang, Z., Zhang, S., Hens, Z., Knopp, D., Niessner, R., Potapkin, D.V., Yu, I., 2014. Biosensors and bioelectronics polymer-coated fl uorescent CdSe-based quantum dots for application in immunoassay. *Biosens. Bioelectron.* 53, 225–231. https://doi.org/10.1016/j.bios.2013.09.045

Sun, Q., Fang, S., Fang, Y., Qian, Z., Feng, H., 2017. Fluorometric detection of cholesterol based on β-cyclodextrin functionalized carbon quantum dots via competitive host-guest recognition. *Talanta.* 167, 513–519. https://doi.org/10.1016/j.talanta.2017.02.060

Sun, Q.Y., Cheng, J.X., Qi, S.Y., Chen, J.F., Hu, D.Y., 2024. Advances in carbon quantum dot technology for food safety: Preparation, modification, applications and mechanisms. *Nano Express.* 5, 1–14. https://doi.org/10.1088/2632-959X/ad15a4

Tian, J., An, M., Zhao, X., Wang, Y., Hasan, M., 2023. Advances in fluorescent sensing carbon dots: An account of food analysis. *ACS Omega* 8, 9031–9039. https://doi.org/10.1021/acsomega.2c07986

V, R., Gujar, V., Pathan, H., Islam, S., Tawre, M., Pardesi, K., Santra, M.K., Ottoor, D., 2019. Bioimaging applications of carbon dots (C. dots) and its cystamine functionalization for the sensitive detection of Cr(VI) in aqueous samples. *J. Fluoresc.* 29, 1381–1392. https://doi.org/10.1007/s10895-019-02448-3

Wang, H., Ai, L., Song, Z., Nie, M., Xiao, J., Li, G., Lu, S., 2023. Surface modification functionalized carbon dots. *Chem. A Eur. J.* 29, e202302383. https://doi.org/10.1002/chem.202302383

Wang, Y., Hu, A., 2014. Carbon quantum dots: Synthesis, properties and applications. *J. Mater. Chem. C* 2, 6921–6939. https://doi.org/10.1039/c4tc00988f

Xiao, F., Miao, J., Tao, H.B., Hung, S., Wang, H., Yang, H. Bin, Chen, J., Chen, R., Liu, B., 2015. One-dimensional hybrid nanostructures for heterogeneous photocatalysis and photoelectrocatalysis. *Small.* 1–17. https://doi.org/10.1002/smll.201402420

Xie, H., Lu, Y., You, R., Qian, W., Lin, S., 2022. Green synthetic nitrogen-doped graphene quantum dot fluorescent probe for the highly sensitive and selective detection of tetracycline in food samples. *RSC Adv.* 12, 8160–8171. https://doi.org/10.1039/d2ra00337f

Xie, R., Qu, Y., Tang, M., Zhao, J., Chua, S., Li, T., Zhang, F., E.H. Wheatley, A., Chai, F., 2021. Carbon dots-magnetic nanocomposites for the detection and removal of Hg2+. *Food Chem.* 364, 130366. https://doi.org/10.1016/j.foodchem.2021.130366

Xing, L.I.U., Yang, L.U.O., 2014. Surface modifications technology of quantum dots based biosensors and their medical applications. *Chinese J. Anal. Chem.* 42, 1061–1069. https://doi.org/10.1016/S1872-2040(14)60753-2

Xiong, J., Zhang, H., Qin, L., Zhang, S., Cao, J., Jiang, H., 2022. Magnetic fluorescent quantum dots nanocomposites in food contaminants analysis: Current challenges and opportunities. *Int. J. Mol. Sci.* 23. https://doi.org/10.3390/ijms23084088

Yan, F., Jiang, Y., Sun, X., Bai, Z., Zhang, Y., Zhou, X., 2018. Surface modification and chemical functionalization of carbon dots: A review. *Mikrochim. Acta* 185, 424. https://doi.org/10.1007/s00604-018-2953-9

Ye, S.-L., Huang, J.-J., Luo, L., Fu, H.-J., Sun, Y.-M., Shen, Y.-D., Lei, H.-T., Xu, Z.-L., 2017. Preparation of carbon dots and their application in food analysis as signal probe. *Chinese J. Anal. Chem.* 45, 1571–1581. https://doi.org/10.1016/S1872-2040(17)61045-4

Zhang, D., Tang, J., Liu, H., 2019. Rapid determination of lambda-cyhalothrin using a fluorescent probe based on ionic-liquid-sensitized carbon dots coated with molecularly imprinted polymers. *Anal. Bioanal. Chem.* 411, 5309–5316. https://doi.org/10.1007/s00216-019-01912-0

Zhang, J., Guo, Y., Pan, G., Wang, P., Li, Y., Zhu, X., Zhang, C., 2020. Injectable drug-conjugated DNA hydrogel for local chemotherapy to prevent tumor recurrence. *ACS Appl. Mater. Interfaces* 12, 21441–21449. https://doi.org/10.1021/acsami.0c03360

Zhang, J., Jing, C., Wang, B., 2022. A label-free fluorescent sensor based on Si, N-codoped carbon quantum dots with enhanced sensitivity for the determination of Cr(VI). *Materials (Basel).* 15. https://doi.org/10.3390/ma15051733

Zhang, Q., Tian, F., Zhou, Q., Zhang, C., Tang, S., Jiang, L., Du, S., 2022. Targeted ginkgo kernel biomass precursor using eco-friendly synthesis of efficient carbon quantum dots for detection of trace nitrite ions and cell imaging. *Inorg. Chem. Commun.* 140, 109442. https://doi.org/10.1016/j.inoche.2022.109442

Zhang, W., Zhong, H., Zhao, P., Shen, A., Li, H., Liu, X., 2022. Carbon quantum dot fluorescent probes for food safety detection: Progress, opportunities and challenges. *Food Control.* 133, 108591. https://doi.org/10.1016/j.foodcont.2021.108591

Zhang, Y., Yuan, X., Jiang, W., Liu, H., 2020. Determination of nereistoxin-related insecticide via quantum-dots-doped covalent organic frameworks in a molecularly imprinted network. *Microchim. Acta.* 187. https://doi.org/10.1007/s00604-020-04435-z

Zhang, Z., Fan, Z., 2021. Application of cerium–nitrogen co-doped carbon quantum dots to the detection of tetracyclines residues and bioimaging. *Microchem. J.* 165, 106139. https://doi.org/10.1016/j.microc.2021.106139

Zhou, C., Zhao, hanqing, Xiong, Z.-L., Zhao, L., 2022. Preparation of carbon quantum dots from tobacco leaves and itsapplication for fluorescence detection of borax in flour and its products. *SSRN Electron. J.* https://doi.org/10.2139/ssrn.4004739

Zhu, W., Zhou, Y., Liu, S., Luo, M., Du, J., Fan, J., Xiong, H., Peng, H., 2021. A novel magnetic fluorescent molecularly imprinted sensor for highly selective and sensitive detection of 4-nitrophenol in food samples through a dual-recognition mechanism. *Food Chem.* 348, 129126. https://doi.org/10.1016/j.foodchem.2021.129126

Zulfajri, M., Gedda, G., Chang, C.J., Chang, Y.P., Huang, G.G., 2019. Cranberry beans derived carbon dots as a potential fluorescence sensor for selective detection of Fe3+ ions in aqueous solution. *ACS Omega.* 4, 15382–15392. https://doi.org/10.1021/acsomega.9b01333

Zuo, P., Lu, X., Sun, Z., Guo, Y., He, H., 2016. A review on syntheses, properties, characterization and bioanalytical applications of fluorescent carbon dots. *Microchim. Acta.* 183, 519–542. https://doi.org/10.1007/s00604-015-1705-3

3 Application of Biosensors in Food Analysis

Radhika Theagarajan, Tanya Kunwar, Saurabh Kharayat, Vivian Singh, Arun Kumar Gupta, and Poonam Mishra

3.1 INTRODUCTION

Food quality and safety surveillance is a significant global concern. Monitoring food safety and quality is essential to ensure that the food reaching consumers is safe to eat (Yumnam et al., 2022). Food is often maintained within specific quality standards to guarantee that its beneficial properties remain intact (Gupta et al., 2022). Additionally, food packaging can provide information about the contents, assuring consumers that the product has been tested and is free from harmful and undesirable substances. Manufacturers who fail to comply with international food standards risk facing severe legal repercussions and economic losses (Han et al., 2018). Common analytical techniques, such as high-performance liquid chromatography (HPLC) and gas chromatography, are widely accepted and regarded as the gold standards for monitoring food quality and safety (Gupta et al., 2022; Medhi et al., 2022).

However, traditional procedures are often time-consuming and cumbersome, requiring expensive instrumentation and skilled operators. In contrast, biosensors offer a highly effective alternative. Biosensors are analytical devices that convert biological responses into measurable electrical signals. Ideally, biosensors should exhibit high specificity, remain unaffected by physical conditions such as pH and temperature, and be reusable (Lv et al., 2018). The term 'biosensor' was first coined by Cammann and later defined by IUPAC (Kaur et al., 2018). A typical biosensor consists of a biological component, such as an enzyme, antibodies, organelles, or transducers, and its high selectivity arises from specific interactions (Senapati et al., 2022).

Once the biological material comes into contact with an appropriate transducer, the biochemical signal is converted into an electrical signal, which can be either continuous or discrete, and is subsequently quantified. The specific analyte or substance to be analyzed determines the biological component and transducer employed. Biosensors have a wide range of applications in food and agriculture, including food adulteration, preservation, processing, medicine, and more (Gupta et al., 2022). In the food industry, biosensors can assess nutrient levels, detect natural toxins and antinutrients, monitor food processing, and identify genetically modified organisms. They are capable of detecting pesticides, antibiotics, proteins, fatty acids, and vitamin

DOI: 10.1201/9781003482611-3

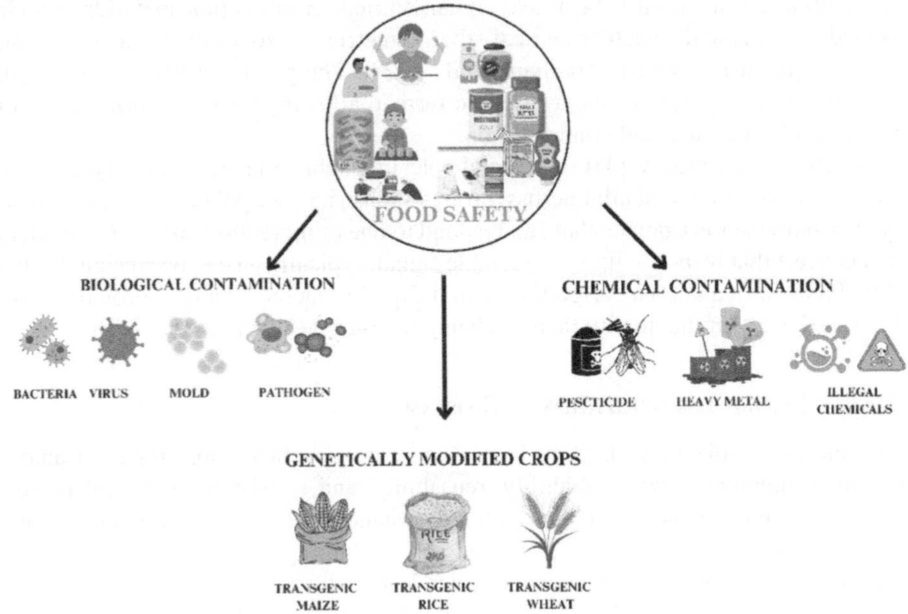

FIGURE 3.1 Factors impacting food safety.

B complex in food through enzymatic and immunogenic reactions (Bhavadharini et al., 2022).

Recent advancements in biosensor technology, including multiplexed detection, smartphone-based biosensors, and the integration of nanotechnology, have significantly improved the monitoring of food quality. Future innovations are anticipated to include enhanced sensing technologies, standardization, regulatory frameworks, and integration with the Internet of Things (IoT) to enable real-time monitoring (Figure 3.1) (Yang et al., 2022).

3.2 PRINCIPLES AND COMPONENTS OF BIOSENSORS

The fundamental concept of biosensor technology involves the use of a bioreceptor, transducer, and processor to convert a physiologically triggered recognition event such as that involving an enzyme or antibody into a detectable signal (Zheng & Zhang, 2021). The end product is a display that indicates both the concentration and presence of the target analyte.

3.2.1 Basic Components of a Biosensor

A bioreceptor is a biomolecule that identifies a target analyte. Various entities, including tissues, microbes, organelles, cells, enzymes, antibodies, nucleic acids, and biomimetics, can function as bioreceptors (Kuri et al., 2020). Transducer converts the recognition event into a measurable signal. Various transduction methods include optical, mechanical, electrochemical, thermometric, piezoelectric, magnetic, and micromechanical systems (Abraham et al., 2024). The processor and signal output system convert the signal into a readable format, allowing for the determination of the analyte's presence and concentration.

Analytical chemistry plays a crucial role in establishing food quality criteria, as quality control is essential across various industries and public services. A food quality biosensor is a device that can respond to one or more food quality parameters and convert that response into a detectable signal, typically an electrical signal. This signal may have a known correlation with the quality factor, or it may provide direct information about the quality factor(s) being measured (Shams et al., 2020).

3.2.2 Enzyme Immobilization in Biosensors

Enzyme immobilization plays a vital role in emergent biosensors for food analysis, as it augments enzyme stability, reusability, and consistent analytical performance. Current advancements have integrated nanomaterials such as metal–organic frameworks, magnetic nanoparticles, and self-healing hydrogels as immobilization matrices, addressing challenges like enzyme leaching and poor mass transfer. These innovations have covered the way for scalable biosensors capable of real-time detection of foodborne pathogens, pesticide residues, spoilage indicators, and allergens, significantly strengthening food safety and quality assurance systems (Tadesse and Liu, 2025; Abdel-Mageed, 2025).

Furthermore, to enhance the performance of biosensors, various enzyme immobilization methods have also been employed, including physical adsorption, that rely on Van der Waals forces between biological components and transducers (Hassan et al., 2021), as well as entrapment techniques, where enzymes are encapsulated within gels, polymers, or printing inks. Additionally, methods such as covalent binding, electrochemical polymerization, and photopolymerization are utilized, offering improved enzyme stability and reusability (Meng et al., 2023).

3.3 CLASSIFICATION OF BIOSENSORS BASED ON TRANSDUCERS

Biosensors can be broadly categorized into two main types:

a) Direct recognition sensors—measure biological interactions directly.
b) Indirect detection sensors—utilize secondary components such as enzymes or fluorescent tags (Herold et al., 2009).

Each category encompasses various types of transducers, including optical, electrochemical, and mechanical transducers. While antibodies are the most commonly used ligands, other molecules, such as aptamers (nucleic acid-binding molecules) and peptides, are also being investigated. The selection of a detector is influenced by several factors, including specificity, sensitivity, and the context of the application

(Barreiros et al., 2022). Currently, the six most significant types of transducers in use are:

1. Enzymatic Biosensors
2. Immunosensors
3. Electrochemical Transducers
4. Optical Transducers
5. Piezoelectricity
6. Colorimetric methods

3.3.1 Enzymatic Biosensors

The first glucose sensor was developed by Clark and Lyons in the early 1960s. It utilized the enzyme glucose oxidase (GOx) as a receptor due to its specificity for glucose (Pohanka, 2022). Enzymes enable the highly specific detection of their respective analytes, even in complex mixtures such as blood and fermentation broth. Examples of such analytes include lactate, glucose, glutamate, and glutamine, all of which are essential to the metabolism of living organisms. Glutamine and glucose support cell growth and function; lactate is produced by cells and serves as an indicator of their amino acid metabolism; and glutamate is an amino acid that cells utilize for energy. Each of these analytes requires a specific set of enzymes for detection (Alhajj et al., 2023). Among many other applications, continuous (online) monitoring of these parameters enables the development, optimization, and control of feeding strategies in cell cultures while simultaneously minimizing patient risks during intensive care and surgical procedures. Figure 3.2 shows the mechanism of an enzyme immunosensor.

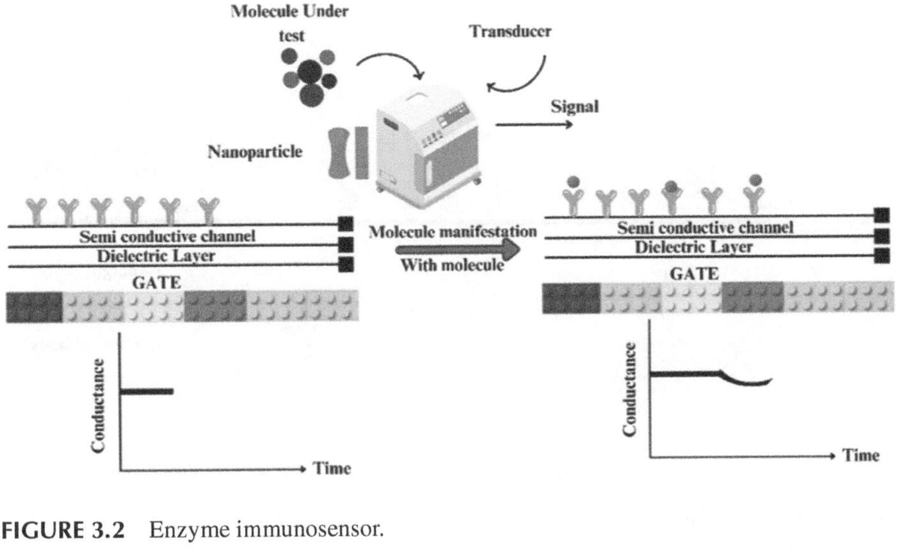

FIGURE 3.2 Enzyme immunosensor.

3.3.2 Immunosensors

The immune system's ability to process information with precision makes it an attractive subject for scientific research. Its primary function is to identify and classify every cell and molecule within the body, distinguishing between harmful and benign entities as part of its defense mechanism. When foreign substances, known as antigens, are detected, specific immune cells produce immunoglobulins (antibodies) that selectively bind to these antigens. This antigen-antibody interaction serves as the foundation for various applications, including the development of biosensors (Megha & Mohanan, 2021). An immunosensor, a type of biosensor that operates on immunological principles, employs an antibody as a bioreceptor to specifically recognize antigens at the molecular level, leading to the formation of a stable immuno-complex. By linking this immunoreaction to the surface of a transducer, such as a thermistor, it becomes possible to detect and quantify the presence of the immuno-complex. The transducer detects the reaction and converts it into an electrical signal, allowing for recording, analysis, and visualization (Prajapati et al., 2025). Transducers can be categorized based on the type of signal they produce (e.g., electrical) or the changes they undergo (e.g., mass) when antigen-antibody complexes form. The design of an immunosensor should ideally incorporate the following requirements: the ability to rapidly identify target antigens, the formation of immuno-complexes without the need for additional reagents, high repeatability in result generation, and ease of detecting the target in real samples (Harshavardhan et al., 2019). In immunosensors, the target analyte can be detected either directly by observing the formation of immune complexes or indirectly by employing a label, such as enzymes or gold nanoparticles, to monitor the binding event (Grieshaber et al., 2008). Figure 3.3 illustrates the fundamental elements of electrochemical immunosensors with enzyme labels where A panel shows principle of detecting antigens and B shows detecting antibodies (Florea et al., 2019)

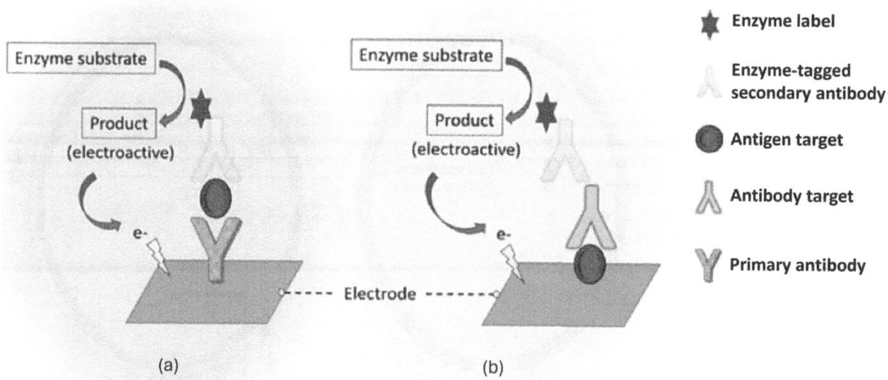

FIGURE 3.3 Fundamental elements of electrochemical immunosensors with enzyme labels and detection process of (a) antigens and (b) antibodies.

3.3.3 Electrochemical Transducers

Biosensors employ various electrochemical techniques, including amperometry, potentiometry, impedance spectroscopy, and conductivity measurements (Lakard, 2020). Among these, amperometry is the most commonly used method, followed by potentiometry. Electrochemical biosensors offer significant advantages, such as miniaturization, cost-effectiveness, high sensitivity, excellent selectivity, rapid detection, and low detection limits, all while requiring minimal sample volumes. These characteristics render them particularly suitable for on-site monitoring of food products.

Depending on the analytes of interest, the selection and optimization of electrode materials, surface modification techniques, electrochemical methodologies, and detection limits are crucial. The findings underscore the potential of electrochemical biosensors to function as sensitive, rapid, precise, and compact sensor platforms for ensuring food safety (Curulli, 2021). Figure 3.4 illustrates the transducer using composite materials to detect chemical compounds (Alhadeff and Bojorge, 2011)

3.3.4 Optical Biosensor

Compared to traditional analytical methods, optical biosensors offer several advantages, including direct, real-time, label-free detection of a wide range of biological and chemical molecules. These advantages encompass cost-effectiveness, compact size, excellent sensitivity, and specificity. New optical biosensors are developed using a variety of innovative concepts and highly interdisciplinary techniques, including chemistry, molecular biology, microelectromechanical systems (MEMS), micro/nanotechnologies, microelectronics, and biotechnology (Yao et al., 2025). Over the

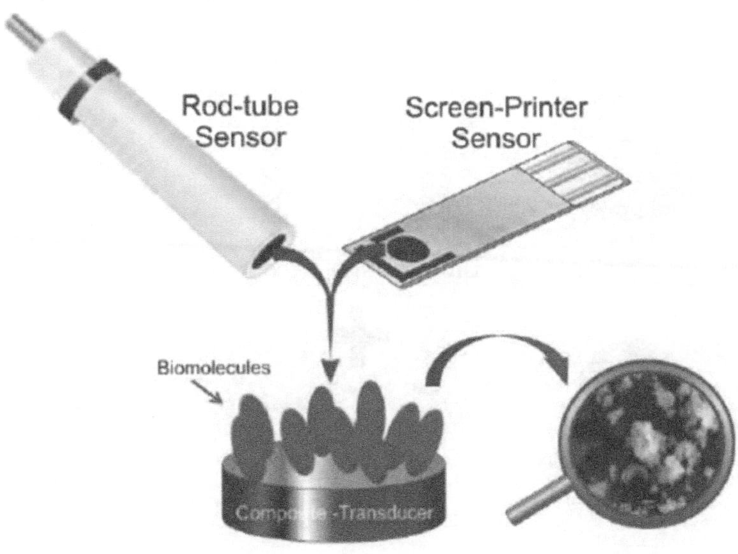

FIGURE 3.4 Electrochemical transducer for analyte detection.

past decade, there has been an exponential increase in research and development related to optical biosensors. The biotechnology sector, environmental applications, and healthcare have been the primary focus of this research and development. There are numerous potential applications for biosensors in biotechnology, medicine, and environmental monitoring, each with specific requirements regarding analyte concentration, output precision, sample concentration, time required for analysis, reusability of the biosensor, and cleaning needs of the system (Haleem et al., 2021).

3.3.5 Piezoelectric Transducers

Piezoelectricity is employed in mass-sensitive transducers, with quartz being the most widely used material. In piezoelectric biosensors, quartz crystals are cut at specific angles (e.g., AT-cut) to optimize their piezoelectric properties (Luong and Guilbault, 2019). The resulting transducer is commonly known as a quartz crystal microbalance (QCM). Piezoelectric biosensors primarily utilize two types of acoustic waveforms: surface acoustic waves (SAWs) and bulk acoustic waves (BAWs). In SAW devices, the placement of two electrodes on one side of the crystal generates a standing surface wave. Although crystals can theoretically achieve very high sensitivity at frequencies ranging from 30 to 200 MHz, BAW devices are more commonly employed in biosensors due to practical limitations in implementing SAW systems at these high frequencies (Leonard et al., 2003). Figure 3.5 presents a schematic overview of the application of optical biosensors for detection (Herrera-Domínguez et al., 2023)

3.3.6 Colorimetric Sensors

Colorimetric sensors utilize visually perceptible color modifications often from pH-responsive dyes or nanoparticle aggregation to indicate the occurrence of food

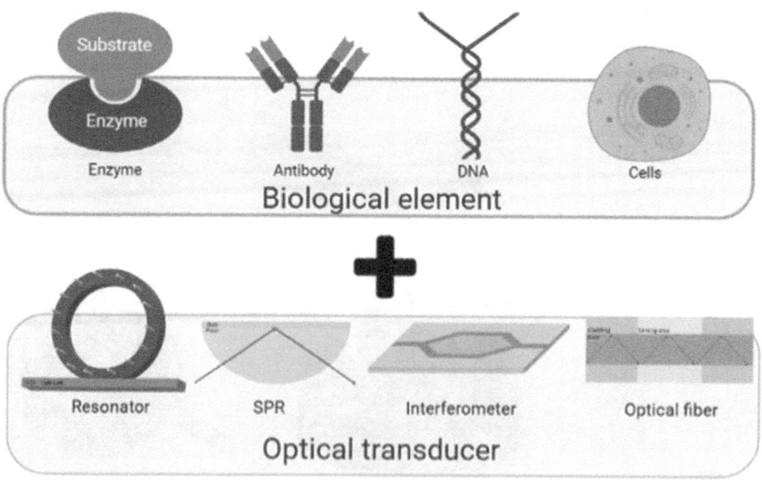

FIGURE 3.5 Mechanism of an optical biosensor.

spoilages, such as volatile amines or bacterial metabolites. In general, this technique provides rapid, cost-effective, and instrument-free detection, more appropriate for on-site food quality monitoring by untrained users. Recent innovations include incorporating natural dyes like anthocyanins into film-based sensors integrated with smartphone-based machine learning for real-time quality tracking, achieving high accuracy under ambient conditions (Kim and Yoo, 2022; Doğan et al., 2024).

Moreover, in food quality applications, colorimetric biosensors are often classified based on their physical format into two key types as described in the next two subsections.

3.3.6.1 Solution-Based Assays

Recently, solid substrate methods have gained popularity for analyzing antioxidants. These solid substrates can be fabricated from various materials, among which paper is one of the most commonly used due to its low cost, portability, and ease of functionalization (Busa et al., 2016). Colorimetric assays are simple straightforward effective tools for measuring antioxidant activity. They are particularly beneficial in fields such as food safety and environmental monitoring due to their ease of use and the advantages offered by substrates and nanoparticle-based materials.

3.3.6.2 Solid Substrate-Based Assays

In recent years, solid substrate methods have gained popularity for analyzing antioxidants. Solid substrates can be fabricated from various materials, with paper being one of the most commonly used due to its affordability, portability, and ease of modification (Busa et al., 2016). Table 3.1 highlights advancements in electrochemical biosensors for detecting antioxidants in food samples and the use of SPR techniques for identifying antibiotic residues.

TABLE 3.1
Development in Electrochemical Biosensors for Determination of Antioxidants in Food Samples and SPR Detection in Antibiotic Residues

Sample	Analyte/Strategy	Biosensor/ Bio-receptor	Limit of Detection (LOD)	Reference
Chicken muscle	Amantadine	Antibody	0.62 μgkg^{-1}	Pan et al., (2022)
Tetracycline	Tetracycline	Aptamer	0.0069 μgkg^{-1}	Wang, Dong, et al. (2018)
Skimmed milk	Glycopeptide antibiotics	Nano molecularly imprinted polymer (MIP)	4.1 $ngmL^{-1}$	Altintas (2018)
Peanut oil, potato chips, and cookies	HRP immobilization on the surface of Au-Pt nanotube/Au graphene-modified carbon electrode	Linear-sweep voltammetry (LSV)	0.046 mgL^{-1}	Wu et al. (2016)

(Continued)

TABLE 3.1 (*Continued*)
Development in Electrochemical Biosensors for Determination of Antioxidants in Food Samples and SPR Detection in Antibiotic Residues

Sample	Analyte/Strategy	Biosensor/ Bio-receptor	Limit of Detection (LOD)	Reference
Blueberry syrup, wine	Laccase immobilization onto AuNPs/graphene nanoplatelets modified screen-printed carbon electrode (AuNPs/ GNPISPCE)	- -	1.5 μM	Zrinski et al. (2020)
Chestnut shell extract	Laccase immobilization on carbon surfaces through potentiostatic deposition of a polydopamine adhesive layer	-	0.29 μM	Almeida et al. (2019)
Food antioxidants	GCE modified with ZIF-9-CoO_x/AgNPs	-	0.0564 pM	Jiang et al. (2019)

3.4 APPLICATIONS OF BIOSENSORS IN FRUITS, VEGETABLES, AND FOOD SAFETY MONITORING

Hazardous compounds and pathogens in food and water pose a serious threat to animal production and the food supply, with significant economic implications. Consuming food products contaminated with bacteria, viruses, parasites, or chemicals that cause foodborne illnesses is a major burden on public health. This issue presents a substantial obstacle to socioeconomic development worldwide (Kuiken et al., 2005). Biosensors represent a novel approach for analyzing food samples and addressing significant challenges. They provide faster and more accurate real-time monitoring compared to standard microbiological methods (Mishra et al., 2018). Additionally, biosensors are particularly effective in detecting bacterial contamination on food and contact surfaces (Ali et al., 2020). Table 3.2 summarizes various research studies on the applications of biosensors in the food sector.

3.4.1 Biosensors in Fruits and Vegetables

With the food industry increasingly transitioning from conventional batch operations to continuous processing, there is a critical need for real-time and robust quality monitoring technologies during both processing and post-production stages. Among the emerging tools, biosensors, including nanosensors, have acquired attention for their ability to provide rapid, sensitive, and selective detection of target analytes throughout the food supply chain.

TABLE 3.2
Various Studies on Biosensors' Applications in Food

Media/Food	Detection/Determination	References
Fish	Determination of K-value and Trimethylamine	Mitsubayashi et al. (2004)
Milk	Presence of urea	Meshram et al. (2018)
	Determination of lactose concentration	
Wine	Detection of iron (III)	Camara-Martos et al. (2016)
	Determination of histamine	Stojanović et al. (2016)
Fresh fruits and Vegetable	Fruit quality	Jawaheer et al. (2003)
	Concentration of total soluble solids	Singh et al. (1996)
	Strawberry quality and fruit composition	Giné Bordonaba and Terry (2009)
Fruit juices	Determination of ascorbic acid	Pisoschi et al. (2010)
Water	Heavy metals detection	Odobašić et al. (2019)

3.4.1.1 Detection of Heavy Metals

Biosensors have emerged as powerful analytical tools for detecting contaminants, particularly heavy metals, in both raw and processed foods (Altug et al., 2022). Monitoring toxic elements is essential due to the stringent food safety standards established by agencies such as the World Health Organization (WHO), which enforces maximum residue limits (MRLs) (Balkrishna et al., 2024). Heavy metal contamination in agricultural produce often arises from industrial emissions, urban runoff, and excessive use of agrochemicals, leading to accumulation in soil, water, and crops (Chen et al., 2020). Although trace elements such as copper (Cu), zinc (Zn), and nickel (Ni) are essential micronutrients, their excessive accumulation can pose significant health risks (Liang et al., 2019).

Enzyme-based biosensors, particularly those utilizing organophosphorus hydrolase, acetylcholinesterase (AChE), alkaline phosphatase, tyrosinase, and laccase, are commonly employed for the detection of heavy metal contaminants (Mukherjee et al., 2021). For instance, laccase-based sensors that are cross-linked with gold electrodes have been effectively used to detect carbamate and formethanate hydrochloride residues in mangoes and grapes (Balkrishna et al., 2024).

Biosensors offer a highly sensitive and efficient elucidation for monitoring heavy metal contamination in food systems, addressing critical food safety challenges. Their continued advancement and integration into regulatory frameworks will be crucial for ensuring global food security and public health protection.

3.4.1.2 Assessment of Fruit Quality

Recent advances in biosensor design now permit targeted quantification of phytochemical constituent's key indicators of nutritional and organoleptic quality in fruits. Sousa et al. (2020) constructed a photoelectrochemical biosensor using a fluorine-doped tin oxide (FTO) electrode modified with a cobalt (II)–polypyrrole complex (CPPI)-sensitized titanium dioxide (TiO_2) and cadmium sulfide (CdS). The sensor demonstrated high sensitivity to naringin, a flavonoid recognized as a marker of bitterness in citrus juices, and it was found to be 97.8–99.6% recoverable.

Following the same experimental paradigm, Das and Sahu (2023) developed a capacitive biosensor utilizing cerium oxide nanoparticles (CeO_2NPs) for the detection of limonin in Citrus limetta. The sensor, prepared through drop-casting, spin-coating, or electrodeposition, exhibited linear capacitive behavior and was analyzed using scanning electron microscopy (SEM), X-ray diffraction (XRD), and Fourier-transform infrared spectroscopy (FTIR). The device demonstrated potential as a cost-effective, rapid, and scalable solution for citrus quality control.

Recent empirical advances indicate the inclusion of electrochemical sensing devices that are applicable for both the detection and reduction of naringin in citrus, thereby enhancing their taste. An interdigitated electrode (IDE) biosensor has been employed to monitor the ripening of pomelo by detecting naringin, demonstrating high selectivity and a sensitivity of 0.008 A for 10 ppb, as well as reusability for three to four cycles. These devices facilitate real-time monitoring, shelf-life management, and quality assessment of citrus products during processing.

3.4.2 Detection of Food Spoilage

The potential for an instantaneous food spoilage index has been explored in food processing, focusing on the accurate detection of spoilage bacteria and metabolites through the capabilities of biosensors (Qu et al., 2007). Due to their remarkable sensitivity and rapid response times, these biosensors serve as effective alternatives to traditional microbiological methods for detecting pathogenic spoilage and microorganisms such as Salmonella spp. and Escherichia coli (Gallo et al., 2020).

Antibody-based detection platforms have become increasingly common methodologies for the general detection of these pathogens, and numerous biosensors have been developed to facilitate the specific binding and identification of these organisms (Ali et al., 2020).

The Food Safety and Standards Authority of India (FSSAI) has been promoting the adoption of rapid and noninvasive technologies for detecting food adulteration, particularly through the use of sensors based on surface plasmon resonance (SPR), which effectively bind to contaminants (Ravindran et al., 2023). Recent technological innovations, including molecularly imprinted polymer (MIP)-based sensors and DNA biosensors, have shown exceptional specificity in differentiating unadulterated products from adulterated counterparts, offering a reliable method for quality assurance.

3.4.3 Immunosensors in Food Analysis

This section explores recent advancements in immunosensor technologies for food analysis, focusing on recent methods for developing and assembling these sensors. It examines examples of food matrices, electrochemical, SPR, and QCM approaches. Immunosensors detect small molecular weight chemicals or high molecular weight molecules using sandwich or competitive tests. Competitive immunosensors can be developed through immobilized antibodies reacting with free antigens or immobilized antigens competing for labeled free antibodies (Ricci et al., 2007).

Among the various types, electrochemical immunosensors have shown great promise in transforming analytical procedures in the food sector due to their specificity, speed, and affordability. Particular focus is placed on recently developed techniques for altering the surfaces of working electrodes. Novel nanomaterials including carbon nanotubes, graphene, and metallic nanoparticles that are frequently combined with polymers or one another have been the subject of numerous advancements. In general, using these materials has resulted in higher sensor sensitivity. To determine whether immunosensors offer solutions in real-world applications, their use in the detection of various infections in actual samples is next examined (Duffy and Moore, 2017).

Recently, the rapid development of biosensors has been fueled by the growing pollution of ecosystems by both biological and non-biological toxins, opening up a new field of study in the analytical sciences. This chapter examine how immunosensors are used in agriculture, water pollution, environmental sciences, human health, and food safety. Pathogens, antigens, antibiotics, pesticides, poisons, allergens, hormones, phytohormones, and cancer indicators are all analyzed. Electrochemical, microgravimetric, optical, and thermometric immunosensors are all types of immunosensors. Improved sensitivity, selectivity, speed, and cost-effectiveness are some benefits of immunosensors. Specifically, when it comes to recognizing and analyzing different chemicals, antibody-based biosensors have done better than conventional techniques. Compounds of nanomolar to picomolar concentrations can be detected by immunosensors (Kumar et al., 2024).

Moreover, immunosensors, a subclass of biosensors that exploit antibody-antigen interactions, offer exceptional selectivity for specific food targets. A comparative overview of recent immunosensor-based studies is provided in Table 3.3.

In conclusion, immunosensors are a powerful form of analysis that has great potential in food analysis. There is ongoing development, especially in regard to the

TABLE 3.3
Comparison of Recent Studies Using Immunosensors

Target Component	Analyte	Type of Immunosensor	Transducer	LOD	Application Matrix	Reference (Year)
Salmonella	Bacterial pathogen	Label-free immunosensor	Electrochemical impedance	3.65×10^2– 3.65×10^8 CFU mL^{-1}	Milk	Li et al. (2024)
Salmonella Enteritidis	Bacterial pathogen	Conductometric immunosensor	PANI/ZnO nanocomposite	6.44 CFU/ mL	Skim milk	Mutlaq et al. (2023)
Pathogenic bacteria	*Salmonella typhimurium*	Label-free immunosensor	Electrochemical impedance	3 ± 7 CFU mL^{-1}	Chicken meat	Soares et al. (2020)
Allergens	Peanut protein (Ara h1)	Sandwich-type immunosensor	Bare screen-printed carbon electrodes (SPCE)	3.5 ng·mL^{-1}	Food products	Freitas et al. (2021)
Veterinary drug residues	Chloramphenicol	piezoelectric immunosensor	-	0.2 ng/mL	Chicken meat, milk, egg, honey	Karaseva and Ermolaeva (2012)

incorporation of nanomaterials and to multiplexing technologies, which will only improve food safety monitoring and quality assurance with greater specificity and efficiency.

3.5 APPLICATIONS OF BIOSENSORS IN FOOD COMPONENT ANALYSIS

3.5.1 Determination of Sugars, Fats, and Proteins

The exact assessment of sugars, fat, and protein is necessary for nutritional and functional food evaluation. Of the components added to food, sugars and artificial sweeteners are added, particularly to sugary drinks, and are associated with health concerns. In this context, a study employing an amperometric biosensor integrated bio-chip-D. Specifically, at a concentration of 250 mM, the study discovered that aspartame, fructose, glucose, and sucrose reduced dissolved oxygen levels by 14.24%, 18.02%, 16.59%, and 20.45%, respectively. This study demonstrated the biosensor's sensitive detection capability for sugars in beverage samples (Umar et al., 2023).

In addition to amperometric detection, colorimetric methods have also been explored. One notable example is the application of colorimetric phenol-sulfuric acid sugar reaction that has been used to produce a biosensor for the detection of DNA and RNA. The target molecule is captured by hybridization after DNA is immobilized in a 96-well microplate. A yellow-orange hue is produced by the reaction in proportion to the desired concentration. This biosensor exhibited a linear response over a broad concentration range, along with excellent sensitivity and selectivity. Additionally, it can identify DNA intercalator chemicals that prevent hybridization. As a result, such biosensors hold promise for applications in drug development, genetic analysis, virus diagnostics, and microbial contamination monitoring (El Aamri et al., 2023).

Moreover, the use of biosensors to detect food peroxidation has advanced significantly. These biosensors offer quick and affordable detection of peroxidation-related analytes by integrating aptamers, metal–organic frameworks, nanomaterials, and enzymes. Therefore, the creation of such novel biosensors is essential, especially for use in food applications (Daci et al., 2024). Another critical application lies in the detection of milk adulteration, a rising global concern that is being addressed with biosensors. Numerous approaches, such as spectroscopic, immunological, and chromatographic procedures, have been developed to identify and detect milk adulterants. Real-time detection is achieved by the employment of electronic detecting equipment such as electronic tongues and noses. These techniques seek to counteract the negative impacts of milk adulteration, which may be detrimental to the health of consumers. Consequently, exploring biosensor-based approaches offers a more efficient and consumer-safe strategy for identifying common milk adulterants (Nagraik et al., 2021).

3.5.2 Analysis of Vitamins and Antioxidants

The assessment of antioxidants is crucial for evaluating the inhibitory effects of beneficial compounds present in the diet. These compounds, which are present in foods produced from plants, have anti-inflammatory, anti-cancer, and antihistaminic effects. Because of their great sensitivity, quick detection time, and user-friendliness, sensors and biosensors are common instruments for this investigation. Accordingly, the development of optical and electrochemical sensors/biosensors for food antioxidant monitoring is well assessed in recent studies (Nejadmansouri et al., 2021).

Antioxidants are indispensable for protecting against oxidative damage triggered by reactive chemical substances. In a biosensor-based study examining the antioxidant activity of teas and vitamin-C-fortified effervescent drinks, it was found that Prussian blue had the greatest mediation effect. The biosensor can identify fake 'antioxidants' or medications that contain less ascorbic acid than stated and demonstrated stability for seven months (Ribeiro et al., 2021).

Further advancing electrochemical methods, another study developed a technique using a pyrolytic graphite sheet to establish an electrochemical technique for detecting Vitamin C concentration. With a sensitivity of 11.7 $\mu Acm^2/mgL^{-1}$ for lower concentrations and 134 $\mu Acm^2/mgL^{-1}$ for higher values, the approach demonstrated a linear response over a broad concentration range. Arugula's vitamin C content and antioxidant activity were considerably raised by the addition of five different kinds of 'green' bio-waste (Škugor Rončević et al., 2022).

Vitamins are necessary micronutrients that are present in trace amounts, and electrochemical sensing techniques have become a practical and affordable way to detect them. When integrated with microfluidic systems, electrochemical microfluidic devices, these methods provide cost-effectiveness, speed, and sensitivity. In order to build portable, compact, and effective analytical systems, these devices are frequently combined with microfluidics. The creation of these instruments addresses issues including selectivity, sensitivity, interference, matrix effects, and calibration and has wide-ranging applications in a variety of scientific, medical, and environmental fields (Draz et al., 2024).

3.5.3 Fermentation Monitoring

Fermentation is a complex biochemical procedure, which requires significant control of metabolites and nutrients to encourage optimal microbial growth and product yield. In general, sensors are important for firm real-time control of fermentation values. In the intricate process of fermentation, which necessitates exact regulation of metabolites and nutrients for maximum growth and product yields, sensors are essential. However, the restricted availability of standard online sensors for bio-reactor monitoring necessitates manual control measures and resource-intensive analysis. To address these challenges, current research is concentrated on creating better online monitoring techniques such as spectroscopic sensors in conjunction with chemometrics, biosensors, and soft sensors (Gargalo et al., 2022).

The application of biosensors in food analysis has drawn a lot of attention in recent years. These analytical tools offer extremely sensitive, cost-effective, and selective cell, chemical or biochemical molecule identification that is pertinent to the food sector. Specifically, these sensors can be used to monitor and regulate biotechnological processes for food transformation and production. With an emphasis on those that may find utility in fermentation production, this review highlights recent developments in electrochemical biosensors with possible applications in food bioprocess technology (Villalonga et al., 2022).

An illustrative example of biosensor application is the use of zymase activity in *Saccharomyces cerevisiae* as a microbiological sensor for environmental monitoring, ecotoxicological evaluation, and water quality assessment. By providing reliable, repeatable, and reasonably priced data, this approach makes it possible to monitor and characterize bio-photonic phenomena, nonchemical water treatments, and environmental stressors from an ecotoxicological perspective (Kernbach et al., 2022). Moreover, the transcription factor (TF)-based whole-cell biosensor approach was created in this work to detect L- and D-lactic acids. It was cross-validated using enzymatic techniques and widely utilized high-performance LC. The use of biosensors for fermentation enhancement and lactic acid determination proved their usefulness as an effective analytical instrument. We investigated the biosynthesis of optically pure L-lactic acid, D-lactic acid, or a combination of both from organic-rich residual fraction (ORRF), a waste product of the production of glucose syrup from wheat starch, by *Lacticasei bacillus paracasei* subsp. *paracasei*, *Lactobacillus delbrueckii* subsp. lactis, and *Lactobacillus amylovorus*. Without further preprocessing, lactic acid could be produced through fermentation of this complicated industrial waste, yielding 0.5 to 0.9 Cmol/Cmol glucose (Augustiniene et al., 2023).

3.5.4 Spoilage Detection and Shelf-Life Estimation

The increasing mass production of agricultural goods, evolving dietary habits, and growing concerns over food safety significantly influence global food security. In order to ensure food safety, biosensors have become a viable solution for detecting pathogens and biomarkers. Due to their ability to provide analytical precision even in complicated food matrices, electrochemical-based biosensors have significantly improved the quantitative detection and screening of food pollutants. For meat, fish, milk, dairy products, and alcoholic beverages, these biosensors are very helpful (Majer-Baranyi et al., 2023).

In response to the persistent challenge of food spoilage, biosensor technologies in the food industry are quickly developing to identify food deterioration, a serious problem that can result in health issues. By identifying volatile organic gases, microbes, and enzymes, these sensors preserve consumer trust and guarantee food safety. The application of biosensors in the food business has addressed issues and opportunities in the field by advancing sensing platforms, identification components, and signal transduction approaches (Kumar et al., 2024a).

Complementing conventional biosensors, nanosensor technology has also gained traction for shelf-life monitoring. Gold, silver, titanium dioxide, and zinc oxide are examples of inorganic nanomaterials that show promise for creating food shelf-life sensors. In addition to biodegradable dietary ingredients like vitamins and antioxidants, these sensors offer quicker and more precise detection of bacteria, poisons, and adulterants. Consequently, the integration of nanotechnology into food monitoring systems enhances both safety assurance and freshness evaluation, supporting sustainable food quality management (Ghosh et al., 2022).

3.6 CHALLENGES AND LIMITATIONS IN FOOD BIOSENSING

Even with major innovations, there are still a number of technical and practical hurdles for biosensors in food analysis. One of the main hurdles is striking the right balance between sensitivity and selectivity when dealing with the intricate combinations found in food, not to mention the disruptions caused by natural food elements and the ongoing need for the sensors to maintain their accuracy and consistency over time (Kulkarni et al., 2022).

Moreover, for instance, the presence of proteins in milk may introduce signal interference, thereby compromising the accuracy of sensor measurements—a type of matrix interference (Péter et al., 2022). Furthermore, enzyme- and DNA-based biosensors frequently exhibit limited operational stability and heightened sensitivity to environmental fluctuations. These constraints underscore the need for real-time monitoring strategies and the development of matrix-compatible platforms to ensure reliable deployment in diverse settings (Neethirajan et al., 2018; Feng et al., 2022; Liu et al., 2020a, 2020b). Figure 3.6 illustrates the challenges and limitations associated with the adoption of biosensors in food quality monitoring.

Current research challenges emphasize the critical requirements for biosensor miniaturization and system integration with wireless technologies for real-time monitoring, raising concerns of feasibility, cost efficiency, scale, and security of data (Patel et al., 2024). In addition to other food-specific innovative calibrators, the absence of uniform protocols for different food matrices impacts biosensor calibration and reliability suppressing reproducibility across biosensors. Moreover, rising contaminants such as microplastics and new food adulterants are difficult to detect because of their low concentration and complex interactions within food systems (Wang et al., 2024). Environmental conditions further affect sensors such as temperature and pH, requiring baseline-controlled standards (Sim et al., 2022). Regulatory compliance and measurement consistency present significant barriers to widespread biosensor adoption, as current limitations in standardization frameworks hinder uniform implementation (Nath and Mahata, 2024). Addressing these challenges

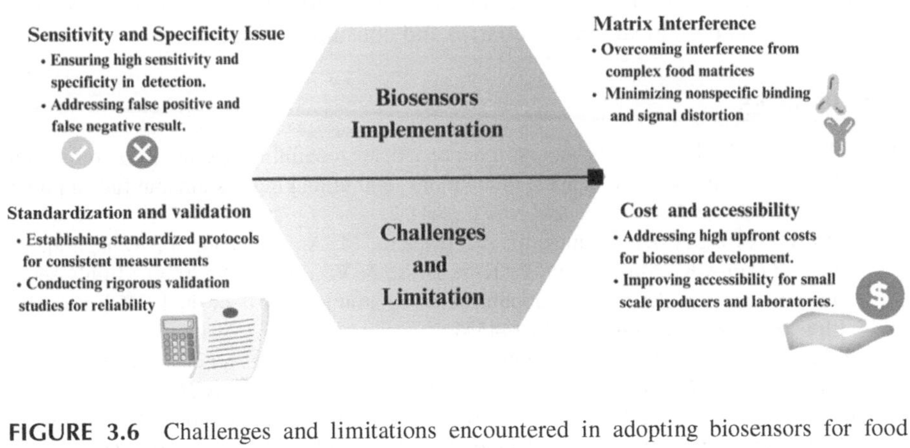

FIGURE 3.6 Challenges and limitations encountered in adopting biosensors for food quality.

requires optimization of biosensor structural design, transduction methodologies, and material composition to enhance firmness, detection accuracy, and operational versatility for food safety monitoring applications.

3.7 RECENT ADVANCEMENTS IN BIOSENSORS FOR FOOD SAFETY

Biosensors have further advanced in sensitivity, specificity, and ability to detect concerns in real time, which is constructive in the food industry. The application of nanotechnology has positively impacted biosensor function by employing nanomaterials like nanoparticles and nanotubes to improve surface chemistry and detection limits. Further, electrochemical biosensors have advanced the detection of pesticides with porous cobalt-based oxides encapsulated in PdAu nanoparticles (Zhao et al., 2023). Enhanced biosensor functionality has allowed for the rapid detection of contaminants, allergens, and quality indicators, surpassing conventional methods (Curulli, 2021). A variety of biosensor platforms, including electrochemical, optical, and piezoelectric sensors, have broadened their scope of application. Moreover, new materials and IoT integration alongside data analytics have miniaturized and multiplexed real-time monitoring (Vasoya, 2023). Yu et al. (2023) demonstrated the real-time monitoring integration with traceability aids in proactive food safety management by monitoring possible risks in the supply chain.

3.8 CONCLUSION

In modern food analysis, biosensors are crucial for the rapid, sensitive, and on-site detection of contaminants, pathogens, and other quality indicators. The performance of biosensors has been enhanced by electrochemical, optical, and nanomaterial-based biosensors allowing real-time food safety monitoring. Traceability along the supply chain can furthermore be improved with IoT integration and data analytics. The continuous efforts to address challenges like matrix interference and sensitivity limitations in biosensors are driving focused design improvements and regulatory standardization. Their widespread adoption promises to transform food safety management, enhancing global quality control and consumer protection.

REFERENCES

Abdel-Mageed, H. M. (2025). Frontiers in nanoparticles redefining enzyme immobilization: A review addressing challenges, innovations, and unlocking sustainable future potentials. *Micro and Nano Systems Letters, 13*(1), 7.

Abraham, D. A., Li, A.-D., Sanmugam, A., Wadaan, M. A., Baabbad, A., Kanagaraj, K., Karuppasamy, K., Maiyalagan, T., Kim, H.-S., & Vikraman. D. (2024). Highly sensitive and selective detection of dopamine using atomic layer deposited HfO_2 ultra-thin films. *Electrochimica Acta, 497,* 144574.

Alhadeff, E., & Bojorge, N. (2011). Graphite-composites alternatives for electrochemical biosensor. In *Metal, Ceramic and Polymeric Composites for Various Uses.* IntechOpen.

Alhajj, M., Zubair, M., & Farhana, A. (2023). Enzyme linked immunosorbent assay. In *StatPearls* [Internet]. Treasure Island (FL): StatPearls Publishing. https://www.ncbi.nlm.nih.gov/sites/books/NBK555922/

Ali, A. A., Altemimi, A. B., Alhelfi, N., & Ibrahim, S. A. (2020). Application of biosensors for detection of pathogenic food bacteria: A review. *Biosensors*, *10*(6), 58.

Almeida, L., Correia, R., Squillaci, G., Morana, A., La Cara, F., Correia, J., & Viana, A. (2019). Electrochemical deposition of bio-inspired laccase-polydopamine films for phenolic sensors. *Electrochimica Acta*, *319*, 462–471.

Altintas, Z. (2018). Surface plasmon resonance based sensor for the detection of glycopeptide antibiotics in milk using rationally designed nanoMIPs. *Scientific Reports*, *8*(1), 11222.

Altug, H., Oh, S. H., Maier, S. A., & Homola, J. (2022). Advances and applications of nanophotonic biosensors. *Nature Nanotechnology*, *17*(1), 5–16.

Augustiniene, E., Jonuskiene, I., Kailiuviene, J., Mazoniene, E., Baltakys, K., & Malys, N. (2023). Application of whole-cell biosensors for analysis and improvement of L-and D-lactic acid fermentation by *Lactobacillus* spp. from the waste of glucose syrup production. *Microbial Cell Factories*, *22*(1), 223.

Balkrishna, A., Kumari, A., Kumar, A., Arya, V., Chauhan, A., Upadhyay, N. K., . . . & Kuca, K. (2024). Biosensors for detection of pesticide residue, mycotoxins and heavy metals in fruits and vegetables: A concise review. *Microchemical Journal*, 111292.

Barreiros dos Santos, M., Rodriguez-Lorenzo, L., Queirós, R., & Espiña, B. (2022). Fundamentals of biosensors and detection methods. In *Microfluidics and Biosensors in Cancer Research: Applications in Cancer Modeling and Theranostics* (pp. 3–29). Cham: Springer International Publishing.

Bhavadharini, B., Kavimughil, M., Malini, B., Vallath, A., Prajapati, H. K., & Sunil, C. K. (2022). Recent advances in biosensors for detection of chemical contaminants in food: A review. *Food Analytical Methods*, *15*(6), 1545–1564.

Busa, L. S. A., Mohammadi, S., Maeki, M., Ishida, A., Tani, H., Tokeshi, M. (2016). Advances in microfluidic paper-based analytical devices for food and water analysis. *Micromachines*, *7*, 86.

Camara-Martos, F., Da Costa, J., Justino, C. I., Cardoso, S., Duarte, A. C., & Rocha-Santos, T. (2016). Disposable biosensor for detection of iron (III) in wines. *Talanta*, *154*, 80–84.

Chen, W. X., Li, Q., Wang, Z., & Sun, Z. J. (2020). Spatial distribution characteristics and pollution evaluation of heavy metals in arable land soil of China. *Huan Jing keXue= HuanjingKexue*, *41*(6), 2822–2833.

Curulli, A. (2021). Electrochemical biosensors in food safety: Challenges and perspectives. *Molecules*, *26*(10), 2940.

Daci, M., Berisha, L., Mercatante, D., Rodriguez-Estrada, M. T., Jin, Z., Huang, Y., & Amorati, R. (2024). Advancements in biosensors for lipid peroxidation and antioxidant protection in food: A critical review. *Antioxidants*, *13*(12), 1484.

Das, S., & Sahu, P. P. (2023). CeO_2 NPs based capacitive sensor for quantification of limonin in citrus limetta juice. *IEEE Sensors Journal*, *23*(7), 7591–7598.

Doğan, V., Evliya, M., Kahyaoglu, L. N., & Kılıç, V. (2024). On-site colorimetric food spoilage monitoring with smartphone embedded machine learning. *Talanta*, *266*, 125021.

Draz, M. U., Yaqub, A., Jafry, A. T., Khan, M., & Ajab, H. (2024). Electrochemical sensing of B-complex vitamins: Current challenges and future prospects with microfluidic integration. *RSC Advances*, *14*(15), 10331–10347.

Duffy, G. F., & Moore, E. J. (2017). Electrochemical immunosensors for food analysis: A review of recent developments. *Analytical Letters*, *50*(1), 1–32.

El Aamri, M., Khalki, Y., Mohammadi, H., & Amine, A. (2023). Development of an innovative colorimetric DNA biosensor based on sugar measurement. *Biosensors*, *13*(9), 853.

Feng, Y., Xu, Y., Liu, S., Wu, D., Su, Z., Chen, G., . . . & Li, G. (2022). Recent advances in enzyme immobilization based on novel porous framework materials and its applications in bio-sensing. *Coordination Chemistry Reviews*, *459*, 214414.

Florea, A., Melinte, G., Simon, I., & Cristea, C. (2019). Electrochemical biosensors as potential diagnostic devices for autoimmune diseases. *Biosensors*, *9*(1), 38.

Freitas, M., Nouws, H. P., & Delerue-Matos, C. (2021). Voltammetricimmunosensor to track a major peanut allergen (Ara h 1) in food products employing quantum dot labels. *Biosensors*, *11*(11), 426.

Gallo, M., Ferrara, L., Calogero, A., Montesano, D., & Naviglio, D. (2020). Relationships between food and diseases: What to know to ensure food safety. *Food Research International*, *137*, 109414.

Gargalo, C. L., Lopez, P. C., Hasanzadeh, A., Udugama, I. A., & Gernaey, K. V. (2022). On-line monitoring of process parameters during fermentation. In *Current Developments in Biotechnology and Bioengineering* (pp. 117–164). Elsevier.

Ghosh, T., Raj, G. B., & Dash, K. K. (2022). A comprehensive review on nanotechnology based sensors for monitoring quality and shelf life of food products. *Measurement: Food*, *7*, 100049.

Giné Bordonaba, J., & Terry, L. A. (2009). Development of a glucose biosensor for rapid assessment of strawberry quality: Relationship between biosensor response and fruit composition. *Journal of Agricultural and Food Chemistry*, *57*(18), 8220–8226.

Grieshaber, D., MacKenzie, R., Vörös, J., & Reimhult, E. (2008). Electrochemical biosensors-sensor principles and architectures. *Sensors*, *8*(3), 1400–1458.

Gupta, A. K., Koch, P., Yumnam, M., Medhi, M., Madufor, N. J., Opara, U. L., & Mishra, P. (2022). Biosensors involved in fruit and vegetable processing industries. In *Biosensors in Food Safety and Quality* (pp. 111–134). CRC Press.

Haleem, A., Javaid, M., Singh, R. P., Suman, R., & Rab, S. (2021). Biosensors applications in medical field: A brief review. *Sensors International*, *2*, 100100.

Han, J. W., Ruiz-Garcia, L., Qian, J. P., & Yang, X. T. (2018). Food packaging: A comprehensive review and future trends. *Comprehensive Reviews in Food Science and Food Safety*, *17*(4), 860–877.

Harshavardhan, S., Rajadas, S. E., Vijayakumar, K. K., Durai, W. A., Ramu, A., & Mariappan, R. (2019). Electrochemical Immunosensors: Working principle, types, scope, applications, and future prospects. *Bio-Electrochemical Interface Engineering*, 343–369.

Hassan, A., Macedo, L. J., Mattioli, I. A., Rubira, R. J., Constantino, C. J., Amorim, R. G., & Crespilho, F. N. (2021). A three component-based van der Waals surface vertically designed for bio-molecular recognition enhancement. *ElectrochimicaActa*, *376*, 138025.

Herold, K. E., Sergeev, N., Matviyenko, A., & Rasooly, A. (2009). Rapid DNA amplification using a battery-powered thin-film resistive thermocycler. *Methods in Molecular Biology*, *504*, 441–458.

Herrera-Domínguez, M., Morales-Luna, G., Mahlknecht, J., Cheng, Q., Aguilar-Hernández, I., & Ornelas-Soto, N. (2023). Optical biosensors and their applications for the detection of water pollutants. *Biosensors*, *13*(3), 370.

Jawaheer, S., White, S. F., Rughooputh, S. D. D. V., & Cullen, D. C. (2003). Development of a common biosensor format for an enzyme based biosensor array to monitor fruit quality. *Biosensors and Bioelectronics*, *18*(12), 1429–1437.

Jiang, X., Liu, X., Wu, T., Li, L., Zhang, R., & Lu, X. (2019). Metal-Organic framework derived carbon-based sensor for monitoring of the oxidative stress of living cell and assessment of antioxidant activity of food extracts. *Talanta*, *194*, 591–597.

Karaseva, N. A., & Ermolaeva, T. N. (2012). A piezoelectric immunosensor for chloramphenicol detection in food. *Talanta*, *93*, 44–48.

Kaur, H., Bhosale, A., & Shrivastav, S. (2018). Biosensors: Classification, fundamental characterization and new trends: A review. *International Journal of Health Sciences and Research*, *8*(6), 315–333.

Kernbach, S., Kernbach, O., Kuksin, I., Kernbach, A., Nepomnyashchiy, Y., Dochow, T., & Bobrov, A. V. (2022). The biosensor based on electrochemical dynamics of fermentation in yeast Saccharomyces cerevisiae. *Environmental Research, 213*, 113535.

Khan, S., Monteiro, J. K., Prasad, A., Filipe, C. D., Li, Y., & Didar, T. F. (2024). Material breakthroughs in smart food monitoring: Intelligent packaging and on-site testing technologies for spoilage and contamination detection. *Advanced Materials, 36*(1), 2300875. MDPI. "Special Issue: Biosensors in Food Analysis and Environmental Detection." Biosensors, https://www.mdpi.com/journal/biosensors/special issues/food environmental.

Kim, D. M., & Yoo, S. M. (2022). Colorimetric systems for the detection of bacterial contamination: Strategy and applications. *Biosensors, 12*(7), 532.

Kuiken, T., Leighton, F. A., Fouchier, R. A., LeDuc, J. W., Peiris, J. S. M., Schudel, A., . . . & Osterhaus, A. D. M. E. (2005). Pathogen surveillance in animals. *Science, 309*(5741), 1680–1681.

Kulkarni, M. B., Ayachit, N. H., & Aminabhavi, T. M. (2022). Biosensors and microfluidic biosensors: From fabrication to application. *Biosensors, 12*(7), 543.

Kumar, A., Kulshreshtha, S., Shrivastava, A., & Saini, A. (2024). Biosensors for food spoilage detection: A comprehensive review of current advances. *Journal of Food Chemistry and Nanotechnology, 10*(S1), S73–S82.

Kumar, H., Dhalaria, R., Guleria, S., Cimler, R., Prerna, P., Dhanjal, D. S., . . . & Kuča, K. (2024). Immunosensors in food, health, environment, and agriculture: A review. *Environmental Chemistry Letters, 22*(5), 2573–2605.

Kuri, P. R., Das, P., & Goswami, P. (2020). Fundamentals of biosensors. In *Advanced Materials and Techniques for Biosensors and Bioanalytical Applications* (pp. 1–28). CRC Press. Boca Raton.

Lakard, B. (2020). Electrochemical biosensors based on conducting polymers: A review. *Applied Sciences, 10*(18), 6614.

Leonard, P., Hearty, S., Brennan, J., Dunne, L., Quinn, J., Chakraborty, T., & O'Kennedy, R. (2003). Advances in biosensors for detection of pathogens in food and water. *Enzyme and Microbial Technology, 32*(1), 3–13.

Li, S., Zhang, Y., Guo, M., Yi, Z., Hu, M., Xiong, C., . . . & Zhang, J. (2024). Rapid detection of Salmonella in milk by labeling-free electrochemical immunosensor based on an Fe3O4–ionic liquid-modified electrode. *Talanta, 270*, 125576.

Liang, G., Gong, W., Li, B., Zuo, J., Pan, L., & Liu, X. (2019). Analysis of heavy metals in foodstuffs and an assessment of the health risks to the general public via consumption in Beijing, China. *International Journal of Environmental Research and Public Health, 16*(6), 909.

Liu, F., Li, K., Zhang, Y., Ding, J., Wen, T., Pei, X., . . . & Li, L. (2020). An electrochemical DNA biosensor based on nitrogen-doped graphene nanosheets decorated with gold nanoparticles for genetically modified maize detection. *MicrochimicaActa, 187*, 1–8.

Liu, S., Kannegulla, A., Kong, X., Sun, R., Liu, Y., Wang, R., . . . & Wang, A. X. (2020). Simultaneous colorimetric and surface-enhanced Raman scattering detection of melamine from milk. *SpectrochimicaActa Part A: Molecular and Bio-molecular Spectroscopy, 231*, 118130.

Luong, J. H., & Guilbault, G. G. (2019). Analytical applications of piezoelectric crystal biosensors. *Biosensor Principles and Applications*, 106–138.

Lv, M., Liu, Y., Geng, J., Kou, X., Xin, Z., & Yang, D. (2018). Engineering nanomaterials-based biosensors for food safety detection. *Biosensors and Bio-electronics, 106*, 122–128.

Majer-Baranyi, K., Székács, A., & Adányi, N. (2023). Application of electrochemical biosensors for determination of food spoilage. *Biosensors, 13*(4), 456.

Megha, K. B., & Mohanan, P. V. (2021). Role of immunoglobulin and antibodies in disease management. *International Journal of Biological Macromolecules*, *169*, 28–38.

Meng, X., O'Hare, D., & Ladame, S. (2023). Surface immobilization strategies for the development of electrochemical nucleic acid sensors. *Biosensors and Bio-Electronics*, *237*, 115440.

Meshram, B. D., Agrawal, A. K., Adil, S., Ranvir, S., & Sande, K. K. (2018). Biosensor and its application in food and dairy industry: A review. *International Journal of Current Microbiology and Applied Sciences*, *7*(2), 3305–3324.

Mishra, G. K., Barfidokht, A., Tehrani, F., & Mishra, R. K. (2018). Food safety analysis using electrochemical biosensors. *Foods*. *7*(2018).

Mitsubayashi, K., Kubotera, Y., Yano, K., Hashimoto, Y., Kon, T., Nakakura, S., . . . & Endo, H. (2004). Trimethylamine biosensor with flavin-containing monooxygenase type 3 (FMO3) for fish-freshness analysis. *Sensors and Actuators B: Chemical*, *103*(1–2), 463–467.

Mukherjee, S., Pal, S., Paria, P., Bhattacharyya, S., Ghosh, K., Pal, A., . . . & Bhattacharyya, N. (2021). On-spot bio-sensing device for organophosphate pesticide residue detection in fruits and vegetables. *Current Research in Bio-Technology*, *3*, 308–316.

Mutlaq, S., Albiss, B., Al-Nabulsi, A. A., Osaili, T., Al-Jaberi, T., Olaimat, A. N., . . . & Ayyash, M. M. (2023). Detection of salmonella enteritidis in milk using conductometric immunosensor coated on polyaniline/zinc oxide nanocomposite. *Foodborne Pathogens and Disease*, *20*(5), 177–185.

Nagraik, R., Sharma, A., Kumar, D., Chawla, P., & Kumar, A. P. (2021). Milk adulterant detection: Conventional and biosensor based approaches: A review. *Sensing and Bio-Sensing Research*, *33*, 100433.

Nath, D., & Mahata, M. K. (2024, May). Performance analysis of dielectrically modulated both sided cavity TFET biosensor using III-V compound semiconductor. In *2024 Parul International Conference on Engineering and Technology (PICET)* (pp. 1–5). IEEE.

Neethirajan, S., Ragavan, V., Weng, X., & Chand, R. (2018). Biosensors for sustainable food engineering: Challenges and perspectives. *Biosensors*, *8*(1), 23.

Nejadmansouri, M., Majdinasab, M., Nunes, G. S., & Marty, J. L. (2021). An overview of optical and electrochemical sensors and biosensors for analysis of antioxidants in food during the last 5 years. *Sensors*, *21*(4), 1176.

Odobašić, A., Šestan, I., & Begić, S. (2019). Biosensors for determination of heavy metals in waters. In *Biosensors for Environmental Monitoring*. IntechOpen.

Pan, Y., Wang, Z., Duan, C., Dou, L., Wen, K., Wang, Z., . . . & Shen, J. (2022). Comparison of two fluorescence quantitative immunochromatographic assays for the detection of amantadine in chicken muscle. *Food Chemistry*, *377*, 131931.

Patel, I., Kulkarni, M., & Mehendale, N. (2024, October). Machine learning-based sensor synchronization for enhanced spatial awareness in assistive technologies. In *2024 International Conference on Intelligent Systems and Advanced Applications (ICISAA)* (pp. 1–6). IEEE.

Péter, B., Farkas, E., Kurunczi, S., Szittner, Z., Bősze, S., Ramsden, J. J., . . . & Horvath, R. (2022). Review of label-free monitoring of bacteria: From challenging practical applications to basic research perspectives. *Biosensors*, *12*(4), 188.

Pisoschi, A. M., Negulescu, G. P., & Pisoschi, A. (2010). Ascorbic acid determination by an amperometricascorbate oxidase-based biosensor. *Revista De Chimie (Bucharest)*, *61*(4), 339–344.

Pohanka, M. (2022). Glucose electrochemical biosensors: The past and current trends. *International Journal of Electrochemical Science*, *16*(7), 210719.

Prajapati, J., Gupta, A., Gautam, R. K., Joshi, J., Kumari, N., & Tiwari, I. (2025). 2D Hexagonal boron nitride nanosheets supported with palladium-doped zinc oxide nanoparticle–based electrochemical sensor for the detection of indole-3-acetic acid. *Ionics*, *31*(6), 6309–6324.

Qu, F., Yang, M., Shen, G., & Yu, R. (2007). Electrochemical biosensing utilizing synergic action of carbon nanotubes and platinum nanowires prepared by template synthesis. *Biosensors and Bioelectronics*, *22*(8), 1749–1755.

Ravindran, N., Kumar, S., CA, M., Thirunavookarasu, S., N., & CK, S. (2023). Recent advances in surface plasmon resonance (SPR) biosensors for food analysis: A review. *Critical Reviews in Food Science and Nutrition*, *63*(8), 1055–1077.

Ribeiro, D. B., Santos Silva, G., Dos Santos, D. R., Castro Costa, A. R., Braga Ribeiro, E., Badea, M., & Nunes, G. S. (2021). Determination of the antioxidant activity of samples of tea and commercial sources of vitamin c, using an enzymatic biosensor. *Antioxidants*, *10*(2), 324.

Ricci, F., Volpe, G., Micheli, L., & Palleschi, G. (2007). A review on novel developments and applications of immunosensors in food analysis. *Analyticachimicaacta*, *605*(2), 111–129.

Senapati, M., Singhal, S., Gupta, A. K., Sonowal, D., Mishra, P., & Sahu, P. P. (2022). Bio/chemical sensors and microsensors involved in meat industry. In *Biosensors In Food Safety and Quality* (pp. 159–175). CRC Press.

Shams, R., Singh, J., Ashraf, S., Manzoor, M., & Dar, A. (2020). Application of biosensors in food quality control. *Journal of Postharvest Technology*, *8*(1), 53–74.

Sim, D., Brothers, M. C., Slocik, J. M., Islam, A. E., Maruyama, B., Grigsby, C. C., . . . & Kim, S. S. (2022). Bio-markers and detection platforms for human health and performance monitoring: A review. *Advanced Science*, *9*(7), 2104426.

Singh, P. C., Bhamidipati, S., Singh, R. K., Smith, R. S., & Nelson, P. E. (1996). Evaluation of in-line sensors for prediction of soluble and total solids/moisture in continuous processing of fruit juices. *Food Control*, *7*(3), 141–148.

Škugor Rončević, I., Skroza, D., Vrca, I., Kondža, A. M., & Vladislavić, N. (2022). Development and optimization of electrochemical method for determination of vitamin C. *Chemosensors*, *10*(7), 283.

Soares, R. R., Hjort, R. G., Pola, C. C., Parate, K., Reis, E. L., Soares, N. F., . . . & Gomes, C. L. (2020). Laser-induced graphene electrochemical immunosensors for rapid and label-free monitoring of *Salmonella enterica* in chicken broth. *ACS Sensors*, *5*(7), 1900–1911.

Sousa, C. S., Lima, K. C., Botelho, C. N., Pereira, N. M., Fernandes, R. N., Silva, G. G., . . . & Luz, R. C. (2020). Photoelectrochemical sensor for determination of naringin at low oxidation potential using a modified FTO electrode with cadmium sulfide and titanium dioxide sensitized with chloroprotoporphyrin IX iron (III). *Journal of Solid State Electrochemistry*, *24*(8), 1715–1726.

Stojanović, Z. S., Mehmeti, E., Kalcher, K., Guzsvány, V., & Stanković, D. M. (2016). SWCNT-modified carbon paste electrode as an electrochemical sensor for histamine determination in alcoholic beverages. *Food Analytical Methods*, *9*(10), 2701–2710.

Tadesse, M., & Liu, Y. (2025). Recent advances in enzyme immobilization: The role of artificial intelligence, novel nanomaterials, and dynamic carrier systems. *Catalysts*, *15*(6), 571.

Umar, L., Rosandi, V. A., Setiadi, R. N., Agustirandi, B., Linda, T. M., & Kuswandi, B. (2023). Amperometric microbial biosensor for sugars and sweetener classification using principal component analysis in beverages. *Journal of Food Science and Technology*, *60*(1), 382–392.

Vasoya, N. H. (2023). Revolutionizing nano materials processing through IoT-AI integration: Opportunities and challenges. *Journal of Materials Science Research and Reviews*, *6*(3), 294–328.

Villalonga, A., Sánchez, A., Mayol, B., Reviejo, J., & Villalonga, R. (2022). Electrochemical biosensors for food bioprocess monitoring. *Current Opinion in Food Science*, *43*, 18–26.

Wang, S., Dong, Y., & Liang, X. (2018). Development of a SPR aptasensor containing oriented aptamer for direct capture and detection of tetracycline in multiple honey samples. *Biosensors and Bioelectronics*, *109*, 1–7.

Wang, Y. Z., Li, Y. R., Zhang, Y. Q., Xiang, Y. M., Bai, R. R., Liu, Y., . . . & Hu, Y. H. (2024). Dual-signal ratiometric electrochemiluminescence biosensor based on Au NPs-induced low-potential emission of PFO Pdots and LSPR-ECL mechanism for ultra-sensitive detection of microRNA-141. *Biosensors and Bioelectronics*, *261*, 116495.

Wu, L., Yin, W., Tang, K., Li, D., Shao, K., Zuo, Y., Ma, J., Liu, J., & Han, H. (2016). Enzymatic biosensor of horseradish peroxidase immobilized on Au-Pt Nanotube/Au-Graphene for the simultaneous determination of antioxidants. *Analytica Chimica Acta*, *933*, 89–96.

Yang, Y., Wang, H., Jiang, R., Guo, X., Cheng, J., & Chen, Y. (2022). A review of IoT-enabled mobile healthcare: Technologies, challenges, and future trends. *IEEE Internet of Things Journal*, *9*(12), 9478–9502.

Yao, J., Ding, T., & Che, L. (2025). A high-sensitivity MEMS fluxgate sensor with double-layer induction coils fabricated by micro-casting. *Journal of Micromechanics and Microengineering*, *35*(4), 045006.

Yu, Q., Wu, T., Tian, B., Li, J., Liu, Y., Wu, Z., . . . & Gu, B. (2023). Recent advances in SERS-based immunochromatographic assay for pathogenic microorganism diagnosis: A review. *AnalyticaChimicaActa*, 341931.

Zhao, Y., Liang, L., Guo, W., Lu, X., Zhao, C., & Gao, F. (2023). The porous hollow cobalt-based oxides encapsulated with bimetallic PdAu Nanoparticles of electrochemical biosensor for highly sensitive pesticides detection. *Nanotechnology*, *34*(28), 285501.

Zheng, L., & Zhang, Y. (2021). Biosensor. *Clinical Molecular Diagnostics*, 345–356.

Zrinski, I., Pungjunun, K., Martinez, S., Zavašnik, J., Stanković, D., Kalcher, K., & Mehmeti, E. (2020). Evaluation of phenolic antioxidant capacity in beverages based on laccase immobilized on screen-printed carbon electrode modified with graphene nanoplatelets and gold nanoparticles. *Microchemical Journal*, *152*, 104282.

4 Laser and Laser Focusing Optics

Principal, Preparation, Analysis, Spectral Acquisition, and Application in Food Quality and Safety

Manpreet Kaur and Yashi Srivastava

4.1 INTRODUCTION

Laser-induced breakdown spectroscopy (LIBS) is a powerful sensor technology used for analyzing a wide range of samples based on atomic emission (Sneddon & Lee, 2004; Gunaratne et al., 2006). Atomic and molecular spectroscopic techniques are commonly used to monitor various types of food and identify adulteration. LIBS is a popular spectroscopic technique for food analysis (Lastra-Mejıas et al., 2019; Juve et al., 2008). LIBS is a modern and promising optical method that can help ensure the quality of food. It has a wide range of applications, including the analysis of mineral composition and various qualitative assessments such as verifying authenticity, distinguishing between samples, detecting bacterial contamination, conducting depth profiling, and mapping elemental surfaces (Kaiser et al., 2012).

4.1.1 Properties of Laser Light

4.1.1.1 Monochromatic

Monochromatic light has a single vibration frequency. Laser light is nearly monochromatic with minimal doppler shift deviation.

4.1.1.2 Coherent

Coherence measures the accuracy of a waveform. For instance, photons emitted by a laser source are in step with stimulating photons.

4.1.1.3 Brightness

Laser technology produces light that is much brighter and more intense than traditional light sources.

DOI: 10.1201/9781003482611-4

4.1.1.4 Unidirectional

Light emitted from laser travels in highly collimated, narrow beam and minimal divergence. (Blutinger et al., 2018).

4.1.2 History

LIBS has a rich history dating back to the invention of the laser in 1960. Just two years later, in 1962, American physicists Frederick Brech and LIyod Cross discovered laser-induced plasma when they focused a laser beam on a target surface. This discovery led to further developments in the 1960s by French and Russian physicists, followed by a period of abandonment until the 1980s. Revitalized by an American research team in the 1980s, LIBS gained traction for its ability to analyze elements in various materials. By 1988, a prototype was developed for in situ analysis of beryllium particles in the air. In the 2000s, LIBS found applications in identifying gases, explosives, and biological agents, showcasing its versatility and potential for various industries (Debras-Guedon, 1963; Korunchikov et al., 1966).

4.1.3 Advantages from Conventional Techniques

LIBS has many advantages over conventional analytic techniques (Senesi et al., 2009). LIBS measurements are performed without the need for sample pretreatment, unlike traditional methods like Atomic Absorption Spectroscopy (AAS) or Inductively coupled plasma atomic emission spectroscopy (ICP-AES) which require digestion of the sample with acid, filtration, dilution, and other steps to prepare the sample for analysis. LIBS requires only a small amount of material, making it minimally destructive and ideal for valuable or limited food samples. The technique uses a focused laser beam to ablate a small amount of the sample's surface. LIBS is equally effective for solid, liquid, and gas samples, providing comprehensive analysis without needing different instruments. LIBS is particularly sensitive to light elements, allowing for a thorough analysis of hydrogen, carbon, and nitrogen. It analyzes samples from a distance, making it suitable for sealed or difficult-to-access containers without sample compromise, and also it can be combined with other techniques like Raman spectroscopy for multi-elemental and molecular surface analysis in food samples.

4.2 PRINCIPLES

A high-energy pulsed laser is directed toward a sample, causing it to ablate or vaporize. A small amount of the sample is vaporized, creating a superheated plasma that ionizes the elements within it. This process happens rapidly and is called breakdown. The excited sample then emits light as it returns to its ground state. This light is collected and dispersed using a spectrometer. The spectrum that is produced can then be analyzed to determine the elements present in the sample. The dispersed light is directed onto a detector, such as a charge-coupled device (CCD) detector, to measure the intensity of the emitted radiation and the frequencies at which the emission is observed. This information can be used to identify and provide information on the number of emitting atoms responsible for the emission (Diedrich, 2007).

4.2.1 LIBS Instrumentation

It consists of a pulsed laser, laser-focusing optics, optics for collecting plasma light, a spectrometer, and a sample chamber.

4.2.1.1 Laser System

The most important part of LIBS instrumentation is the laser system. Q-switched Nd: YAG lasers are commonly used due to their ability to generate short, intense pulses of light. These lasers can produce pulses with durations ranging from ns to ps or even fs. The laser pulse energy, pulse width, and repetition rate are crucial parameters that affect the LIBS analysis (Rakovsky et al., 2014).

4.2.1.2 Optics

Focusing optics are used to concentrate the laser beam onto the sample surface. This includes lenses, mirrors, and optical fibers. The spot size of the laser on the sample is an important factor impacting the intensity of the plasma generated. A smaller spot size usually results in higher irradiance (power per unit area) on the sample (Sneddon & Lee, 1998).

4.2.1.3 Light Collection Optics

After the laser pulses hit the sample and create plasma, the light emitted by the plasma needs to be collected and directed toward the spectrometer for analysis. Lenses and mirrors are used to collect and guide the light. In some cases, fiber optics are used to transmit the collected light to the spectrometer (Santos et al., 2012).

4.2.1.4 Spectrometer

The spectrometer is a crucial component for analyzing the emitted light from the plasma. It consists of a spectrograph and a detector. The spectrograph separates the emitted light into its different wavelengths (colors), creating a spectrum. The detector then measures the intensity of each wavelength. Spectral range, resolution, and attainment time are important parameters for the spectrometer's performance. Different types of spectrometers, such as Echelle spectrometers or CCD-based spectrometers, may be used depending on the application and desired spectral range (Singh & Thakur, 2007).

4.2.1.5 Detector

The detector, often a CCD or Intensified Charge-Coupled Device (ICCD), captures the spectrum produced by the spectrograph. It converts the light into electronic signals that can be processed to determine the intensity of each wavelength. The number of photoelectrons produced by the detector is proportional to the intensity of the emitted light, providing quantitative information about the elements present in the sample (Rakovsky et al., 2014).

4.2.1.6 Sample Chamber

LIBS instruments often have a specially designed sample chamber where the sample is placed. The chamber is designed to hold the sample securely and protect the optics from damage. It may also have mechanisms to rotate the sample, ensuring that fresh material is exposed to the laser for each analysis (Shah et al., 2019).

4.2.2 Physical Interactions

LIBS is a technique that uses a laser pulse to create high-temperature plasma on the sample being analyzed. The laser pulse is so intense that it causes the sample to vaporize, atomize, and undergo excitation and de-excitation. When the plasma emits light, that light is captured and analyzed to identify the components present in the sample (Anabitarte et al., 2012; Costa et al., 2019).

4.3 PARAMETERS INFLUENCING THE SIGNAL

The LIBS instrumentation utilizes a laser to generate plasma by providing the necessary energy. The pulse energy, pulse time, length, and number of pulses are key laser-related parameters (Anabitarte et al., 2012). The target material absorbs photons of laser radiation, which depends on the laser's wavelength and the optical properties of the sample. The pulse energy alters the material's state (Anabitarte et al., 2012). Wavelength is a crucial factor in laser-matter interactions as it affects the ablation process by influencing material absorbency. The absorption mechanism varies based on material properties (Anabitarte et al., 2012). It has been found that an increase in delay time results in a decrease in emission intensity, plasma temperature, and electron number density. This is due to the conversion of ion thermal energy into kinetic energy with plasma expansion (Hanif et al., 2011).

4.4 DIFFERENT METHODS FOR VARIOUS PHYSICAL STATES OF MATTER

Fresh food products, such as vegetables, are typically washed, dried to a constant weight, ground, and homogenized before analysis using LIBS (Kim et al., 2012). Smaller particle sizes result in improved signal quality. Analyzing powders can be problematic and can lead to defocusing of the focal spot. Therefore, the powder is usually pressed into pallets using a hydraulic press and pressure to provide a harder and more uniform sample for LIBS analysis. Sometimes, a binder is required when the powder does not form cohesive pallets. Analyzing liquid samples can be challenging due to the disruption caused by laser pulses on the liquid surface. To overcome this issue, a thin layer of the liquid is applied onto a rotating stainless-steel wheel, and the liquid volume is constantly refilled. To enable LIBS analysis, the liquid can be solidified by freezing it or depositing a small volume of solution onto a solid substrate. The choice of substrate is critical in this process (Jantzi et al., 2016). It has been demonstrated in the field of environmental science that LIBS is feasible for gases and aerosols, but no research has been conducted on its application to food products (Dutouquot et al., 2014).

4.5 SPECTRAL ACQUISITION

LIBS measurements are typically conducted under ambient atmospheric conditions; however, conducting these measurements under vacuum or low-pressure inert gas conditions may expand the spectral range (Hahn & Omenetto, 2012). To obtain a LIBS spectrum, a high-powered laser is applied onto the surface of the sample, resulting in the collection of emitter plasma light. A small crater is formed as a result, with its dimensions depending on the LTSD. It is important to maintain a constant LTSD (Santos et al., 2012). For homogenous samples, ten shots may be sufficient, while heterogeneous samples require more (Tognoni & Cristoforetti, 2016). LIBS spectra consist of a large number of spectral lines. To identify the emission lines, the NIST provides an online database that is commonly used. The intensity of the lines in a LIBS spectrum differs for different elements (Thakur, 2007). During LIBS analysis, multiple spectra are examined for quantitative and qualitative analysis.

4.6 APPLICATIONS

Few applications of LIBs are shown in Table 4.1.

4.6.1 Adulterant Detection in Various Foods

4.6.1.1 In Milk

Milk adulteration refers to the act of adding any substance to milk, either intentionally or unintentionally, to improve its appearance or increase its volume, which changes its composition. This can involve adding water, skim milk, fat replacement, and modification of protein content. LIBS can be used to detect milk adulteration. To do that, the milk sample is prepared by lyophilization for 24 hours to remove moisture. Then, a compact pellet is created from milk powder (250 mg) using a hydraulic press. The pellets can be directly analyzed without requiring additional preparation. To ensure accuracy, samples are positioned precisely using a micro-metric positioner. The laser is directed toward the surface of the pellet, creating a spot. Plasma radiation is then emitted and collected through an optical fiber. The radiation goes through a fused silica collimator and is analyzed with the help of a spectrometer. It is worth noting that all samples can be analyzed at room temperature. Qualitative analysis involves collecting LIBS spectra for each sample. To standardize the spectral variation, all spectra need to be normalized. After normalization, the spectra for each sample are compiled into a database. For data analysis, chemometric methods such as PCA (principle component analysis) can be used. In qualitative analysis of milk mixture detection, spectra for different milk species are recorded within the wavelength range of 190–450 nm. In quantitative analysis of adulteration with melamine, the CN band is analyzed using spectroscopy in the presence of increasing concentrations of melamine in the milk sample. A corresponding amplification in the CN molecule signal is observed. The spectral range of CN emission within the wavelength 382–389 nm can be observed. The emission band of CN is associated with both the molecular nature and concentration of melamine. NN can be used for qualitative classification and quantitative approaches. A study was conducted that used PCA and NN to analyze LIBS spectra. The results showed a correct classification rate of 98%, with a level of robustness of 100%. To quantify melamine, two methodologies were developed. The first was a univariate analysis using the CN emission band, while the second was a multivariate calibration NN model. The R^2 values obtained were 0.982 and 0.999, respectively (Moncayo et al., 2017) (Figure 4.1).

TABLE 4.1
Applications and Accuracy of LIBS in Food Analysis

Sr. No.	Food Sample	Component Detection	Sample Preparation	LIBS Setup Specifications	Data Analysis Methods/Models	Elements Emission Lines (nm)	Accuracy	References
1.	Milk	Adulteration (mixture of milk from different species and melamine in toddler milk sample)	Samples were freeze-dried, powdered, and compressed for pallet formation	Q-switched Nd: YAG laser, 1,064 nm, pulse duration of 4 ns, 100 mm focal-distance lens, laser pulse energy 100 mJ. repetition rate 1 Hz	PCA and NN	C (247.4–248.4), Mg (278.5–281.3), Ca (314.7–320.1, 368.6–376.1, 389. 49–401.2, 442.6–446.5), Sr (405.5–409.0, 420.5–423.4), CN (382.1–390.0)	100% of correction classification, R^2-NN–0.999 MPE-5%	Moncayo et al. (2017)
2.	Tea	Toxic elements (Br, Cr, Si)	Tea grains were pressed using a hydraulic press to form pallets	Q-switch Nd: YAG, wavelength 266 nm, repetition rate of 20 Hz, maximum output energy of 30 mJ, 8 ns pulses, UV convex lens having focal length 30 mm, 500 nm spectrograph	The spectral assignment was carried out by using the NIST Database and fingerprint wavelengths of each element	Cr (286.5), Br (793.9), Si (784.9), limit of detection (LOD) (mg L^{-1})–Cr (12), Br (11), Si (1)		Gondal et al., 2016a
3.	Honey	Adulterants- High Fructose Corn Syrup (HFSC), sugarcane syrup	The honey sample was placed in a polythene flask (1mL capacity), and the flask placed in the sample holder	1,064 nm Q-switched Nd:YAG laser,50 mJ maximum power energy with pulse duration less than 8 ns, spectral range from 200 nm to 630 nm	PLS-DA model, PLS regression model	HFSC- emission lines at 394.86 nm (Ca I), 396.91 nm (Ca II), 397.38 nm (Ca I), and 398.2 nm (Ca I) Sugarcane syrup- emission lines at 251.75 nm (Fe I), 252.15 nm (Fe I), 253.43 nm (Fe II), and 395.58 nm (Fe I)	100%	Nespeca et al. (2020)

4.	Beef	Offal adulteration (copper content in minced beef)	Fresh beef steaks are minced, dried and ground into powder; pallets are formed using a hydraulic press	Q-switched Nd: YAG laser, 1,064 nm, pulse energy 150 mJ and pulse duration of 5 ns, spectrophotometers wavelength 185–904 nm, focal length 80 mm, scanning 100 different, 10×10 grid pattern, repetition rate of 1 Hz.	PLSR modelling	C I (247.9 nm), Mg II (279.6 nm), Zn I (330.3 nm), Ca II (393.34 nm), Ca I (422.7 nm), Mg I (518.4 nm), Na I (588.9 nm and 589.6 nm), H I (656.3 nm), N I (742.4 nm, 744.3 nm and 746.9 nm), K I (766.5 nm and 769.9 nm), O I (777.2 nm), Rb I (780.0 nm) and Na I (818.3 nm and 819.4 nm), Cu I (324.7 nm and 327.4 nm) LOD of Cu—1 ppm	R^2_{CV}—0.85, RMSECV—43.5 ppm, R^2_p—0.85, RMSEP—36.8ppm	Casado-Gavalda et al. (2017)
	Brown rice flour	Adulteration (talc powder, corn flour, gypsum powder, sorghum flour, buckwheat flour)	Pallets are formed from mixed powder by using electric tablets press	Q-switched Nd: YAG laser, wavelength: 532 nm; pulse duration: 8 ns, laser energy:40mJ, repetition frequency:1Hz,	Linear Discrimination Analysis NB-naïve Bayes (LDA), NB, SVM, and CNN model	60 lines of N, H, O, C, Ca, Na, K according to ICP-MS result and NIST atomic spectral database	TR-LIBS Training set (%); LDA—89.63, NB—91.11, SVN—94.34, and 1D CNN—90.71. Prediction set (%)— LDA—90.59, NB—91.27, SVM—95.40 and 1D CNN—85.66	Ma et al. (2022)
6.	Pistachio	Adulteration (green pea and spinach)	Samples were dried, ground, and pallets were prepared using pallet press	50 mJNd:YAG laser, 1,064 nm, repetition rate—8 Hz, focal length of lens—10 cm, laser operate at 30 mJ^{-1}. Scanned region— 150	PCA and PLC	Ranged between 186 and 900 nm for Fe, Ca, Mg, K, Na, C H, O, Zn, and Na	Pistachio-R^2—0.995, LOD—2.04% Spinach—R^2—0.993, LOD—1.64%	Sezer et al. (2019)

(Continued)

TABLE 4.1 (*Continued*)
Applications and Accuracy of LIBS in Food Analysis

Sr. No.	Food Sample	Component Detection	Sample Preparation	LIBS Setup Specifications	Data Analysis Methods/Models	Elements Emission Lines (nm)	Accuracy	References
7.	Meat	Adulteration (identification of different meat species)	Fat removed from the sample, grounded, and protein extracted, then drying of meat, removal of fat and milling to collect protein, and myofibrillar protein was collected, dried, milled, and turned into pallets using a hydraulic press	1064 nm Nd:YAGlaser, lens-10 cm focal distance,8 Hz repetition rate and 40 mJ/pulse Scanned region- 150	PCA and PLC	Ranged between 213 nm and 822 nm for Fe, C, Mg, Zn, Ca, Na, H, K, N, and O	LOD- beef adulteration with chicken-2.84%, R^2—0.962 and pork meat—3.89%, R^2—0.967	Sezer et al., 2021
8.	Green leafy vegetable	Minerals' detection	Sample were sorted, washed, vacuum dried, powdered, and pallets were formed	Nd:YAG laser, max. pulse energy—425 mJ, repetition rate—10 Hz, convex lens—15 cm, spectrometer—200–900 nm wavelength range		Si (288.1, 251.4 nm), Mg (285.1, 279.5, 279.8 nm), Ti (334.9, 336.1, 337.2 nm), Fe (344.1, 238.3, 240.4 nm), Ca (422.6, 393.3, 396.8, 443.5, 445.5 nm), C (247.8, 229.6 nm), and Al (308.2, 309.2 nm)		Shukla et al. (2016)

9.	Poultry meat	Calcium	Samples were frozen after collection until analysis, for analysis, sample were thawed, stirred, covered with plastic wrap, and pressed flat	Nd:YAG laser, pulse energy— 230 mJ,15 cm focal length lens, spectrum range—200–800 nm	PLS	Ca II (392.10–394.60 nm and 395.41–398.20 nm), Ca I (421.11–424.29 nm), Na I (587.71–590.78), and K I (764.81–768.2 nm and 768.50–771.28 nm)	R^2_{CV} of 0.82 and RMSECV of 17 mg/100 g [very low (90 mg/100 g Ca), intermediate (20–90 mg/100 g Ca), and high (>90 mg/100 g Ca)]	Andersen et al. (2016)
10.	Liquid infant formula	Calcium	Rotating wheel of liquid chamber submerged into liquid, creating a thin layer which is presented to laser beam	150 mJ Q-switched Nd:YAG laser, pulse duration—5ns, spectrometer wavelength—181–904 nm, repetition rate—2 Hz, rotatory wheel speed—12 rpm	PLSR	Ca II 393.37, 396.85 nm; Ca I 422.67 nm; H I 656.29 nm; N I 744.23, 746.83, 821.63 nm; K I 766.49, 769.90 nm; O I 777.19 nm; and Na I 589.05, 589.59 nm.	R^2_C—0.96, R^2_{CV}—0.89, RMSECV—10.23 mg 100 mL^{-1}, RMSEP—6.45 mg 100 mL^{-1}	Cama-moncunill et al. (2020)
11.	Maize tissues	Inorganic nutrients	Sample dried, ground, and compressed to make pallets	Nd: YAG laser,200 mJ at 532 nm, 5 ns pulse width,10 Hz repetition rate, spectrometers' wavelength—200 nm and 720 nm	NIST database	Spectral range— 240–290 nm, Si-I (288.16 nm, 263.13 nm 243.51 nm) C-I (247.85 nm), Mg-II (279.55 nm, 280.27 nm), K (691.11 nm, 693.88 nm), Al (394.40 nm, 396.15 nm), Ca (643.90, 649.38 nm)		Jabbar et al. (2020)
12.	Wheat flour	Lead	Sample was pressed into pallet form	LIBS–LIF system, Q-switched Nd:YAG laser, 266 nm, spherical quartz lens—f = 150 mm)		Lead (405.78 nm)	LOD-73.8 ppb RMSECV—1.24 mg/kg LOQ—0.246 ppm, less than 4.07% RE between LIBS-LIF and ICP-OES/MS methods	Wang et al. (2024)

(Continued)

TABLE 4.1 (*Continued*)
Applications and Accuracy of LIBS in Food Analysis

Sr. No.	Food Sample	Component Detection	Sample Preparation	LIBS Setup Specifications	Data Analysis Methods/Models	Elements Emission Lines (nm)	Accuracy	References
13.	Bakery products	Toxic and dietic element	Sample vacuum dried, ground to make powder, pallets formed using hydraulic press	Q-switched Nd:YAG laser, 5 ns pulse duration, 400 mJ laser energy at 532 nm, 10 Hz repetition rate, focusing lens—15 cm	NIST database	Fe I (250.33 nm, 251.17 nm, 255.88 nm, 259.93 nm, 263.13 nm, 613.73 nm), Fe II (273.95 nm, 274.69 nm, 274.91 nm, 275.57 nm), Mg (279.55 nm, 280.27 nm 285.21 nm), Zn (32,823 nm, 330.25 nm, 334.50 nm 336.23 nm), Na (588.99 nm, 589.59 nm), K (766.49 nm, 769.89), Ti (307.29 nm, 307.52 nm)		Ahmed et al. (2021)
1.	Rice	Toxic metals (Cr, Pb, Cu, and Cd)	Sample ground, sieved and pressed to make pallets	Nd:YAG laser, pulse energy— 150 mJ, pulse width—9 ns, wavelength—1,064 nm, spectrometer range—200 nm– 600 nm	Multivariate analysis based on PCA	Pb (373.99nm), Cd (346.6nm), Cr (534.5nm), and Cu (450.59nm)	Relative errors in the range of 1.25–11.8% when compared with the AAS method	Salloom et al. (2021)
2.	Edible oil	Heavy metals	Liquid drops placed on substrate and transferred to disk under LIBS	Nd:YAG laser at 1,064 nm, laser energy 165 mJ, two quartz lens focal length—75 mm and 150 mm, spectrometer range—between 200 and 794 nm	NIST database	C I (247.86nm), Si I (243.52nm), Si I (288.16nm), Zn I (213.86nm), Cd I (228.80nm), Cu I (324.75nm), Cr I (425.43nm)	Detection limits for Cd, Cu, and Cr in oils are 0.49–0.08 ppm, Zn—3.9 ppm, R^2 for Cd— 0.9939, Zn—0.9972, Cu—0.9969, Cr—0.9929	Vinic et al. (2020)

3.	Honey	Geographical origin (unifloral honey and multifloral honey)	Samples were added to plate, no other sample preparation needed.	Laser energy—80 mJ, wavelength—532 nm, frequency—1 Hz, lens' focal length—100 mm	PCA, LDA and SVM	Mg I 285.23 nm, Ca II 393.37 nm, Na I589.03 nm, K I766.57nm	Accuracy rates (acacia honey-LDA—74.1%, SVM— 82.6%), multifloral honey- LDA—98.6%, SVM—98.7%	Zhao et al. (2020)
4.	Olive oil	Authenticity and geographical discrimination	Sample placed in petri dish and subjected to laser beam	Q-switched Nd: YAG laser, at 1,064 nm, laser energy 90 mJ, lens of 10 cm focal length	LDA, ERTC, RFC, XGBoost	H (656.3 nm, 486.1 nm), O I (777.2, 777.4, and 777.5 nm), N I (742.4, 744.2, 746.8 nm, and 818.8, 821.6, 824.2 nm), C I (247.9, 795.2,833.5 nm, 906.2, 940.6 nm)	LDA— 100%, ERTC— 100%, RFC— 99.8 ± 0.3%, XGBoost—99.8±0.2%	Gyftokostas et al. (2021)
5.	Grape seeds	Geographical origin	Seeds cleaned, dried, ground, sieved, and powder compressed to make pallets using tablet press machine	Q-switched Nd:YAG nanosecond pulsed laser (532 nm), max. energy— 200 mJ, wavelength range—230–880 nm	PCA, SVM, RBFNN, ELM, and CNN	Spectral range—380.01–860.04 nm	SVM, RBFNN, and ELM- calibration accuracy—96.6%, prediction accuracy—86.7%, CNN—100%, 96.7%	He et al., 2020
6.	Ber	Microbial assessment	Round pallets of sample were prepared using hydraulic press	Q-switched Nd:YAG laser beam (1,064 nm), pulse duration—10 ns, pulse energy—17 mJ/pulse, repetition rate—10 Hz	PLS-DA	Spectral range—187–887 nm to assess elemental compositional changes	89%	Sharma et al. (2022)
7.	Soyabean oil	Toxins (Fenthion)	Using a micropipette, sample oil is sucked and dispersed in ceramic sheet cavity, which undergoes LIBS analysis	Nd:YAG laser (1,064 nm), max. energy—200 mJ, pulse width—8 ns, operating frequency—1 Hz, focusing lens—100 mm	ELM, Boruta-ELM, GA-Boruta-ELM	Spectra in the band ranges 210~219 nm and 250~258 nm	GA-Boruta-ELM model has a superior prediction ability with R_P^2—0.9962, $RMSE_P$—11.005, and $MAPE_P$—1.66%	Ding et al. (2024)

(*Continued*)

TABLE 4.1 (*Continued*)
Applications and Accuracy of LIBS in Food Analysis

Sr. No.	Food Sample	Component Detection	Sample Preparation	LIBS Setup Specifications	Data Analysis Methods/Models	Elements Emission Lines (nm)	Accuracy	References
8.	Chard	Pesticides	Chard leaves are washed, dried, pulverized, and sieved, and pallets are prepared	Nd:YAG laser (1,064 nm, 100 mJ/pulse), repetition rate—2 Hz, focal length lens—200 mm	PCA, LDA, and MANOVA	C—415.3 nm, P—253.6 nm, S—416.26 nm, Cl—725.66 nm	MANOVA—95% confidence level, LDA— 9.5% classification error, PCA—77%	Martino et al., 2021
9.	Apple	Pesticide residue (Chlorpyrifos)	Apple pieces are air-dried	Q-Switched Nd:YAG laser (1,064 nm), pulse energy—200 mJ, pulse width 3–5 ns, repetition rate—20 Hz	PCA	P (213.62 and 214.91 nm), S (393.33 and 396.89 nm), and Cl (837.594 nm)		Ma & Dong, 2014
10.	Sheep colostrum	Protein evaluation	Sample frozen at −20°C till analysis	Q-switched Nd: YAG (1,064 nm) laser, 5 ns laser pulse, 150 mJ energy, repetition rate—1 Hz	PCA, ANOVA	CN— 386.16 nm, 387.14 nm, C_2—516.52 nm, Ca—422.67 nm	PCA—95.8% PC1 and PC2—80.8%	Abdel-Salam et al. (2019)
11.	Canned tuna fish	Protein evaluation	Content of can drained, homogenized, and pressed into tablets	Q-switched Nd: YAG (1,064 nm) laser, 5 ns laser pulse, 50 mJ energy, repetition rate—1 Hz	Conventional method of protein estimations validates the results	CN—388.34 nm, C_2—516.52 nm	$R^2 = 0.99$	Abdel-Salam et al. (2018)
12.	Rice grain	Grains quality	Sample cryogenically ground and pressed into pallets	Q-switched Nd: YAG (1,064 nm) laser, 8 ns laser pulse, 50 mJ energy, lens focal distance—10cm	PCs, PCA, K-NN	300–600 nm range	94% prediction accuracy	Perez-Rodriguez et al. (2023)

13.	Maize kernels	Genetically modified	Sample dried, ground, and pressed to form pallets	Q-switched Nd: YAG (532 nm) laser, 8 ns laser pulse, 200 mJ energy, repetition rate—1–10 Hz	PCA, PLS-DA, SIMCA, ELM	Range—235–850 nm	ELM—100%	Liu et al. (2018)

Abbreviations: MPE: mean prediction error; NIST: National Institute of Standards and Technology; PLSR: partial least square regression; SVM: support vector machine; ERTC: extremely randomized trees classifier; RFC: random forest classifier; XGBoost: extreme gradient boosting classifier; RBFNN: radial basis function neural network; ELM: extreme learning machine; Boruta-ELM: Boruta extreme learning machine; GA-Boruta-ELM: genetic algorithm Boruta extreme learning machine; MANOVA: multivariate analysis of variance; K-NN: k-nearest neighbors; SIMCA: soft independent modeling of class analogy; Tukey HSD: Tukey's honest significant difference; R^2_p: coefficients of determination in prediction; ICP-OES/MS: inductively coupled plasma optical emission spectroscopy/mass spectrometry; LIBS-LIF: laser-induced fluorescence combined with laser-induced breakdown spectroscopy; TR-LIBS: time-resolved laser-induced breakdown spectroscopy

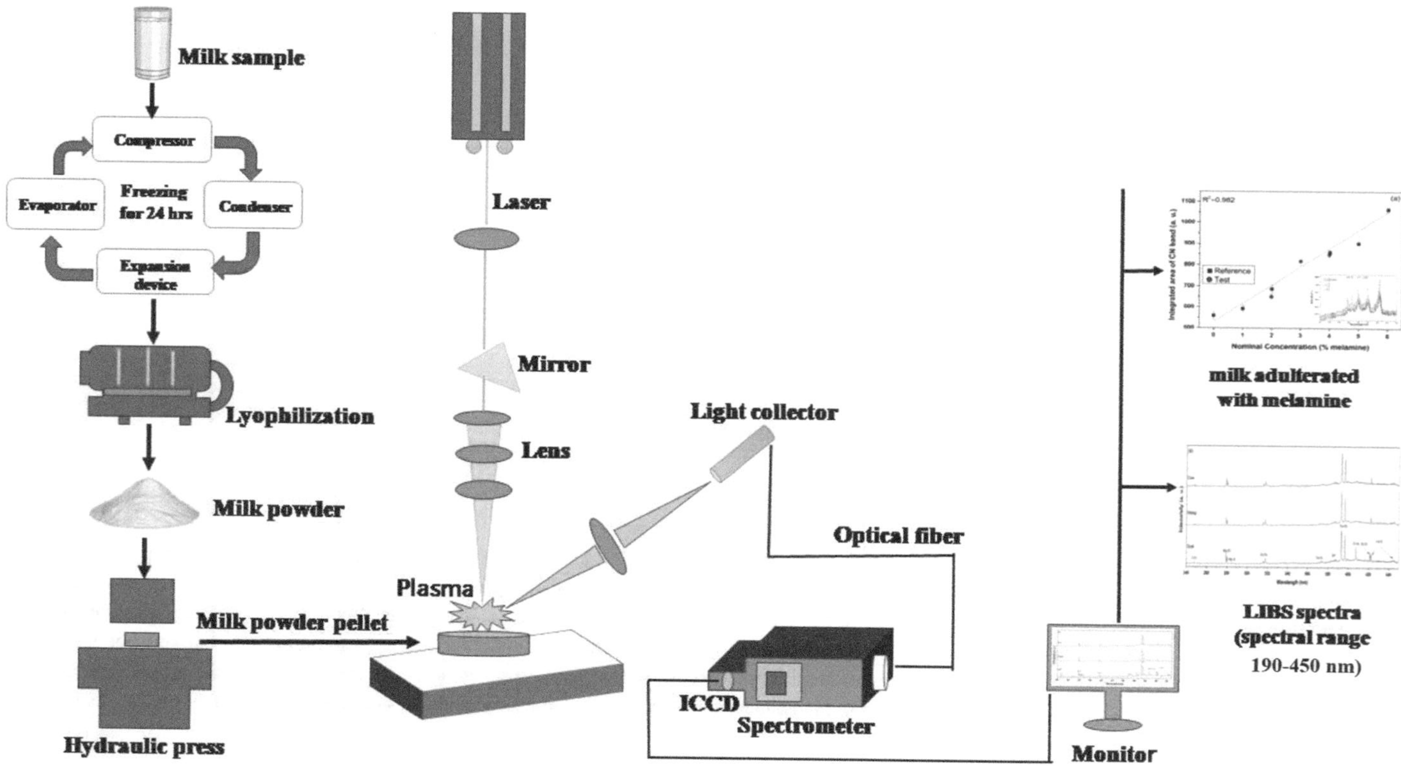

FIGURE 4.1 Diagrammatic representation of milk sample preparation for LIBS analysis and LIBS setup.

Source: Adopted from Moncayo et al. (2017)

4.6.1.2 Tea

Tea is the second most consumed beverage worldwide, following water. Tea leaves contain several chemical species, including Cu, Zn, Fe, Mg, and others. Cr, Br, Cd, and Mn are hazardous, even at low concentrations. The LIBS spectra for the tea samples are within wavelength range of 250–350 nm. Atomic and ionic emission lines identified within this range are Fe, Cr, Ca, and Si. The atomic emission lines identified within the wavelength range of 760–800 nm are K, Br, Cu, and Si. Variation in the emission line intensities across different tea samples indicates different concentration of elements. Quantitative LIBS analysis involves using the intensities of ionic and atomic emission lines to measure species in the sample. This is done by creating a calibration curve based on a set of samples with known concentrations, allowing for accurate quantification. A study found that tea samples contain Fe, Cr, K, Br, Cu, Si, and Ca in concentrations in ranges of 378–656, 96–124, 1421–6785, 99–1476, 17–36, 2–11, and 92–130 mg L^{-1} respectively. Detection limits were estimated to be 22, 12, 14, 11, 6, 1, and 12 mg L^{-1} for these elements, respectively (Gondal et al., 2016).

4.6.1.3 Honey

Honey is a value-added product, it is susceptible to adulteration with less expensive substances like HFSC or sugarcane syrup for financial gain (Pilizota & Tiban, 2009). The spectra could be obtained using a solid analysis. LIBS system experiments shall be conducted in air under ambient pressure, with a spark discharge system to reheat the plasma. Adulterant detection in honey through LIBS analysis was assessed using PLS-DA classification models. A study found that adulterated honey samples showed positive values in specific emission lines. The lines at 394.86 nm (Ca I), 396.91 nm (Ca II), and 397.38 nm (Ca I) indicated the presence of HFCS, while the lines at 251.75 nm (Fe I), 252.15 nm (Fe I), 253.43 nm (Fe II), and 395.58 nm (Fe I) revealed the presence of sugarcane syrup. The study concluded that higher Ca content in HFCS and higher Fe content in sugarcane syrup allowed for the identification of the adulterant in honey (Nespeca et al., 2020).

4.6.1.4 Brown Rice Flour

Products with low nutritional value are often mixed with products with high nutritional value. As a kind of food with high nutritional value, brown rice flour is always mixed with some low-priced food to obtain additional economic benefits, such as sorghum flour, talc powder, and corn flour. To facilitate LIBS analysis, the powdered sample should be compressed into pallets using controlled pressure. In the LIBS experimental setup, a laser pulse is generated from the laser source and directed onto the sample surface using a lens, resulting in the formation of plasma for spectral analysis. The plasma emission is collected by the spectrometer through a light collector and fiber optic system, allowing for the elemental composition of the sample to be analyzed in detail. LDA, NB, and support vector machine (SVM) models are effective for qualitative analysis of adulterants in the sample. The accuracy of the classification models generally follows an initial increase, then decreases with an increase in delay time. PSLR model provides a quantitative analysis of the adulterant ratio, with an LOD of 7.930%. based in LIBS. According to a study, LOD based on LIBS ranged from 8% to 51%. However, the LOD of the five adulterants based on TR-LIBS decreased from about 4% to 19%. This implies that TR-LIBS significantly enhances the detection performance (Ma et al., 2022).

4.6.1.5 In Meat

LIBS analysis is a valuable method for species differentiation in meat through protein examination. Sample preparation begins with fat removal, grinding, and subsequent protein extraction. Following this, the meat undergoes drying to a constant weight, fat removal using Soxhlet extraction, and milling for bulk protein extraction. For myofibrillar protein isolation, homogenization in a phosphate buffer, centrifugation, and dissolution of resulting pellets in KCl phosphate buffer are employed. Further centrifugation yields supernatant rich in myofibrillar protein, which undergoes isoelectric point precipitation to separate actin and myosin proteins. The bulk protein, actin, and myosin were washed, dried, ground using a mill, and then compressed into pellets using a hydraulic press. The setup for LIBS involves a laser and a lens to focus the laser beam on the sample. The plasma emission is collected and recorded by the spectrometer. Multivariate techniques, PCA and PLS, were used for data analysis. The recorded LIBS spectra were in the range of 188–900 nm, and PLS analysis calculated LOD values for beef adulteration with chicken and pork meat as 2.84% and 3.89%, respectively, using bulk proteins (Sezer et al., 2020).

4.6.2 Minerals' Detection in Various Food

4.6.2.1 In Green Leafy Vegetables

LIBS technique is useful to carry out the mineral analysis of green leafy vegetables. To prepare a sample for LIBS analysis, green leafy vegetables are washed and sorted. The sample is vacuum-dried and powdered using a grinder, and pallets are formed using a hydraulic press that applies pressure. The LIBS setup consists of a laser that delivers pulse energy. The laser induces plasma on the sample surface, and a lens is used to collect the emission from the plasma. The collected light is then transferred to a spectrometer using a fiber cable. In a study, the spectra of fresh green leaves were recorded. The spectral lines of Si, Mg, Ti, Fe, and Ca were present, confirming the presence of these minerals. The mineral content in samples of fresh leaves from chenopodium, spinach, fenugreek, chickpea, and mustard was studied by recording their LIBS spectra. All samples contained persistent spectral lines of Si (288.1, 251.4 nm), Mg (285.1, 279.5, 279.8 nm), Ti (334.9, 336.1, 337.2 nm), Fe (344.1, 238.3, 240.4 nm), Ca (422.6, 393.3, 396.8, 443.5, 445.5 nm), C (247.8, 229.6 nm), and Al (308.2, 309.2 nm), confirming the presence of these minerals in them. The LIBS spectra were recorded within the wavelength range of 200 to 500 nm. The presence of Na, K, N, and O is confirmed by LIBS spectra from 700 to 825 nm (Shukla et al., 2016) (Figure 4.2).

4.6.2.2 In Poultry Meat

It is important to assess the calcium content in MSM to ensure the quality of poultry products. Calcium is a crucial factor in determining the quality of MSM, as it

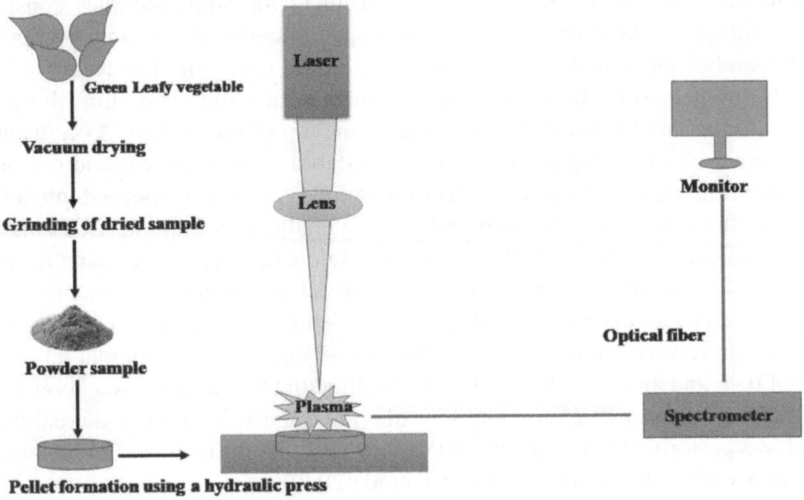

FIGURE 4.2 Diagrammatic representation of green leafy vegetable sample preparation for mineral analysis and LIBS setup.

Source: Adopted from Shukla et al. (2016)

serves as an indicator of bone residue present in the meat. According to international standards, the calcium content in fresh MSM should not exceed 1,000 ppm. LIBS is an emerging technique that shows promise due to its speed and its ability to directly detect mineral elements with minimal sample preparation.

The sample must be immediately frozen upon collection at −20°C and kept frozen until analysis. The sample was gently thawed in a controlled environment at 5°C to preserve its integrity. After thawing, the sample was stirred to mix meat juice back into the sample to ensure an even distribution of moisture. Subsequently, the samples were carefully transferred into square weighing dishes. To maintain uniformity, the samples were covered with plastic wrap and then pressed flat upside down on a table surface. Finally, the samples were carefully stored in ice to preserve their integrity until the time of measurement. The LIBS setup comprises a laser and a spectrometer equipped with a detector that needs to be cooled to −67°C during measurement. The laser is focused on the sample using a lens with a focal length, which creates a spot. The light emitted from the plasma is collected using a lens with a specific focal length positioned at an angle away from the target, relative to the laser axis. The spectrum is recorded in the range between 200 and 800 nm. The focusing lens and collection lens must be cleaned with deionized water and a cleaning agent after each sample to remove any meat juice. Data can be analyzed utilizing a PLS calibration model (Andersen et al., 2016).

4.6.2.3 In Liquid Infant Formula

LIBS technique offers the potential to directly analyze liquid food products. Specifically, it can be employed as a predictive tool for determining the calcium content in ready-to-eat infant formula. The LIBS instrument for liquid samples consists of a laser with pulse duration, coupled with a spectrometer. The system comprises a liquid chamber that has a rotary wheel and a liquid reservoir. The rotary wheel is partially submerged in the liquid, which creates a thin film. This thin film is then exposed to the laser beam that is focused on the top of the system. As a result, the laser interacts with the liquid. It is possible to stabilize the thickness and uniformity of a liquid film. Approximately 40 ml of the samples needs to be poured into a liquid reservoir after rinsing the reservoir and wheel with distilled water for LIBS analysis. The data was analyzed using PSLR modeling. Before that, a preprocessing technique was applied to the spectra. In the study, the correct arrangement of calcium content was indicated by an examination of the calcium emission lines spectra at 422.67 nm (Ca I). A study was conducted, and it was found that the PLSR model had an R^2_C value of 0.96 and an R^2_{cv} value of 0.89, indicating that the model was a good fit. The RMSECV value was 10.23 mg per 100 mL^{-1}, and the validation of the calibration model was performed to test the model's robustness, resulting in an RMSEP value of 6.45 mg per 100 mL^{-1} (Cama-Moncunill et al., 2020).

4.6.2.4 In Wheat Flour

Accurate methods for evaluating micronutrients are essential for determining food fortification needs and understanding its nutritional impact. LIBS technique has been effectively employed in the assessment of the elemental composition of wheat grain tissues. To prepare a sample for LIBS analysis, it is essential to ensure that the sample is thoroughly mixed with cellulose in a cryogenic mill equipped with a self-contained liquid nitrogen bath, followed by a 5-minute pre-cooling period proceeded to homogenize the samples for 30 minutes. The preparation of pallets (test sample) involves the transfer of powdered material to a die set followed by the application of force. Subsequently, the pallets must be stored in a desiccator before analysis. The LIBS instrumentation comprises a laser that generates pulses. The laser is directed onto the pallet surface using a convergent lens with a specific focal length. The sample is positioned within a sample holder, and the plasma emission is directed through a telescope with a silica lens. Subsequently, the emission is channeled into an optical fiber which is then linked to a spectrometer. The data processing involves averaging spectra. A study found that the correlation coefficients of P, K, Ca, Mg, S, Fe, Mn, and Zn determined by LIBS and/or EDXRF varied from 0.9705 for Zn to 0.9990 for Mg using LIBS and from 0.9306 for S to 0.9974 for K using EDXRF. The measurements' coefficients of variation ranged from 1.2 to 20% for LIBS and from 0.3 to 24% for EDXRF (Peruchi et al., 2014).

4.6.3 Identification of the Geographical Origin of Food

4.6.3.1 Honey

The chemical composition of honey is known to vary regionally and is influenced by environmental factors such as climate and altitude. To identify the geographical origins of honey, LIBS combined with multivariate methods has been employed. This analytical approach has shown promise in accurately discriminating the source of honey based on its unique chemical fingerprint. The use of LIBS and multivariate analysis can provide a valuable tool for the authentication and quality control of honey. It is noteworthy that no sample preparation is requisite for the analysis of honey samples through LIBS. The honey samples are accurately weighed and subsequently deposited onto plates for the LIBS analysis. This procedure ensures the elimination of any extraneous variables that may otherwise interfere with the analysis process. The experiment involves utilizing a laser to ablate a sample, followed by the collection of plasma light through the achromatic lens. The collected light is subsequently conveyed to the spectrograph and ultimately detected by a charged-coupled device. The multivariate analysis involved the utilization of spectra from the same origin for representing regional characteristics. PCA was utilized for quantitatively visualizing the distribution of honey (the calibration samples) with score plots. LDA and SVM were employed for the quantitative classification of geographical origins. LDA and SVM are two widely recognized multivariate analysis algorithms. Additionally, a one-way ANOVA test was conducted to examine the regional difference in LIBS emissions. Each peak wavelength in the LIBS spectrum corresponds to a specific element present in the NIST database. The peak intensity is proportional to the elemental concentration. According to a previously conducted study, the accuracy rates of LDA and SVM models for acacia honey were 74.1% and 82.6%, respectively. Meanwhile, the accuracy rates of LDA and SVM models for multifloral honey were 98.6% and 99.7%, respectively (Zhao et al., 2020) (Figure 4.3).

4.6.3.2 Olive Oil

Ensuring the quality of olive oil is crucial as adding cheaper or lower-quality oil is a common adulteration practice that raises ethical, health, and financial concerns. The designations of olive oil as Protected Designation of Origin (PDO) and Protected Geographical Indication (PGI) serve as crucial indicators of the product's authenticity and quality and also function as a guarantee of its brand identity. These designations are legally recognized and regulated by the European Union and are granted only to olive oils that meet strict criteria related to the geographic origin of the olives, the production process, and the sensory characteristics of the oil. For businesses operating in the olive oil industry, obtaining PDO/PGI certification can provide a competitive advantage by enhancing consumer trust and loyalty and by enabling differentiation from other products in the market. LIBS technique represents a valuable tool for the identification of the geographical origin of olive oil. Through the analysis of the elemental composition of the sample, this technique allows for accurate and reliable determination of the area where the olives are grown. The olive oil samples are carefully collected and stored in opaque glass bottles to preserve their quality. The bottles are maintained at a temperature ranging between 2°C and 4°C throughout the storage period. Before conducting the laser measurements, the oil samples are left to equilibrate at room temperature for approximately 4 hours. This step ensures that the samples are at a consistent temperature before the analyses are conducted. To conduct the LIBS experiment, a small glass Petri dish is used to hold around 2 ml of the sample. The sample is then subjected to a laser beam that is focused on it through a lens to generate plasma. The plasma radiation is collected

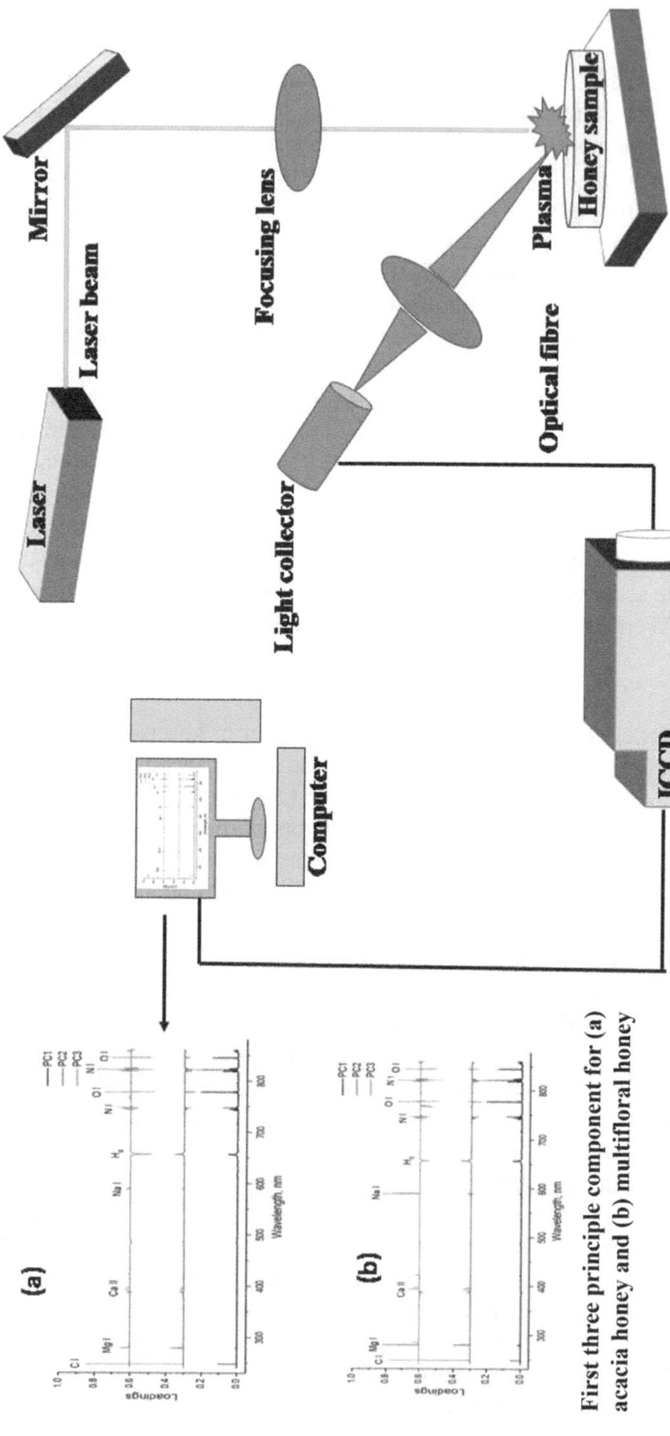

FIGURE 4.3 Diagrammatic representation of LIBS analysis of honey. First three principle component for (a) acacia honey and (b) multifloral honey

Source: Adopted from Zhao et al. (2020)

by another lens and transferred to an optical fiber that is coupled to a spectrometer for spectral analysis. The spectrometer is equipped with a detector to measure the results accurately. The utilization of machine learning techniques in the analysis of LIBS spectroscopic data has been implemented. The spectrum shows atomic hydrogen emissions at 656.3 and 486.1 nm, oxygen emissions at 777.2–777.5 nm and 615.7, 715.7, and 926.4 nm, as well as nitrogen emissions at 742.4, 744.2, 746.8 and 818.8, 821.6, 824.2 nm. Carbon lines are also present at 247.9, 795.2, 833.5, 906.2, and 940.6 nm (Gyftokostas et al, 2021).

4.6.4 Microbial Assessment

The LIBS technique is a valuable method for identifying various types of bacteria. For LIBS analysis, it is essential to choose ferroalloy plates as the substrate. Before conducting any experiments, all the instruments used must be sterilized. To transfer bacteria from the slanted solid medium, an inoculation ring can be used to inoculate it into another medium. The medium should then be placed inside an incubator and purified. To grow bacteria, the bacterial colony can also be inoculated in a liquid medium and placed inside a shaking box with constant temperature and rotating speed. The bacteria can then be placed into a visible light spectrometer. It is important to maintain a relatively constant bacteria concentration each time. For the experiment, the bacteria solution should be pipetted onto the surface of the metal substrate, spread evenly, and then dried. The experimental setup for LIBS includes a pulse laser, a lens that focuses the laser beam onto the sample surface, a vibrational mirror, a field mirror, and a spectrometer. Spectral pretreatment is necessary for data analysis. PCA and SVM models can be used to identify bacteria with an accuracy rate of 98.93% (Liang et al., 2023).

4.6.4.1 Ber

The pathogenic bacterium *Agrobacterium tumefaciens* causes crown gall disease in *Ziziphusmauritiana* (Indian Ber) plants. To investigate this disease, the LIBS technique could be employed to detect chemical changes. To identify the bacteria, Gall samples from diseased *Ziziphusmauritiana* plants are collected, cut into small segments, and sterilized before washing with sterile water. These segments are then suspended in distilled water, gently squeezed, and the suspension streaked onto nutrient agar Petri dishes. The plates are incubated to allow bacterial growth for further analysis. In the LIBS analysis process, the samples are ground, and pellets are prepared using a hydraulic press. The LIBS spectra for these palletized *Ziziphusmauritiana* (Indian ber) samples are then recorded. The optical emission from the laser-induced plasma is collected using a lens and directed to a CCD spectrometer. All LIBS spectra are recorded in an open-air environment for analysis. The LIBS spectra were recorded in the range of 187–887 nm to analyze changes in elemental composition. The results from LIBS revealed decreasing trends in Ca and increasing trends in Mg, K, Si, Mn, and Na in the diseased samples of *Ziziphusmauritiana* (Indian ber) (Sharma et al., 2022) (Figure 4.4).

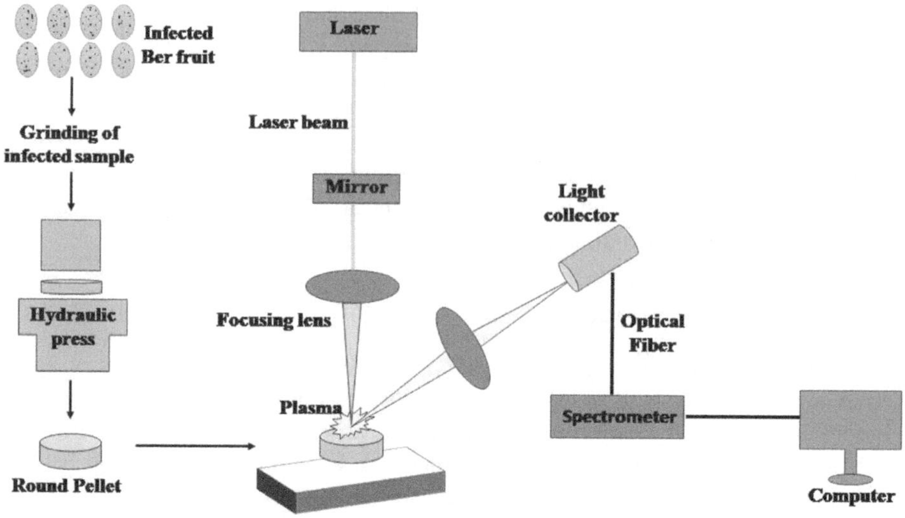

FIGURE 4.4 Diagrammatic representation of ber sample preparation for LIBS analysis. *Source:* Adopted from Sharma et al. (2022)

4.6.5 Detection of Toxins/Pesticide Residue

4.6.5.1 Determination of Fenthion in Soybean Oil

Analyzing organophosphorus pesticides in vegetable oil involves two main steps: extracting the pesticide and instrumental detection. A drop of oil is deposited on a ceramic sheet surface. The ceramic sheet, coated with the oil film, is then allowed to stand to ensure even distribution. The LIBS experimental setup includes a laser that is focused onto the sample surface through a lens to create plasma. The spectral data from the plasma is collected by a fiber optic and transferred to a spectrometer. This combination of substrate-assisted LIBS technique and the GA-Boruta-ELM model shows promise for quantitatively determining pesticide residues in edible vegetable oils. The model achieved R_P^2 of 0.9962, with an $RMSE_P$ of 11.005 and $MAPE_P$ of 1.66% (Ding et al, 2024).

4.6.5.2 Detection of Pesticide Residue on Apple Surface

Chlorpyrifos residue on apple surfaces was detected using the LIBS technique, which identifies pesticide contamination. To prepare the sample, apple pieces were air-dried under controlled conditions. The experimental setup included a laser and spectrometer. The laser was focused on the sample surface, and the resulting laser-induced plasma signal was collected via fiber and transferred to the spectrometer. Pretreatment was required before data analysis. A study revealed that the spectral characteristics of phosphorus (213.62 and 214.91 nm), sulfur (393.33 and 396.89 nm), and chlorine (837.594 nm) in chlorpyrifos were detected by LIBS. Chemometric methods revealed distinct differences in spectra between clean apples and those sprayed with various concentrations of chlorpyrifos (Ma & Dong, 2014).

4.6.6 Quality Evaluation of Different Food Products

4.6.6.1 Protein Evaluation

4.6.6.1.1 In Sheep Colostrum

To conduct LIBS analysis on colostrum, a sample must first be collected and sent for microbiological analysis. After completion of the analysis, the sample is frozen at a temperature of −20°C until spectroscopic analysis can be conducted. Before analysis, the frozen sample is thawed, and a droplet is carefully distributed onto an ashless filter paper. The sample is then allowed to partially dry and spread evenly across the filter paper for analysis. LIBS setup consists laser which produces laser pulses that are focused onto the sample using a lens. Fiber optic cables are used to collect and guide the emission light of the laser-produced plasma to the spectrometer. A spectrometer, coupled with an ICCD camera, is used to disperse and detect the collected light. Finally, the spectrum is analyzed using LIBS software. The normalization of spectra is a critical aspect of analysis that must be considered to avoid fluctuations. In an experiment, it was found that the intensities of CN and C_2 increase with higher protein content. The results demonstrate that higher CN intensity corresponds to lower total bacterial count in colostrum (Abdel-Salam et al., 2019).

4.6.6.1.2 In Canned Tuna Fish

Tuna is a popular fish that provides an excellent source of high-quality protein. To prepare the sample for LIBS analysis, the contents of the can are drained and thoroughly homogenized before a sample is taken and pressed into a tablet. For LIBS measurement, a laser created short pulses of light focused on a sample of tuna fish. Emitted light was collected by an optical fiber, directed to a spectrometer, and detected by a camera controlled with specialized software. In an experiment, the spectral analysis of the obtained data revealed that the intensities of CN and C_2 molecular bands, which appear at 388.34 nm and 516.52 nm, respectively, are indicative of the protein content and thus the quality of different tuna fish (Abdel-Salam et al., 2018).

4.6.7 Cereal Quality

4.6.7.1 Rice Grain

The labeling of rice according to its quality features depends on the inspection of images and/or visual examination. To prepare the rice sample for LIBS analysis, cryogenic grinding was used, which involves pre-freezing, grinding, and then freezing between two milling steps. Powdered samples were then turned into pellets by applying pressure. Rice pallets were analyzed using a LIBS setup which includes a laser, lens for focalization, and an optical fiber for transmission of plasma emission to the spectrometer. Preprocessing is performed before data analysis to eliminate unwanted variations and ensure reliable results. The recorded emission lines were identified within the range of 300–600 nm (Perez-Rodriguez et al., 2023) (Figure 4.5).

4.6.7.2 Detection of Genetically Modified Food

The LIBS technique can be used to identify GM maize kernels. To prepare the sample, pretreatment is carried out that involves drying the maize kernels in an

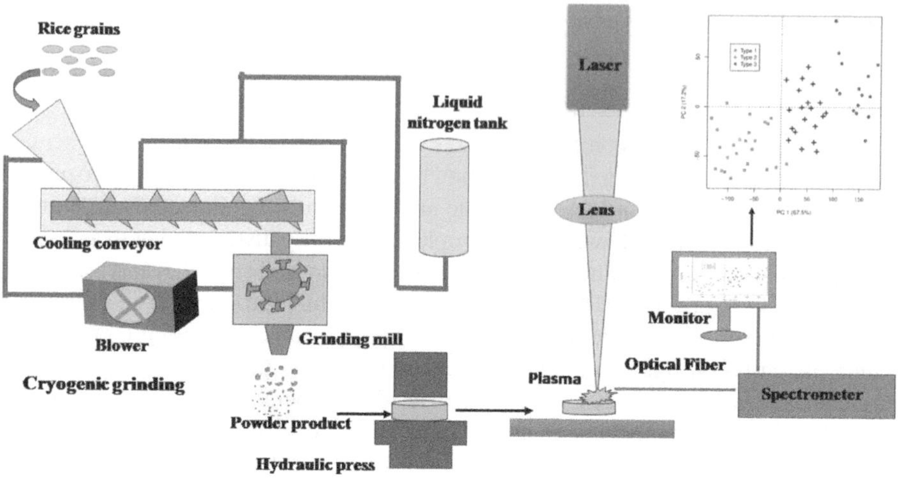

FIGURE 4.5 Diagrammatic representation of sample preparation of rice for LIBS analysis. *Source:* Adopted from Perez-Rodriguez et al. (2023)

oven at 60°C for 6 hours. Once dried, the kernels are ground into a fine powder using a grinder. The powder sample is then pressed into a tablet using a pressing machine that applies pressure. The resulting GM maize tablet can then be collected as the experimental sample for LIBS analysis. The LIBS system is composed of a pulsed laser that generates laser pulses to ablate the sample. Additionally, the system includes a spectrometer equipped with an ICCD camera to detect optical signals, along with other accessories such as lenses, mirrors, and an energy meter. To prepare for an experiment, the system must be preheated, and the experiment must be conducted in an air environment. Preprocessing of data is crucial before conducting data analysis to eliminate any environmental noises present in the raw data. Chemometric methods can be utilized to distinguish between GM and non-GM crops. PCA can serve as a selection method. Among various models, the PCA loading-ELM model is most suitable for practical application in GM maize identification. In a study, it was found that mineral element content in GM maize was lower than in non-GM maize. It has demonstrated 100% identification accuracy in both calibration and prediction sets (Liu et al., 2018).

4.7 CHALLENGES AND LIMITATIONS

The technology of LIBS has several challenges and limitations that need to be addressed (Yang et al., 2022; Yu et al., 2020). The complexity of elements in food products poses a challenge for detecting heavy metal elements using LIBS. The technique may struggle to accurately detect and quantify these elements due to their presence in varying concentrations and interferences from matrix components. Enhancing the LIBS signal is crucial for improving detection limits and accuracy. Various methods such as nanoparticle-assisted enhancement, laser pulse additions, and auxiliary devices have been explored. However, achieving consistent and effective signal enhancement remains a challenge. The presence of a wide variety of elements in food product samples can lead to interference in LIBS spectra. Matrix effects can affect the accuracy and precision of measurements, especially when analyzing multiple elements simultaneously. Foods have diverse compositions including carbohydrates, lipids, proteins, and additives. This complexity can lead to varied responses to laser interaction, impacting the reproducibility of LIBS signals due to small focused beam sizes and sample mass vaporized. In powdered materials like flour, differences in particle size distribution and composition can impact laser-sample interaction and signal quality. Pelletizing powdered samples is common, but choosing suitable binders for pellets can be challenging, as some may not be food-safe. High moisture or fat content in foods can hinder LIBS signal quality, particularly for trace element detection. Sample drying or defatting is a solution, but these preparation steps can be time-consuming and may eliminate the advantages of LIBS. Certain foods require specific sample preparations, such as forming dry sheets or wafers, which must maintain sample composition. However, temperature-sensitive samples must be carefully handled to avoid altering their composition during preparation. Achieving adequate sensitivity for trace element detection in high-moisture or high-fat foods poses challenges. Using higher energy levels or specialized system designs like double pulse or plasma spatial confinement may be necessary, but these methods can introduce complexity and may not always be efficient. Optimizing LIBS systems for food analysis, including beam focus, energy levels, and plasma control, is crucial to overcome matrix effects and achieve reliable results.

4.8 FUTURE SCOPE

The technology used by LIBS is still in need of further development and improvement. This applies particularly to the following areas (Chavan et al., 2023). Currently, the use of LIBS applications in food processing is limited to the laboratory or pilot phase, which means further research and development are required for industrial-scale implementation. Overcoming challenges such as regulating heat damage and understanding the production mechanism of laser-induced components are crucial to successfully implementing this technology. Additionally, assessing consumer acceptance of laser-processed foods may require new regulations to guide the growth of this industry. Extensive research is required to understand the effects and mechanisms of laser-induced changes in food. Optimization of laser parameters for specific applications, development of mathematical models for predictive analysis, and exploration of new research avenues are necessary to enhance understanding. Further research is needed to develop optimized equipment, integrate it with other technologies for better performance, and explore its applications in emerging fields like precision agriculture. To promote research and development in LIBS technology for food processing, it is necessary to have more government support, including funding and regulatory frameworks. Clear regulations and standards can help guide industry adoption, while incentives can encourage widespread use for food safety, quality, and sustainability.

4.9 CONCLUSION

LIBS has proven to be a valuable and versatile tool for elemental analysis in food products. It offers numerous advantages such as rapid analysis, minimal sample usage, and sensitivity to light elements, making it well suited for applications in food safety and quality control. However, challenges such as lower limits of detection and sample matrix interference must be addressed for optimal use. The study highlights LIBS' effectiveness in detecting adulteration, verifying authenticity, and assessing the quality of various food matrices including beef, rice, vegetables, oils, and more. Despite its potential, further research is needed to overcome challenges related to heavy metal detection, signal consistency, and reproducibility in complex food matrices. Additionally, efforts toward commercialization and broader implementation of LIBS in food processing require government support and continued exploration of its applications. Overall, LIBS emerges as a promising technology with the potential to revolutionize food analysis and ensure consumer safety and product quality.

ACKNOWLEDGMENT

The authors are thankful to Prof. Raghavendra Prasad Tiwari, Vice Chancellor of Central University of Punjab, Bathinda, India, and University of Allahabad for giving facilities to write this book chapter. Special thanks to Dr. Poonam Mishra for providing the opportunity to contribute to this book chapter.

REFERENCES

Abdel-Salam, Z.A., Abdel-Salam, S.A.M., Abdel-Mageed, I.I., & Harith, M.A. (2019). Evaluation of proteins in sheep colostrum via laser-induced breakdown spectroscopy and multivariate analysis. *Journal of Advanced Research*, 15, 19–25.

Abdel-Salam, Z.A., Alexeree, S.M.I., & Harith, M.A. (2018). Utilizing biosynthesized nano-enhanced laser-induced breakdown spectroscopy for proteins estimation in canned tuna. *SpectrochimicaActa Part B. Atomic Spectroscopy,* 149, 112–117.

Ahmed, N., Farooq, K., Shahida, S., ul Haq, K., Abbasi, S.A., Umar, Z.A., & Baig, M.A. (2021). Spectrochemical analysis of Pakistani bakery breads using laser induced breakdown spectroscopy. *Optik*, 226, 165743.

Anabitarte, F., Cobo, A., & Lopez-Higuera, J.M. (2012). Laser-induced breakdown spectroscopy: Fundamentals, applications, and challenges. *ISRN Spectroscopy*, 1–12.

Andersen, M.B.S., Frydenvang, J., Henckel, P., & Rinnan, A. (2016). The potential of laser-induced breakdown spectroscopy for industrial at-line monitoring of calcium content in comminuted poultry meat. *Food Control*, 64, 226–233.

Blutinger, J.D., Meijers, Y., Chen, P.Y., Zheng, C., Grinspun, E., & Lipson, H. (2018). Characterization of dough baked via blue laser. *Journal of Food Process Engineering*, 232, 56–64.

Cama-Moncunill, X., Markiewicz-Keszycka, M., Cullen, P.J., Sullivan, C., & Casado-Gavalda, M.P. (2020). Direct analysis of calcium in liquid infant formula via laser-induced breakdown spectroscopy (LIBS). *Food Chemistry,* 309, 125754.

Casado-Gavalda, M.P., Dixit, Y., Geulen, D., Cama-Moncunill, R., Cama-Moncunill, X., Markiewicz-Keszycka, M., Cullen, P.J., & Sullivan, C. (2017). Quantification of copper content with laser induced breakdown spectroscopy as a potential indicator of offal adulteration in beef. *Talanta*, 169, 123–129.

Chavan, P., Yadav, R., Sharma, P., & Jaiswal, A.K. (2023). Laser light as an emerging method for sustainable food processing, packaging, and testing. *Foods*, 12, 2983.

Costa, V.C., Augusto, A.S., Castro, J.P., Machado, R.C., Andrade, D.F., Babos, D.V., Speranca, M.A., Gamela, R.R., & Pereira-Filho, E.R. (2019). Laser induced-breakdown spectroscopy (LIBS): History, fundamentals, applications and potentials. *Quimica Nova*, 42, 527–545.

Debras-Guedon, J., & Liodec, N. (1963). The use of a ruby laser as an excitation source in emission spectroscopy. *ComptesRendus de l' Académie des Sciences*, 257, 3336–3339.

Diedrich, J. (2007). *Laser-Induced Breakdown Spectroscopy on Bacterial Samples*, Wayne State University, Detroit, MI, pp. 6–7 (MS. thesis).

Ding, Y., Wang, Y., Chen, J., Chen, W., Hu, A., Shu, Y., & Zhao, M. (2024). Substrate-assisted laser-induced breakdown spectroscopy combined with variable selection and extreme learning machine for quantitative determination of fenthion in soybean oil. *Photonics*, 11, 129.

Dutouquet, C., Gallou, G., Le Bihan, O., Sirven, J.B., Dermigny, A., Torralba, B., et al. (2014). Monitoring of heavy metal particle emission in the exhaust duct of a foundry using LIBS. *Talanta*, 127, 75–81.

Gondal, M.A., Habibullah, Y.B., Baig, U., & Oloore, L.E. (2016a). Direct spectral analysis of tea samples using 266 nm UV pulsed laser-induced breakdown spectroscopy and cross validation of LIBS results with ICP-MS. *Talanta*, 152, 341–352.

Gondal, M.A., Shemis, M.A., Khalil, A.A., Nasr, M.M., & Gondal, B. (2016b). Laser produced plasma diagnosis of carcinogenic heavy metals in gallstones. *Journal of Analytical Atomic Spectrometry*, 31, 506–514.

Gunaratne, T., Kangas, M., Singh, S., Gross, A., & Dantus, M. (2006). Influence of bandwidth and phase shaping on laser induced breakdown spectroscopy with ultrashort laser pulses. *Chemical Physics Letters*, 423, 197–201.

Gyftokostas, N., Stefas, D., Kokkinos, V., Bouras, C., & Couris, S. (2021). Laser-induced breakdown spectroscopy coupled with machine learning as a tool for olive oil authenticity and geographic discrimination. *Scientific Reports*, 11, 5360.

Hahn, D.W., & Omenetto, N. (2012). Laser-induced breakdown spectroscopy (LIBS), part II: Review of instrumental and methodological approaches to material analysis and applications to different fields. *Applied Spectroscopy*, 66(4), 347–419.

Hanif, M., Salik, M., & Baig, M. (2011). Quantitative studies of copper plasma using laser induced breakdown spectroscopy. *Optics and Lasers in Engineering*, 49, 1456–1461.

Harhira, A., Bouchard, P., Rifai, K., Haddad, J. E., Sabsabi, M., Blouin, A., & Laflamme, M. (2017). Advanced laser-induced breakdown spectroscopy (LIBS) sensor for gold mining, In *The 56th Conference of Metallurgists COM*. 13.

He, Y., Zhao, Y., Zhang, C., Li, Y., Bao, Y., & Liu, F. (2020). Discrimination of grape seeds using laser-induced breakdown spectroscopy in combination with region selection and supervised classification methods. *Foods*, 9(2), 199.

Jabbar, A., Akhtar, M., Mahmood, S., Ahmed, R., & Baig, M.A. (2020). Determination of major inorganic nutrients in maize tissues by calibration-free laser induced breakdown spectroscopy. *Analytical Letters*, 53(8), 1328–1341.

Jantzi, S.C., Motto-Ros, V., Trichard, F., Markushin, Y., Melikechi, N., & De Giacomo, A. (2016). Sample treatment and preparation for laser-induced breakdown spectroscopy. *SpectrochimicaActa Part B: Atomic Spectroscopy*, 115, 52–63.

Juve, V., Portelli, R., Boueri, M., Baudelet, M., & Yu, J. (2008). Space-resolved analysis of trace elements in fresh vegetables using ultraviolet nanosecond laser-induced breakdown spectroscopy. *Spectrochimica Acta Part B*, 63, 1047–1053.

Kaiser, J., Novotny, K., Martin, M.Z., Hrdlicka, A., Malina, R., Hartl, M., et al. (2012). Trace elemental analysis by laser-induced breakdown spectroscopy—biological applications. *Surface Science Reports*, 67(11–12), 233–243.

Kim, G., Kwak, J., Choi, J., & Park, K. (2012). Detection of nutrient elements and contamination by pesticides in spinach and rice samples using laser-induced breakdown spectroscopy (LIBS). *Journal of Agricultural and Food Chemistry*, 60(3), 718–724.

Korunchikov, A.I., & Yankovskii, A.A. (1966). Some aspects of the plasma uptake and spectrum excitation produced by laser radiation. *Journal of Applied Spectroscopy*, 5, 429–434.

Lastra-Mejıas, M., Aroca-Santos, R., Izquierdo, M., Cancilla, J.C., Mena, M.L., & Torrecilla, J.S. (2019). Chaotic parameters from fluorescence spectra to resolve fraudulent mixtures of fresh and expired protected designation of origin extra virgin olive oils. *Talanta*, 195, 1–7.

Liang, J., Wang, S., Luo, X., Zhang, Y., Chen, F., Mi, Z., Zhang, L., Wang, G., Zhang, W., Liu, Z., Ma, X., Ye, Z., Zhu, Z., Yin, W., & Jia, S. (2023). Non-contact bacterial identification and decontamination based on laser-induced breakdown spectroscopy. *Journal of Photochemistry and Photobiology B: Biology*, 244, 112719.

Liu, X., Feng, X., Liu, F., Peng, J., & He, Y. (2018). Rapid identification of genetically modified maize using laser-induced breakdown spectroscopy. *Food and Bioprocess Technology*, 12, 347–357.

Ma, F., & Dong, D. (2014). A measurement method on pesticide residues of apple surface based on laser-induced breakdown spectroscopy. *Food Analytical Methods*, 7, 1858–1865.

Ma, H., Shi, S., Zhang, D., Deng, N., Hu, Z., Liu, J., & Guo, L. (2022) Time-resolved laser-induced breakdown spectroscopy for accurate qualitative and quantitative analysis of brown rice flour adulteration. *Foods*, 11, 3398.

Martino, L. J.,D'Angelo, C.A., Marinelli, C., & Cepeda, R. (2021). Identification and detection of pesticide in chard samples by laser-induced breakdown spectroscopy using chemometric methods. *Spectrochimica Acta Part B: Atomic Spectroscopy*, 177, 106031.

Moncayo, S., Manzoor, S., Rosales, J.D., Anzano, J., & Caceres, J.O. (2017). Qualitative and quantitative analysis of milk for the detection of adulteration by Laser Induced Breakdown Spectroscopy (LIBS). *Food Chemistry*. 232, 322–328.

Nespeca, M.G., Vieira, A.L., Júnior, D.S., Neto, J.A.G., & Ferreira, E.C. (2020). Detection and quantification of adulterants in honey by LIBS. *Food Chemistry*, 311, 125886.

Perez-Rodriguez, M., Mendoza, A., González, L.T., Lima Vieira, A., Pellerano, R.G., Gomes Neto, J.A., & Ferreira, E.C. (2023). Rice labeling according to grain quality features using laser-induced breakdown spectroscopy. *Foods*, 12, 365.

Peruchi, L.C., Nunes, L.C., Carvalho, G.G.A. de, Guerra, M.B.B., Almeida, E., Rufini, I.A., Santos Jr., D., & Krug, F.J. (2014). Determination of inorganic nutrients in wheat flour by laser-induced breakdown spectroscopy and energy dispersive X-ray fluorescence spectrometry. *SpectrochimicaActa Part B*, 100, 129–136.

Pilizota, V., & Tiban, N.N. (2009). Category: Natural products advances in honey adulteration detection. *Food Safety Magazine*, 60–64.

Rakovsky, J., Cermak, P., Musset, O., & Veis, P. (2014). A review of the development of portable laser induced breakdown spectroscopy and its applications. *SpectrochimicaActa Part B: Atomic Spectroscopy*, 101, 269–287.

Ropodi, A.I., Panagou, E.Z., & Nychas, G.-J.E. (2016). Multispectral imaging (MSI): A promising method for the detection of minced beef adulteration with horsemeat. *Food Control,* 73, 57–63.

Salloom, H.T., Hamad, T.K., & Mohammed, A.H. (2021). Detection of harmful metal elements in rice using laser induced breakdown spectroscopy. *Journal of Physics: Conference Series*, 1818, 012178.

Santos Jr., D., Nunes, L.C., de Carvalho, G.G.A., da Silva Gomes, M., de Souza, P.F., de Oliveira Leme, F., dos Santos, L.G.C., & Krug, F.J. (2012). Laser-induced breakdown spectroscopy for analysis of plant materials: A review. *Spectrochimica Acta Part B: Atomic Spectroscopy*, 71, 3–13.

Senesi, G., Dell'Aglio, M., Gaudiuso, R., De Giacomo, A., Zaccone, C., De Pascale, O., Miano, T., Capitelli, M., 2009. Heavy metal concentrations in soils as determined by laser-induced breakdown spectroscopy (LIBS), with special emphasis on chromium. *Environmental Research*, 109, 413–420.

Sezer, B., Apaydin, H., Bilge, G., & Boyaci, I.H. (2019). Detection of Pistacia vera adulteration by using laser-induced breakdown spectroscopy. *Food Chemistry*, 99(5), 2236–2242.

Sezer, B., Bjelak, A., Velioglu, H.M., & Boyaci, I.H. (2021). Protein-based evaluation of meat species by using laser-induced breakdown spectroscopy. *Meat Science*, 172, 108361.

Shah, S.K.H., Iqbal, J., Ahmada, P., Khandaker, M.U., Haq, S., & Naeem, M. (2019). Laser induced breakdown spectroscopy methods and applications: A comprehensive review. *SpectrochimicaActa Part B: Atomic Spectroscopy*, 150, 49–79.

Sharma, N., Kumar, S., Lee, Y., Singh, V.K., & Singh, V.K. (2022). Spectroscopic investigations of healthy and diseased ber (ziziphusmauritiana) fruits using laser-induced breakdown spectroscopy in combination with partial least squares-discriminant analysis. *Arabian Journal for Science and Engineering*, 47, 7519–7529.

Shukla, P., Kumar, R., & Kumar Rai, A. (2016). Detection of minerals in green leafy vegetables using laser-induced breakdown spectroscopy. *Journal of Applied Spectroscopy*, 83(5), 815–820.

Singh, J.P., & Thakur, S.N. (2007). *Laser-Induced Breakdown Spectroscopy.* Elsevier.

Sneddon, J., & Lee, Y.-I. (1998). Laser-induced breakdown spectrometry. *Chemistry Educator*, 3, 1–7.

Sneddon, J., & Lee, Y.-I. (2004). Laser-induced breakdown spectrometry. *Current Topics in Analytical Chemistry*, 4, 111–118.

Thakur, S.N. (2007). Atomic emission spectroscopy. In J.P. Singh & S.N. Thakur (Eds.), *Laser-Induced Breakdown Spectroscopy* (1st ed., pp. 23–48). Amsterdam: Elsevier.

Tognoni, E., & Cristoforetti, G. (2016). Signal and noise in laser induced breakdown spectroscopy: An introductory review. *Optics & Laser Technology*, 79, 164–172.

Vinic, M., Aruffo, E., Andreoli, F., Ivkovic, M., & Lazic, V. (2020). Quantification of heavy metals in oils with μL volume by laser induced breakdown spectroscopy and minimization of the matrix effect. *SpectrochimicaActa Part B*, 164, 105765.

Wang, Y., Chen, Y., & Li, R. (2024). Direct and sensitive lead analysis of wheat flour using laser-induced fluorescence combined with laser-induced breakdown spectroscopy under low-energy ultraviolet laser ablation. *Journal of Applied Spectroscopy*, 1–8.

Yang, Z., Ren, J., Du, M., Zhao, Y., & Yu, K. (2022). Enhanced laser-induced breakdown spectroscopy for heavy metal detection in agriculture: A review. *Sensors*, 22(15), 5679.

Yu, K., Ren, J., & Zhao, Y. (2020). Principles, developments and applications of laser-induced breakdown spectroscopy in agriculture: A review. *Artificial Intelligence in Agriculture*, 4, 127–139.

Zhao, Z., Chen, L., Liu, F., Zhou, F., Peng, J., & Sun, M. (2020). Fast classification of geographical origins of honey based on laser-induced breakdown spectroscopy and multivariate analysis. *Sensors*, 20, 1878.

5 Nanocomposites
Recent Trends of Nanocomposites in Food Analysis and Safety

Rajamanickam R., Mahesh Jaiwar, Anbarasan R., Sajeevkumar V.A., and Kumar R.

5.1 INTRODUCTION

The branch of studying and manipulating materials at the nanoscale conceptually began by Richard Feynman's landmark speech, 'There's plenty of Room at the bottom' delivered in 1959 at the California Institute of Technology (Feynman, 1960). In the context of science, 'nano-' is a prefix used to indicate an extremely small scale, specifically one-billionth of a meter. The concept of nanotechnology was officially introduced in 1974 by Professor Norio Taniguchi of Tokyo Science University, who provided a detailed definition emphasizing the manipulation of materials at the atomic and molecular scales through processes such as separation, consolidation, and deformation involving individual atoms or molecules. Taniguchi's definition underscored the significance of precision engineering techniques applied to thin films and semiconductor procedures, highlighting the potential for advancement in these areas through nanoscale manipulation.

5.2 APPLICATIONS OF NANOSCIENCE AND TECHNOLOGY

Nanoscience is the fundamental science that aims to observe, understand, and explain the unique behaviors and properties of materials at the nanoscale; whereas nanotechnology involves the manipulation and utilization of nanoscale phenomena for practical applications. At this scale, materials exhibit significantly different physical and chemical properties compared to their macroscale counterparts (Duncan, 2011).

Nanoscale materials exhibit diverse properties based on their morphological characteristics such as size, shape (e.g., spheres, rods, tubes, wires, cubes), surface area, aspect ratio, crystallinity, core-shell structure, porosity, surface morphology, and surface area-to-volume ratio. These features strongly influence their functional properties, including quantum behavior, as well as thermal (Savage & Rao, 2004; Lozano-Steinmetz et al., 2022), conductive (Kane Jacob et al., 2020), electronic, optical (Lance Kelly et al., 2003), catalytic (Gawande et al., 2016) antimicrobial

DOI: 10.1201/9781003482611-5

(Omerović et al., 2021) properties. For instance, quantum effects, such as quantum confinement, become significant at very small sizes and can alter a material's electrical conductivity or optical emission. A well-known example is gold at nanoscale, which appears red or purple due to these effects, as demonstrated in the historical Lycurgus Cup. These distinctive characteristics have revolutionized multiple industries, driving advancements across various fields, including electronics, catalysis, healthcare, food technology, aerospace, automotive industry, among others.

5.3 FUNDAMENTALS OF NANOCOMPOSITES

A composite is a material made up of two or more distinct substances that combine to create a new material with enhanced properties. In most cases, the individual components retain their own physical and chemical characteristics, while contributing superior overall performance compared to each component alone. Composites consist of two main components: a matrix and reinforcing materials embedded in the matrix. Based on the matrix, composites are classified into polymer matrix composites (Jaber et al., 2023), metal matrix composites, and ceramic matrix composites (Das & Das, 2020). When the reinforcement material is in the nanoscale, the composite is referred to as nanocomposite (Newnham & Giniewicz, 2000). Nanoscale additives are instrumental in enhancing the mechanical, thermal, barrier, electrical, and antimicrobial characteristics of composites, rendering them highly advantageous across diverse sectors such as food packaging, pharmaceuticals, electronics, and construction. Notably, within these applications, polymer-based nanocomposites hold significant importance in food packaging as a result of their adaptable mechanical properties, customizable barrier features, and cost-effectiveness.

The categorization of polymer-based nanocomposites can be based on several factors, including the nature of the polymer matrix, the fillers or reinforcement substances employed, their structural configuration within the polymer matrix, and the scale of the nanomaterials utilized (Karak, 2018) (Figure 5.1). One key classification is based on the state of dispersion of the nanomaterials or structural arrangements inside the polymer matrix (Kubišová et al., 2010). But this classification mostly applies to nanoclay-based composition. The classifications are as follows:

1) Intercalated nanocomposite: In this type of polymer nanocomposite, the nanomaterials retain their original crystallographic structures. However, the spacing between interlayers, planes, or sheets increases compared to their initial state. This occurs as polymer chains intercalate between nanoclays forming a sandwich-like structure.
2) Exfoliated nanocomposite: These nanocomposites are formed when nanomaterials are completely dispersed within the polymer matrix. During this process, nanoclays lose their original layered structure, resulting in a disordered or randomly distributed arrangement throughout the polymer. This structural transformation enhances mechanical and barrier properties, making exfoliated nanocomposites superior to intercalated ones.

FIGURE 5.1 Classifications of polymer-based nanocomposites.

3) Phase-separated nanocomposite: It is called phase-separated nanocomposite as one can find two phases—polymer and nanoclay. In this, polymer does not break through the clay platelets. This leads to poor interaction between the two components. Hence, improvements in mechanical and barrier properties are less when compared to exfoliated and intercalated nanocomposites.

Another way of classification is based on the type of polymer matrix used, which includes:

1. Thermoplastic nanocomposites: If the matrix is a thermoplastic polymer such as polyethylene and polypropylene, the nanocomposite is called thermoplastic nanocomposite.
2. Thermosetting nanocomposites: If the matrix used is thermosetting polymers like polyester and polyurethane, it is known as thermosetting nanocomposite.
3. Elastomeric nanocomposites: They are composed of elastometric materials such as silicone rubber and polyurea (Meng et al., 2022).

Polymer-based nanocomposites can also be classified on the basis of the reinforcing materials used such as (i) nanoclay-based nanocomposite, (ii) carbon-based nanocomposites, (iii) metal/metal oxide nanoparticles incorporated, and (iv) biopolymer-based nanocomposites. Another way of classification is based on the dimension of the nanomaterials used like zero-dimensional, one-dimensional, and two-dimensional nanocomposites.

5.4 ROLE OF NANOTECHNOLOGY IN REVOLUTIONIZING FOOD SECTOR

5.4.1 Importance of Food Safety and Quality Preservation

With changing lifestyles, shifting food consumption patterns, and the growing challenge of food loss, conventional packaging alone is no longer sufficient to ensure food security and quality. Hence, to provide safe and sustainable food for all, advanced packaging concepts such as modified atmospheric packaging, controlled atmospheric packaging, and smart packaging have been introduced. These innovations help extend shelf life, reduce waste, and enhance food preservation, addressing global challenges in food availability and safety. While traditional materials and production methods have contributed to smart packaging advancements, nanotechnology offers unparalleled potential for developing innovative materials with superior properties.

5.4.2 Role of Nanotechnology in Revolutionizing Food Sector

Nanotechnology pertains to the deliberate control and alteration of substances at dimensions of the nanometer scale, wherein materials manifest distinct physical and chemical attributes from their larger-scale versions. This transformative manipulation, as elucidated by Duncan (2011), holds the capacity to revolutionize various industries, including the realm of food, by fostering advancements in food safety protocols, processing techniques, packaging methods, and the creation of functional food products. The contributions of nanotechnology to the food sector have been underscored by researchers such as Singh et al. (2023) and He and Hwang (2016), who have underscored its significance in driving innovation and progress in diverse aspects of the food industry.

Applications can be broadly categorized into two main groups, viz., food nanostructured ingredients and food nanosensing (Hossain et al., 2021). These food nanostructured ingredients can be further divided into three subclasses as shown in Figure 5.2. They can be used as food additives to enhance texture (Mu et al., 2019), smart nano-carriers for supplements (He et al., 2019; Ayala-Fuentes & Chavez-Santoscoy, 2021), antimicrobial agents (Raja et al., 2023), and fillers in improving the mechanical strength and the barrier properties (Khanzada et al., 2022). In contrast, nanosensing focuses on ensuring quality (Lv et al., 2024) and food products' safety (Manafi et al., 2025).

5.4.2.1 Techniques for the Synthesis of Nanocomposites

The creation of nanocomposites with carefully regulated characteristics is essential for food analysis applications, and a variety of synthesis techniques are available to accomplish this goal. These techniques vary from chemical processes to physical and biological methods, each influencing the structure, size, and composition of the resultant nanocomposites (Basumatary et al., 2020).

Various techniques are employed in the synthesis of nanocomposites, each offering unique advantages and applicability depending on the desired properties and applications. The techniques used for the synthesis of the nanocomposites can be broadly classified into three methods which are physical methods, chemical methods, and the biological or green methods. Techniques of synthesis of nanocomposites are beyond the scope of this chapter.

Physical methods comprise various techniques like the melt blending/extrusion, electrospinning, layer-by-layer assembly, and ultrasonication. While chemical methods include techniques such as in-situ polymerization, sol-gel synthesis, co-precipitation,

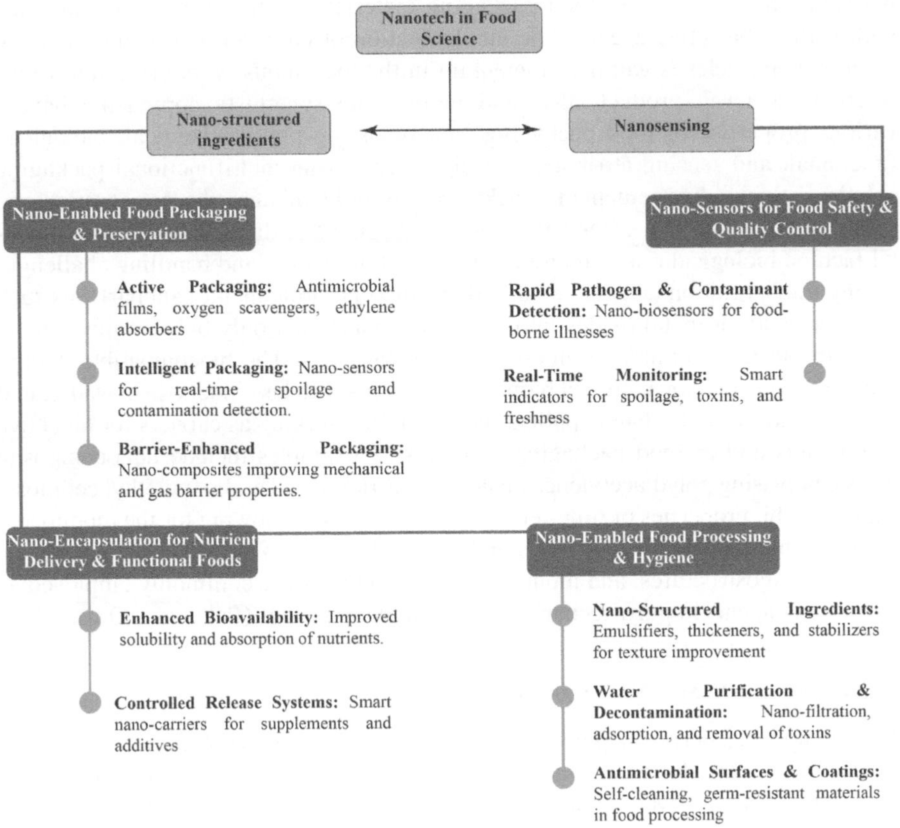

FIGURE 5.2 Applications of nanotechnology in food science.

and micro emulsion, biological methods include the use of microorganisms such as bacteria, fungi, and plants for the synthesis of the nano materials.

5.5 NANOCOMPOSITES IN FOOD ANALYSIS

Nanocomposites utilized in food analysis encompass a diverse array of raw materials including polysaccharides, metals, and metal oxides. These innovative materials hold great promise in bolstering food analysis and safety, offering tailored solutions for various applications. Of particular interest are bio-based nanocomposites, characterized by their biodegradability and biocompatibility, presenting a sustainable alternative to conventional non-degradable polymers within food-related contexts. Derived from renewable sources, these nanocomposites mitigate environmental impact. Notably, polysaccharide-based nanocomposites, employing natural polymers like starch, cellulose, and chitosan, have garnered attention due to their biodegradable nature, biocompatibility, and nontoxic properties (Pal et al., 2021). Chitosan, a derivative of sugars renowned for its inherent antimicrobial properties, is frequently integrated into specialized food packaging materials, particularly when combined with natural bioactive agents. The amalgamation of chitosan with metal or metal oxide nanoparticles is gaining momentum in the food industry, offering innovative solutions to elevate product safety and quality. This synergistic approach enhances various properties of food packaging, encompassing physicochemical, biological, functional, and sensing attributes, thereby engendering multifunctional packaging solutions. Such advancements in packaging not only enhance the preservation and safety of food but also extend its shelf life by safeguarding against environmental factors, biological risks, microorganism contamination, and handling challenges during transportation (Jogaiah et al., 2025). Incorporating metal nanoparticles such as silver, gold, or metal oxides into a chitosan matrix not only bolsters mechanical strength but also confers antimicrobial characteristics. The biocompatible, biodegradable, and nontoxic nature of biopolymeric nanoparticles renders them well-suited for a myriad of food-related applications, including serving as carriers for bioactive compounds and in food packaging. Noteworthy examples include nanocomposite films comprising polydiacetylene, silver nanoparticles, and carboxymethyl cellulose, which exhibit properties of time-temperature indicators, allowing for the monitoring of perishable food items like fruits and vegetables. Moreover, silver nanoparticles, cellulose nanostructures, and montmorillonite nanoclay are commonly employed as reinforcing agents in polysaccharide-based nanocomposites (Pal et al., 2021).

5.5.1 Metal-Based Nanocomposites

Nanocomposites derived from metals have emerged as a popular choice in the field of food packaging due to their superior mechanical strength, thermal stability, and barrier properties. Silver nanoparticles incorporated into polymer matrices are well known for imparting broad-spectrum antimicrobial attributes to these nanocomposites. The synergistic antimicrobial effects achieved by combining chitosan and zinc oxide nanoparticles in composite films present a promising avenue for the development of innovative food packaging solutions (Pal et al., 2021). The capacity of chitosan to chelate ions facilitates the creation of stable complexes with transition and post-transition metal ions when combined with metal nanoparticles within chitosan matrices, yielding nanocomposites that exhibit significantly improved bactericidal efficacy, flexibility, thermal stability, and structural integrity compared to unmodified chitosan films (Jogaiah et al., 2025). Moreover, the utilization of natural plant extracts like pomegranate seeds, lemon juice, and guava leaves in the synthesis of metal nanoparticles provides an environmentally friendly approach to producing nanocomposites for food packaging). Intelligent active packaging systems have been developed utilizing silver nanoparticles and anthocyanin-rich purple corn extract. Additionally, cellulose nanocomposite films containing silver nanoparticles have shown remarkable mechanical, thermal, and optical properties, along with antimicrobial activity against *E. coli* and *Staphylococcus aureus* (Pal et al., 2021).

5.5.2 Metal Oxide-Based Nanocomposites

Metal oxide-based nanocomposites, such as those containing nanoparticles of titanium dioxide or zinc oxide, are valued for their UV-blocking and antimicrobial properties in various applications. These nanocomposites, when integrated into composite materials, have been found to decrease the water solubility, moisture content, and water vapor permeability of films made from a blend of cassava starch and agar (Pal et al., 2021). Notably, the antimicrobial attributes of metal nanoparticles, in comparison to essential oils derived from plants, are characterized by their broad-spectrum effectiveness, stability, and efficacy in eradicating bacteria without necessitating specific receptor interactions. This feature is believed to reduce the likelihood of bacterial resistance development (Jogaiah et al., 2025). The addition of metal oxide nanoparticles with the food packaging materials increases mechanical strength, barrier properties, UV-blocking capacity, thermal resistance, and optical characteristics. Specifically, when metal and metal oxide nanoparticles are introduced into active packaging films based on chitosan, they demonstrate notable antioxidant properties that can effectively combat oxidative degradation and delay lipid oxidation in high-fat food products (Jogaiah et al., 2025).

5.5.3 Bio-Based Nanocomposites

Bio-based nanocomposites have emerged as environmentally friendly alternatives to conventional polymers, garnering significant attention in various fields. A recent study by Pal et al. (2021) demonstrated the enhanced antimicrobial properties of polyvinyl alcohol/cassava starch nanocomposites through the incorporation of oregano essential oil and silver nanoparticles, showcasing synergistic effects. The application of bionanocomposites in active packaging extends to the development of controlled-release systems designed to deliver beneficial additives such as nutrients, vitamins, and probiotics within food products. By incorporating components like antioxidants, antibacterial agents, enzymes, and plant extracts, bionanocomposites can bolster packaging properties of conventional food packaging materials, thereby enhancing their overall performance (Biswas et al., 2022). These nanocomposites can serve diverse functions in food preservation, acting as gas absorbers, antimicrobial agents, and carriers for enzyme immobilization, ultimately contributing to improved food stability. Moreover, the utilization of nanolaminates in food packaging offers a versatile approach to coating various food items like meats, cheeses, vegetables, fruits, and baked goods. Studies have indicated that the inclusion of nanoclay and silicate components in nanocomposites can elevate the levels of volatiles, antioxidant capacity, and organic acids, consequently enhancing their antibacterial properties (Biswas et al., 2022). This multifaceted approach underscores the potential of bio-based nanocomposites in revolutionizing the field of active food packaging through their diverse functionalities and enhanced performance attributes.

5.5.4 Carbon-Based Nanocomposites

Carbon-based nanocomposites, such as carbon nanotubes and graphene, have been extensively utilized in the food industry to improve the mechanical strength and barrier properties of packaging materials (Biswas et al., 2022). The incorporation of carbon nanotubes and nanocellulose into polymer matrices has shown to enhance the mechanical, thermal, and barrier properties of these nanocomposites, thereby expanding their potential applications in food packaging. Despite the exceptional electrical and mechanical properties of carbon nanotubes and nanofibers that make them favorable for food packaging, their high cost and processing challenges limit their widespread use (Biswas et al., 2022). Recent developments in materials like carbon nanodots and graphene quantum dots have positioned them as promising components for biosensing and bioimaging applications due to their unique optical properties. For example, a nanocomposite composed of graphene oxide and copper oxide has exhibited significant antibacterial activity against *S. aureus* and *E. coli.* This antimicrobial effect of graphene-based nanostructures is attributed to their ability to cause cell membrane disruption and oxidative stress, inhibiting bacterial growth and spore formation. The modification of graphene derivatives with metal compounds such as Ag, ZnO, or TiO_2 has been demonstrated to enhance their inherent antimicrobial properties (Barra et al., 2020). Furthermore, the inclusion of cellulose nanocrystals has been found to improve the physical properties of films, resulting in enhanced water resistance and mechanical strength. The combination of graphene derivatives with biopolymer-based composites presents a promising approach for developing active food packaging solutions that can prolong the shelf life of food products (Barra et al., 2020). Cellulose nanocrystals, specifically, have played a key role in enhancing the physical characteristics of films, including strengthening mechanical properties and water resistance (Pal et al., 2021). Furthermore, polymer nanocomposites have been deployed as gas barriers to mitigate carbon dioxide leakage from carbonated beverage containers. The use of copper-based nanocomposites is prevalent for impeding the growth of bacteria, viruses, and fungi due to their extensive surface area, which interacts with cell membranes, thereby intensifying their antibacterial properties (Biswas et al., 2022). The efficacy of nanocomposite films loaded with plant extracts in preventing microbial spoilage, such as inhibiting the growth of *E. coli* and *Staphylococcus aureus*, underscores their potential in extending the shelf life of food products (Pal et al., 2021). Nanomaterials, with their enhanced antibacterial capabilities, can be harnessed for detecting food spoilage using nanosensors.

5.6 TECHNIQUES IN FOOD ANALYSIS USING NANOCOMPOSITES

The incorporation of nanocomposites into food analysis has revolutionized the field, providing enhanced sensitivity, selectivity, and efficiency in detecting various food components and contaminants. Nanocomposites offer unique advantages due to their high surface area, tunable properties, and ability to interact with target analytes at the nanoscale (Mousa & Dong, 2020). Nanomaterials are now used in the food industry as nanosensors, delivery vehicles, and antimicrobial agents. In the food industry, nanosensors can be employed to detect pathogens and toxins, to monitor food quality, and to track food throughout the supply chain (Biswas et al., 2022). Nanocomposites have emerged as powerful tools for food analysis, providing enhanced sensitivity, selectivity, and efficiency in detecting various food components and contaminants (Biswas et al., 2022).

5.6.1 Nanosensors for Food Analysis and Safety

Nanosensors are intricate devices that utilize nanomaterials to detect and quantify physical, chemical, or biological attributes with exceptional sensitivity and precision. In the realm of food analysis, nanosensors play a pivotal role in identifying a wide array of elements such as pathogens, toxins, pesticides, and various contaminants. They exhibit significant potential in recognizing spoilage organisms, pathogens, contaminants, and toxins in food packaging and can be tailored for applications in food analysis, assessments, drinking water, aromas, as well as medical diagnostics (Biswas et al., 2022). Specifically, nanoparticle-based sensors integrated within packaging materials have the capability to identify particular pathogenic microorganisms and indicate food spoilage to consumers through alterations in color (Ramkumar et al., 2019). These sensors offer real-time monitoring functionalities, facilitating prompt identification and management of potential food safety risks (Gómez-Llorente et al., 2022). Moreover, surface-enhanced Raman spectroscopy (SERS)-based nanosensor-shave emerged as a robust instrument for food analysis, enabling swift and precise detection of various analytes, encompassing pesticides, antibiotics, and food additives (Ramkumar et al., 2019). These sensors leverage the enhancement of Raman signals of the target analyte upon interaction with plasmon-resonant nanoparticles, like gold or silver nanoparticles. The utilization of smart biosensors and biomimetic sensors supports in the detection and eradication of infections by functioning as synthetic cell surfaces. Innovative technologies such as biosensor arrays, nano-test strips, electronic noses, and nanocantilevers are currently under development, with the ability to alter color upon detecting any indication of food spoilage (Biswas et al., 2022). Electrochemical sensors are increasingly being adopted in the agri-food sector for quality and safety assessments owing to their advantages, including ease of use, high sensitivity, and rapid response times, particularly for direct detection purposes.

5.6.2 Applications of Nanocomposites in Food Safety and Quality

Nanocomposites are used for detecting microorganisms that cause spoilage as well as diseases transmitted by food and revealing carcinogens in food items (Biswas et al., 2022). Nanocomposites can be used to assess the freshness of food.

5.6.3 Detection of Contaminants Using Nanocomposites

Using nanosensors, a wide variety of toxins, including bacteria, fungal poisons, and chemical pollutants, can be detected. The integration of nanotechnology plays a crucial role in guaranteeing the safety and quality of food products. Within the food industry, significant progress has been made in the detection of pathogens through advanced technological methods. Nanotechnology contributes to the enhancement of food preservation techniques, the refinement of food quality and safety standards, and the surveillance of food supply chains through the deployment of nanosensors and nano-barcodes. Foodborne illnesses, which affect millions of individuals globally, can have serious repercussions for public health and economies. Early detection of foodborne pathogens is essential for preventing outbreaks and minimizing their impact. Nanomaterials, including nanoparticles, nanotubes, and nanowires, have demonstrated great potential for pathogen detection due to their high surface area, unique optical and electrical properties, and ability to interact with biological molecules (Vasilescu et al., 2017). Nanosensors integrated into food packaging can detect volatile organic compounds released by spoiling food, providing real-time information about the freshness and safety of the product.

5.6.4 Use of Nanocomposites in Food Packaging

Nanocomposites have attracted considerable attention within the food packaging sector due to their capacity to enhance barrier properties, extend product shelf life, and mitigate microbial contamination. The incorporation of antimicrobial nanoparticles, such as silver or zinc oxide nanoparticles, into packaging materials has proven effective in impeding the growth of spoilage microorganisms and pathogens on food surfaces, consequently prolonging the preservation of food items. Additionally, the implementation of nanocomposite coatings on packaging substrates has been found to enhance their barrier properties, thereby reducing the permeability to gases, moisture, and organic vapors. This decline in oxygen permeability offers a significant advantage by preventing the oxidation of food components, including lipids and vitamins, thereby preserving the nutritional quality of food products. Forecasts from the U.S. Department of Energy indicate that the integration of nano-enabled sustainable packaging solutions could potentially lead to a 20% reduction in the carbon footprint of the food industry by the year 2033. The increasing focus on sustainable food packaging solutions, bolstered by governmental initiatives and evolving consumer preferences, is expected to propel advancements in nanocomposite materials further. Metal nanoparticles possess broad-spectrum antimicrobial properties, as well as enhanced barrier and sensing functionalities. The method by which they function entails eliminating bacteria without requiring specific receptor interactions, thereby reducing the likelihood of resistance emergence and expanding their range of antibacterial effects. The incorporation of metal and metal oxide nanoparticles into nanocomposite films based on chitosan represents a significant advancement in the realm of functional food packaging, effectively converting these materials into dynamic and responsive solutions. These nanocomposites not only improve food preservation and quality control but also act as active shields within the domain of food packaging.

5.6.5 Spoilage Detection in Food Using Nanocomposites

Spoilage in food products primarily arises from microbial proliferation, enzymatic processes, and chemical transformations. The assessment of food freshness can be facilitated through the deployment of nanosensors capable of identifying alterations in volatile organic compounds, pH levels, and temperature changes. These nanosensors offer instantaneous feedback on the condition of food items, thereby diminishing the likelihood of spoilage occurrences and the onset of foodborne illnesses. In actively contributing to the maintenance of food quality, these sensing devices transcend their conventional role as passive protective barriers. The capacity of specific metal nanoparticles to adsorb ethylene gas, a critical agent in the maturation and aging of fruits and vegetables, aids in regulating respiration rates, prolonging the shelf life of fresh produce, and delivering real-time insights into spoilage progression, thereby broadening the scope of intelligent packaging solutions. By employing sophisticated sensors to meticulously monitor food quality, this approach guarantees reduced wastage and enhanced safety in food consumption. Packaging mechanisms assume a pivotal function in safeguarding the integrity of food products across the supply chain, ensuring their safety and optimizing their longevity by furnishing robust defenses against environmental stressors and contamination risks. Active packaging, integrating elements that interact with the food or its surrounding environment, represents a noteworthy advancement in packaging technologies (Jogaiah et al., 2025). These systems are capable of dispensing preservatives, sequestering ethylene, or managing moisture content, actively contributing to the extension of shelf life and the preservation of food quality (Soares et al., 2023). Intelligent packaging builds upon this principle by integrating sensors and indicators that oversee diverse parameters, such as temperature, pH levels, or the presence of specific gases, thereby supplying real-time updates regarding the status of the food product (Jogaiah et al., 2025).

5.7 CHALLENGES AND LIMITATIONS

Despite the promising applications of nanocomposites in food analysis and safety, several challenges and limitations need to be addressed. Concerns regarding the potential toxicity of nanomaterials and their migration from packaging materials into food products have been raised (Biswas et al., 2022). Due to the possible migration of nanoparticles and their effects on human health and the environment, the broad use of chitosan nanocomposites with metal NPs is hampered by a number of issues, including the absence of standardized mass production procedures. It is essential to conduct thorough risk assessments to evaluate the potential health and environmental impacts of nanocomposites before their commercialization in the food sector (Jogaiah et al., 2025). The acceptance of nanotechnology-based food products by consumers represents a pivotal consideration in the food industry. One of the primary obstacles faced is the need for standardized procedures for large-scale production. The transition from experimental laboratory conditions to industrial-level manufacturing without compromising the enhanced characteristics of the resulting nanocomposites is a multifaceted endeavor. Transparent labeling and effective communication strategies are indispensable to foster consumer trust and approval, enabling them to be informed about the incorporation of nanomaterials in food items and their associated advantages and drawbacks. The absence of comprehensive regulatory frameworks and consistent testing methodologies for nanocomposites within the food sector presents a significant hurdle. This deficiency is exacerbated by the fact that existing regulations may not sufficiently account for the distinctive properties and potential hazards linked with nanomaterials, generating ambiguity for both producers and regulators. The establishment of unambiguous guidelines and procedures for the assessment of safety, characterization, and oversight of nanocomposites is crucial to ensure their judicious and sustainable implementation in the food industry (Odobašić et al., 2015). Challenges and limitations of nanocomposites are shown in Figure 5.3.

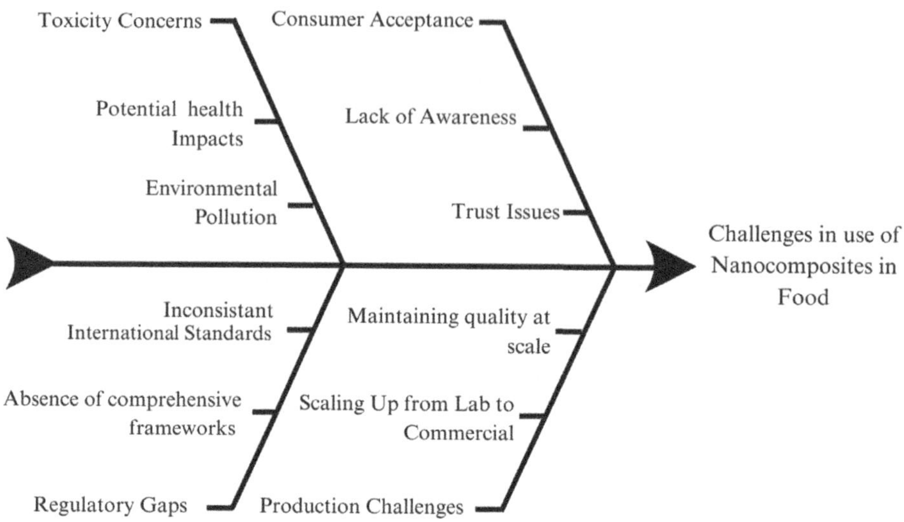

FIGURE 5.3 Challenges and limitations of using nanocomposites in food.

5.7.1 Nanocomposites' Safety and Toxicity in Food

The meticulous evaluation of risks associated with nanocomposites is essential to ensure their safe utilization in food packaging, mandating in-depth toxicity investigations to assess potential health implications. The mitigation of these concerns necessitates a thorough toxicological characterization and enhanced migration assessments (Jogaiah et al., 2025). The utilization of diverse chemical methods in nanoparticle synthesis can yield adverse outcomes leading to the generation of deleterious non-environmentally friendly by-products that contaminate the surroundings (Biswas et al., 2022). Therefore, the adoption of green synthesis techniques is imperative to address these challenges. Despite the biocompatibility and film-forming properties of chitosan, which could aid in regulating nanoparticle migration, there remains a notable absence of extensive ecotoxicity investigations encompassing prolonged trophic transfer assessments, microbial diversity evaluations, and standardized analyses of nanomaterial risks (Jogaiah et al., 2025).

5.7.2 Nanocomposite Regulations

The management of nanoparticles in food and food packaging materials, along with associated health and safety concerns, necessitates the establishment of appropriate regulatory frameworks to govern the utilization of nanotechnology (Ramkumar et al., 2019). Standardization is essential for the sourcing of raw materials, intermediary products, and final food items utilized in the production process. Regulatory measures should encompass the physical and chemical properties inherent to nanoparticles (Ramkumar et al., 2019). It is imperative to standardize the identification and quantification of nanoparticles present in food commodities, devise regulatory protocols, stipulate maximum consumer exposure thresholds, formulate dietary pharmacokinetic models, and implement standardized analytical methodologies (Biswas et al., 2022). It is noteworthy that regulatory frameworks and viewpoints vary between nations. Determinations on whether specific compounds are suitable for inclusion in food products, their safety for consumption, among other factors, are crucial. The endorsement and clearance of nanostructured compounds in food are obligatory before any product is introduced to the market. These evaluations regarding the incorporation of nanostructured compounds are indispensable for mitigating the toxicological risks posed by such substances (Ramkumar et al., 2019). The European Food Safety Authority (EFSA) and the Food and Drug Administration (FDA) stand as prominent regulatory bodies engaged in overseeing nanomaterials. The development of standardized methodologies for synthesizing, characterizing, and utilizing nanocomposites is pivotal for ensuring reproducibility and comparability across different research studies and applications. Both the FDA and EFSA advocate for conducting risk assessments of nanomaterials on a case-by-case basis (Gottardo et al., 2021). This approach involves evaluating the potential hazards linked with specific nanomaterials and assessing the probability of dietary exposure to these materials.

5.7.3 Consumer Acceptance and Environmental Impact

The acceptance of food products utilizing nanotechnology by consumers plays a crucial role, as highlighted by Soares et al. (2023). The intricate long-term environmental implications of nanocomposites, encompassing their behavior in the environment, transport mechanisms, and potential repercussions on ecosystems, remain partially understood. It is imperative to establish transparent labeling practices and effective communication strategies to educate consumers about the incorporation of nanomaterials in food items, elucidating both their advantages and drawbacks to foster consumer trust and acceptance. The absence of comprehensive regulatory frameworks and standardized evaluation protocols for nanocomposites within the food sector presents a notable obstacle, as noted by Ramkumar et al. (2019). While some research projects present optimistic outlooks regarding the integration of nanotechnology in the food industry, concerns regarding ingestion risks are on the rise, as highlighted by Soares et al. (2023). Furthermore, the volume of information accessible to consumers significantly influences their perceptions of applications of nanotechnology, as emphasized by Ramkumar et al. (2019).

5.8 FUTURE SCOPE OF NANOCOMPOSITE IN FOOD ANALYSIS AND SAFETY

The future of functional nanocomposites in food analysis holds immense promise for revolutionizing food safety, quality control, and nutritional monitoring. Nanotechnology applications, integrated with artificial intelligence (AI) and IoT, offer promising prospects for market expansion in food systems. Nanomaterial-based sensors can detect pathogens, toxins, and spoilage indicators with high sensitivity and speed. Nanocomposites are actively being researched for their utilization in active and intelligent packaging systems, aimed at prolonging the shelf life of food items, monitoring their freshness, and furnishing immediate information regarding their quality and safety attributes. Additional investigation is imperative to fully explore the potential of nanocomposites for targeted delivery of nutrients and bioactive compounds, alongside their application in pioneering food processing methodologies that can augment food quality and safety standards. The advancement of bio-based and biodegradable nanocomposites stands as a promising research domain that could contribute significantly to the mitigation of the environmental repercussions associated with food packaging materials. The amalgamation of nanotechnology with cutting-edge analytical methodologies is instigating the creation of highly sensitive and selective sensing platforms for food analysis. Notably, materials like carbon nanotubes, metallic nanoparticles, and nanostructured composites are progressively pivotal in the development of sophisticated sensing and biosensing systems tailored for comprehensive food analysis, facilitating prompt and precise identification of contaminants, pathogens, and quality indicators (Biswas et al., 2022; Soares et al., 2023). Ongoing research endeavors are concentrated on overcoming the constraints of current nanocomposite-based food analysis techniques and broadening their scopes to encompass novel realms within food safety and quality assessment. Advancements in material science, nanotechnology, and analytical chemistry are poised to propel the creation of more sophisticated and versatile nanocomposites for food analysis, thereby laying the groundwork for a safer, healthier, and more sustainable food supply chain (Biswas et al., 2022; Ramkumar et al., 2019).

5.9 CONCLUSION

Nanocomposites have emerged as a revolutionary material with significant implications in the realm of food analysis and safety. Their capacity to amalgamate distinctive attributes such as antimicrobial properties, enhanced barrier efficacy, and mechanical robustness renders them exceptionally beneficial in prolonging shelf life, detecting contaminants, and enhancing packaging efficacy within the food industry. Whether originating from metallic elements, metal oxides, carbon-based configurations, or biodegradable polymers, these amalgams can be customized to fulfill specific requirements within the food sector. The choice of base components and the method of amalgamation—whether it be through physical, chemical, or environmentally conscious means—play a pivotal role in determining the efficiency and safety of the end product. Recent progressions in synthesis methodologies have facilitated the creation of environmentally sustainable and expansible nanocomposites that are wellsuited for practical implementation. Furthermore, advancements in biosensors and intelligent packaging are ushering in novel avenues for real-time monitoring of food quality. Despite the encouraging advancements, obstacles persist. Issues surrounding regulations, expenses, material integrity, and scalability necessitate resolution prior to widespread integration. Nevertheless, through persistent research efforts, enhanced manufacturing techniques, and interdisciplinary cooperation, it is anticipated that nanocomposites will assume an even more substantial role in upholding food safety and quality standards in the future.

REFERENCES

Ayala-Fuentes, J. C., & Chavez-Santoscoy, R. A. (2021). Nanotechnology as a Key to Enhance the Benefits and Improve the Bioavailability of Flavonoids in the Food Industry. *Foods*, *10*(11), 2701. https://doi.org/10.3390/foods10112701

Barra, A., Santos, J. D. C., Silva, M. R. F., Nunes, C., Ruiz-Hitzky, E., Gonçalves, I., Yildirim, S., Ferreira, P., & Marques, P. A. A. P. (2020). Graphene Derivatives in Biopolymer-Based Composites for Food Packaging Applications. *Nanomaterials*, *10*(10), 2077. https://doi.org/10.3390/nano10102077

Basumatary, I. B., Mukherjee, A., Katiyar, V., & Kumar, S. (2020). Biopolymer-Based Nanocomposite Films and Coatings: Recent Advances in Shelf-Life Improvement of Fruits and Vegetables [Review of Biopolymer-Based Nanocomposite Films and Coatings: Recent Advances in Shelf-Life Improvement of Fruits and Vegetables]. *Critical Reviews in Food Science and Nutrition*, *62*(7), 1912.

Biswas, R., Alam, M., Sarkar, A., Haque, M. I., Hasan, Md. M., & Hoque, M. (2022). Application of Nanotechnology in Food: Processing, Preservation, Packaging and Safety Assessment [Review of Application of Nanotechnology in Food: Processing, Preservation, Packaging and Safety Assessment]. *Heliyon*, *8*(11).

Das, S., & Das, S. (2020). Properties for Polymer, Metal and Ceramic Based Composite Materials. In *Elsevier eBooks* (pp. 815–821). https://doi.org/10.1016/B978-0-12-803581-8.11897-1

Duncan, T. V. (2011). Applications of Nanotechnology in Food Packaging and Food Safety: Barrier Materials, Antimicrobials and Sensors. *Journal of Colloid and Interface Science*, *363*, 1–24.

Feynman, R. P. (1960). There's Plenty of Room at the Bottom. *Engineering and Science*, *23*(5), 22–36. https://calteches.library.caltech.edu/1976/1/1960Bottom.pdf

Gawande, M. B., Goswami, A., Felpin, F.-X., Asefa, T., Huang, X., Silva, R., Zou, X., Zboril, R., & Varma, R. S. (2016). Cu and Cu-Based Nanoparticles: Synthesis and Applications in Catalysis. *Chemical Reviews*, *116*(6), 3722–3811. https://doi.org/10.1021/acs.chemrev.5b00482

Gómez-Llorente, H., Hervás, P., Pérez-Esteve, É., Baviera, J. M. B., & Fernández-Segovia, I. (2022). Nanotechnology in the Agri-Food Sector: Consumer Perceptions [Review of Nanotechnology in the Agri-Food Sector: Consumer Perceptions]. *NanoImpact*, *26*, 100399.

Gottardo, S., Mech, A., Drbohlavová, J., Małyska, A., Bøwadt, S., Sintes, J. R., & Rauscher, H. (2021). Towards Safe and Sustainable Innovation in Nanotechnology: State-of-Play for Smart Nanomaterials [Review of Towards Safe and Sustainable Innovation in Nanotechnology: State-of-Play for Smart Nanomaterials]. *NanoImpact*, *21*, 100297.

He, X., Deng, H., & Hwang, H. M. (2019). The Current Application of Nanotechnology in Food and Agriculture. *Journal of Food and Drug Analysis*, *27*(1), 1–21. https://doi.org/10.1016/j.jfda.2018.12.002

He, X., & Hwang, H. (2016). Nanotechnology in Food Science: Functionality, Applicability, and Safety Assessment. *Journal of Food and Drug Analysis*, *24*(4), 671–681. https://doi.org/10.1016/j.jfda.2016.06.001

Hossain, A., Skalicky, M., Brestic, M., Mahari, S., Kerry, R. G., Maitra, S., Sarkar, S., Saha, S., Bhadra, P., Popov, M., Islam, M. T., Hejnak, V., Vachova, P., Gaber, A., & Islam, T. (2021). Application of Nanomaterials to Ensure Quality and Nutritional Safety of Food. *Journal of Nanomaterials*, *2021*, 1–19. https://doi.org/10.1155/2021/9336082

Jaber, A. A., Abbas, S. A., Farah, A. A., Kopeć, K. K., Alsalik, Y. M., Tayeb, M. A., & Verghese, N. (2023). Effect of Fiber Sizing Levels on the Mechanical Properties of Carbon Fiber-Reinforced Thermoset Composites. *Polymers*, *15*(24), 4678. https://doi.org/10.3390/polym15244678

Jacob, K., de Caro, D., Faulmann, C., & Valade, L. (2020). Nanoparticles of Molecular Conductorsand Superconductors: Progress Over the Last Ten Years. *European Journal of Inorganic Chemistry, 45*, 4237–4246. https://doi.org/10.1002/ejic.202000761.

Jogaiah, S., Günyaktı, A., Mujtaba, M., Thakur, A., Britto, S. D., Nagaraja, G., Belorkar, S. A., & Shetty, H. S. (2025). Chitosan-Metal and Metal Oxide Nanocomposites for Active and Intelligent Food Packaging; A Comprehensive Review of Emerging Trends and Associated Challenges [Review of Chitosan-Metal and Metal Oxide Nanocomposites for Active and Intelligent Food Packaging; A Comprehensive Review of Emerging Trends and Associated Challenges]. *Carbohydrate Polymers, 357*, 123459.

Karak, N. (2018). Fundamentals of Nanomaterials and Polymer Nanocomposites. In *Elsevier eBooks* (pp. 1–45). https://doi.org/10.1016/b978-0-12-814615-6.00001-1

Khanzada, B., Mirza, B., & Ullah, A. (2022). Chitosan Based Bio-Nanocomposites Packaging Films with Unique Mechanical and Barrier Properties. *Food Packaging and Shelf Life, 35*, 101016. https://doi.org/10.1016/j.fpsl.2022.101016

Kubišová, H., Měřínská, D., & Svoboda, P. (2010). PP/Clay Nanocomposite: Optimization of Mixing Conditions with Respect to Mechanical Properties. *Polymer Bulletin, 65*(5), 533–541. https://doi.org/10.1007/s00289-010-0269-y

Lance Kelly, K., Coronado, E., Zhao, L. L., & Schatz, G. C. (2003). The Optical Properties of Metal Nanoparticles: The Influence of Size, Shape, and Dielectric Environment. *The Journal of Physical Chemistry B, 107*(3), 668–677.

Lozano-Steinmetz, F., Ramírez-Navarro, M. P., Vivas, L., Vasco, D. A., Singh, D. P., & Zambra-Sazo, C. (2022). Thermal and Rheological Characterization of Aqueous Nanofluids Based on Reduced Graphene Oxide (rGO) with Manganese Dioxide Nanocomposites (MnO_2). *Nanomaterials, 12*(17), 3042. https://doi.org/10.3390/nano12173042

Lv, Y., Shi, G., Ji, Z., Shi, C., Luo, Y., Hong, H., Tan, Y., & Yang, X. (2024). Development and Application of Visual Fish Freshness Indicator Film Incorporated with Anthocyanins Encapsulation by Whey Protein-Propylene Glycol Alginate Nanoparticles. *International Journal of Biological Macromolecules*, 137054. https://doi.org/10.1016/j.ijbiomac.2024.137054

Manafi, L., Mahmoudi, R., Pakbin, B., Ezati, P., Mousavi, S., Mehrabi, A., & Javadi, A. (2025). Synergistic Effects of Black Pepper Nano-emulsion and Lactobacillus Acidophilus Against Salmonella Enteritidis in Food Systems. *LWT*, 117335. https://doi.org/10.1016/j.lwt.2025.117335

Meng, Q., Guo, G., Qin, X., Zhang, Y., Wang, X., & Zhang, L. (2022). Smart Multifunctional Elastomeric Nanocomposite Materials Containing Graphene Nanoplatelets. *Smart Materials in Manufacturing, 1*, 100006. https://doi.org/10.1016/j.smmf.2022.100006

Mousa, M., & Dong, Y. (2020). The Role of Nanoparticle Shapes and Structures in Material Characterisation of Polyvinyl Alcohol (PVA) Bionanocomposite Films. *Polymers, 12*(2), 264.

Mu, R., Hong, X., Ni, Y., Li, Y., Pang, J., Wang, Q., Xiao, J., & Zheng, Y. (2019). Recent Trends and Applications of Cellulose Nanocrystals in Food Industry. *Trends in Food Science & Technology, 93*, 136–144. https://doi.org/10.1016/j.tifs.2019.09.013

Newnham, R. E., &Giniewicz, J. R. (2000). Nonmechanical Properties of Composites. In *Elsevier eBooks* (pp. 431–463). https://doi.org/10.1016/b0-08-042993-9/00052-8

Odobašić, A., Ćatić, S., Šestan, I., &Bratovčić, A. (2015). Application of Polymer Nanocomposite Materials in Food Packaging. *Croatian Journal of Food Science and Technology, 7*(2), 86.

Omerović, N., Djisalov, M., Živojević, K., Mladenović, M., Vunduk, J., Milenković, I., Knežević, N. Ž., Gadjanski, I., & Vidić, J. (2021). Antimicrobial Nanoparticles and Biodegradable Polymer Composites for Active Food Packaging Applications. *Comprehensive Reviews in Food Science and Food Safety, 20*(3), 2428–2454. https://doi.org/10.1111/1541-4337.12727.

Pal, K., Sarkar, P., Anis, A., Wiszumirska, K., & Jarzębski, M. (2021). Polysaccharide-Based Nanocomposites for Food Packaging Applications [Review of Polysaccharide-Based Nanocomposites for Food Packaging Applications]. *Materials, 14*(19), 5549. Multidisciplinary Digital Publishing Institute.

Raja, F. N. S., Worthington, T., & Martin, R. A. (2023). The Antimicrobial Efficacy of Copper, Cobalt, Zinc and Silver Nanoparticles: Alone and in Combination. *Biomedical Materials, 18*(4), 045003. https://doi.org/10.1088/1748-605x/acd03f

Ramkumar, C., Vishwanatha, A., & Saini, R. (2019). Regulatory Aspects of Nanotechnology for Food Industry. In Dasarahally-Huligowda, L.K., Goyal, M.R., & Suleria, H.A.R. (eds) *Nanotechnology Applications in Dairy Science: Packaging, Processing, and Preservation* (pp. 169–184). Apple Academic Press.

Savage, T., & Rao, A.M. (2004). Thermal Properties of Nanomaterials and Nanocomposites. In Tritt, T.M. (eds) *Thermal Conductivity. Physics of Solids and Liquids.* Boston, MA: Springer. https://doi.org/10.1007/0-387-26017-X_12

Singh, R., Dutt, S., Sharma, P., Sundramoorthy, A. K., Dubey, A., Singh, A.,& Arya, S. (2023). Future of Nanotechnology in Food Industry: Challenges in Processing, Packaging, and Food Safety. *Global Challenges, 7,* 2200209. https://doi.org/10.1002/gch2.202200209

Soares, M. G., Bevilaqua, G. C., & Lima, M. de. (2023). Potential Applications of Environmentally Friendly Nanoparticles in Food Matrices: A Review [Review of Potential Applications of Environmentally Friendly Nanoparticles in Food Matrices: A Review]. *Food and Bioprocess Technology, 16*(12), 2742.

Taniguchi, N. (1974). On the Basic Concept of 'Nano-Technology'. In *Proceedings of the International Conference on Production Engineering, Tokyo, Part II* (1974). Japan Society of Precision Engineering.

Vasilescu, A., Hayat, A., Gáspár, S., & Marty, J. (2017). Advantages of Carbon Nanomaterials in Electrochemical Aptasensors for Food Analysis. *Electroanalysis, 30*(1), 2.

6 Smartphone-Based Techniques for Food Analysis

Application of Computer Vision and Artificial Intelligence as Rapid Detection Techniques in the Food Sector

Monica Yumnam, Keshram Dulait, and Poonam Mishra

6.1 INTRODUCTION

Smartphone sensors are integral components that detect and measure physical, environmental, and biometric data, significantly enhancing smartphone functionality across diverse domains, including environmental science, food analysis, healthcare, navigation, and augmented reality. The ability of smartphones to assess various physical attributes has transformed their utility beyond communication, enabling advanced applications in real-time data collection, analysis, and decision-making (McGonigle et al., 2018). Smartphone-based sensors can be broadly categorized into two types: built-in sensors and add-on sensors. Built-in sensors are seamlessly integrated within smartphones and include cameras, accelerometers, pedometers, thermometers, barometers, humidity probes, light detectors, magnetometers, proximity sensors, and biometric sensors, all commonly found in iPhones and Android devices. On the other hand, add-on sensors involve integrating external, more advanced sensors with smartphones, transforming them into versatile analytical tools. These external sensors utilize various transducer technologies, including colorimetric, electrochemical, fluorescence, surface-enhanced Raman spectroscopy (SERS), and impedance-based sensors, expanding the range of detectable analytes and enhancing detection reliability, sensitivity, and selectivity. Incorporating cutting-edge technologies such as nanotechnology, microfluidics, and 3D printing further improves the analytical

DOI: 10.1201/9781003482611-6

performance of these sensors, enabling cost-effective and miniaturized solutions for rapid diagnostics (Gao & Wu, 2016).

Given the widespread availability of smartphone, these tools offer cost-effective, real-time testing solutions, particularly beneficial in resource-limited settings. Traditional food safety assessments rely on laboratory-based techniques for detecting chemical contaminants and microbial pathogens, which can be time-consuming, costly, and logistically challenging, especially in remote areas. Smartphone-based analytical tools bridge this gap by enabling rapid, on-site detection and remote confirmation of results, democratizing access to advanced analytical technologies (Ozcan, 2014). As technological advancements continue to refine smartphone-based sensor systems, their integration into food safety, public health, and environmental monitoring will make it possible for more accessible, rapid, and cost-effective analytical solutions.

This chapter explores the principles, applications, and recent developments in smartphone-based techniques for food analysis, highlighting the application of computer vision and artificial intelligence (AI) as rapid detection techniques in the food sector.

6.2 DIFFERENT TECHNIQUES UTILIZED FOR FOOD ANALYSIS USING SMARTPHONES

Various analytical techniques have been applied to smartphone-based systems, allowing for rapid and on-site food quality assessment. Among the most important approaches are colorimetric analysis, fluorescence detection, microscopic imaging, electrochemical sensing, and electro-chemiluminescence. Figure 6.1 illustrates a schematic diagram of the utilization of smartphones using different techniques. Each of these strategies uses distinct detection concepts and smartphone components, such as the camera, sensors, and computational capacity, to produce trustworthy analytical results in the food sector.

6.2.1 Colorimetric Techniques

Colorimetric analysis is widely used in food analysis to determine the concentration of specific compounds based on their light absorption properties. It relies on changes in color intensity or hue to indicate chemical composition. It utilizes a smartphone's built-in camera and processing capabilities to assess color changes in food samples. These color variations serve as indicators of key food properties, such as freshness, ripeness, chemical composition, or the presence of contaminants. The approach typically involves placing a test strip or sensor in contact with the food sample, triggering a reaction that results in a visible color change. The smartphone camera captures this change, and specialized image-processing software analyzes the color intensity or hue to extract quantitative or qualitative information such as pH levels, sugar content, or the presence of allergens, toxins, or microbial contaminants (Geng et al., 2023). In smartphone-based colorimetry techniques, color models like CIELAB (International Commission on Illumination L*a*b*), RGB (Red, Green, Blue), and HSV (Hue, Saturation, Value) are commonly used to analyze and quantify color. Each model offers a distinct approach to representing color with its application depending on the specific requirements of the analysis.

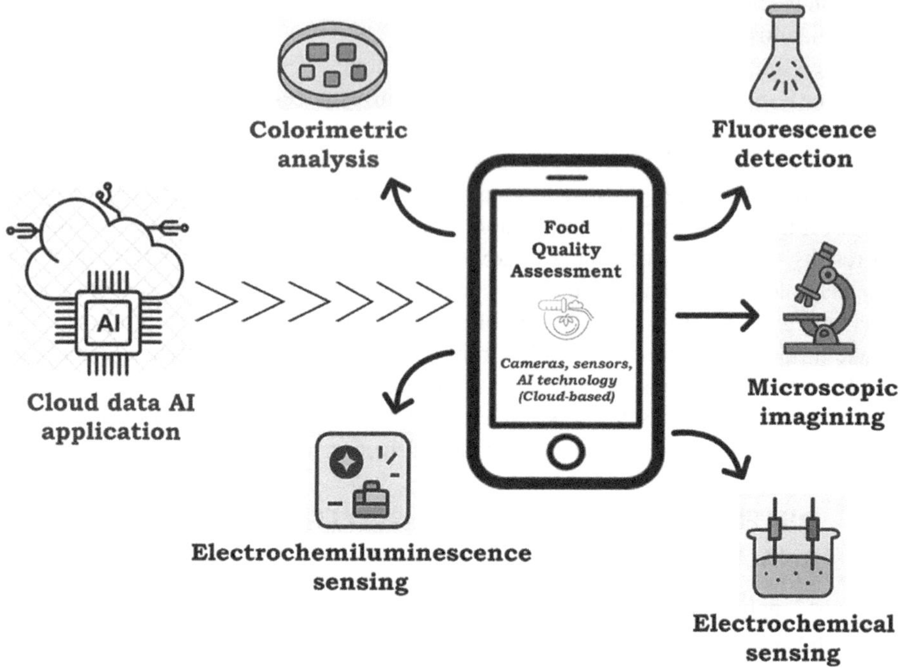

FIGURE 6.1 Schematic diagram of the utilization of smartphones integrated with AI using different techniques.

Commonly used software for extracting color parameters include Image J, Adobe Photoshop and Illustrator, free smartphone applications, and custom software developed by researchers. Additionally, the integration of AI technologies, particularly machine learning algorithms, is enhancing data interpretation from colorimetric sensors, improving accuracy and efficiency in analysis (Yu et al., 2022). The captured images undergo processing either through dedicated mobile applications or by transferring the data to a computer for advanced analysis.

Smartphone colorimeters measure sample concentrations based on color changes over time and are employed in both solid and liquid sample analyses. These devices offer rapid, cost-effective, and portable solutions for qualitative and quantitative analyses in remote or resource-limited areas. Despite their advantages, smartphones face challenges in precise food analysis due to limited control over imaging settings, automatic image processing, and changes in lighting conditions. These factors can affect color accuracy and pixel intensity, making it harder to get consistent and reliable measurements. However, advancements in camera technology, sensors, and computational power have improved their performance. Smartphones have been successfully used for colorimetric analysis as a standalone tool or equipped with external sensing components such as water monitoring with simple holders, ELISA assays, and fish spoilage detection via 3D-printed optical components (Kılıç et al., 2018; Yumnam et al., 2023). Custom cradles, fiber optics, and external light sources have enhanced their capabilities for biomarker detection and food quality assessment. New designs, such as surface plasmon resonance-based sensors and compact spectrometers using built-in flash and compact disks, further demonstrate their potential as portable, low-cost alternatives to sophisticated lab equipment.

6.2.2 Fluorescent Technique

Fluorescence techniques work by exciting specific molecules, called fluorophores, with light at a specific wavelength, causing them to absorb energy and jump to an excited state. When these excited molecules return to their ground state, they emit light at a longer wavelength (lower energy), which can be detected and analyzed to identify or quantify the target molecule, and this process is characterized by the absorption of excitation light and subsequent emission of fluorescence light at a different wavelength, allowing for high sensitivity and specificity in various applications like microscopy, analysis, and diagnostics (Radotić et al., 2023). The efficiency of fluorescence detection depends on excitation-emission properties, quantum yield, and the specificity of the interaction with the target analyte. Modern smartphones equipped with an external part can serve as efficient portable fluorometric devices. Key components of smartphone-based fluorescence systems include an excitation source, such as an LED or laser diode, to induce fluorescence in the sample. Optical filters are incorporated to selectively capture emitted fluorescence while minimizing background noise. The smartphone camera, coupled with dedicated software, enables image acquisition, processing, and quantification of fluorescence intensity. Additionally, the integration of microfluidic and paper-based platforms enhances portability, making these systems well suited for field applications and on-site analysis (Wang et al, 2023).

Fluorescence techniques in food analysis utilize the ability of certain organic compounds in food to absorb light and then re-emit it at longer wavelengths. These techniques are widely used because they offer rapid, nondestructive analysis with minimal sample preparation (Mishra, 2022).

Fluorescence spectroscopy has various applications in food analysis, including assessing food quality, safety, identification, and freshness. It is commonly used to evaluate the quality of dairy products like milk, cheese, and yogurt, as well as olive oil, by detecting chemical changes that occur during spoilage or degradation, such as oxidative fat deterioration. In food safety, this technique helps detect contaminants such as microbial growth (bacteria, molds), mycotoxins, and hazardous chemicals by identifying fluorescent substances under specific wavelengths (SádeCká et al. 2007). Additionally, fluorescence fingerprinting aids in food authentication by collecting unique fluorescence spectra, ensuring products are genuine, and determining their geographical origin, such as wines or spices. The technique is also highly effective in assessing food freshness, particularly for fish and other perishable products, as the degradation of proteins and lipids alters fluorescence properties, allowing for freshness evaluation (Deshpande & Kanungo, 2023; Radotić et al., 2023). The article "Glowing colors of foods: application of fluorescence and chemometrics in food studies" by Ewa Sikorska and Igor Khmelinskii (2016) reviews the applications of fluorescence spectroscopy in qualitative and quantitative analysis of foods. It highlights fluorescence as a sensitive and selective technique for detecting chemical changes in food products. Fluorescent probes further enable the detection of antioxidant properties in amino acids, contributing to a deeper understanding of food composition. Smartphone-integrated fluorescence sensors have also emerged as powerful tools for real-time analysis, facilitating applications in freshness assessment and bioactive compound measurement in food and beverages (Radotić et al., 2023).

6.2.3 Microscopic Imaging Technique

Advancements in smartphone technology have enabled their use in microscopic imaging, offering a portable and cost-effective alternative to traditional laboratory microscopes. Smartphone-based microscopic imaging techniques utilize the device's built-in camera and processing capabilities to capture and analyze high-resolution microscopic images. However, additional components are often incorporated to achieve the required magnification and image clarity, enhancing the smartphone's imaging capabilities (Banik et al., 2021).

One of the key components in smartphone-based microscopy is the use of external optical lenses, such as macro or microscope lenses, which can be attached to the smartphone camera. These lenses provide the necessary magnification for detailed structural analysis of food samples. Proper illumination is also essential for clear and accurate imaging, with LED lights or other external light sources commonly used to enhance visibility and contrast. To ensure stable and consistent imaging conditions, custom-designed 3D-printed holders or adapters are often employed. These accessories securely position the smartphone and the sample, minimizing movement and ensuring precise focus. Additionally, dedicated software applications play a crucial role in processing and analyzing images. These applications enhance image quality, enable accurate measurements, and generate quantitative data, making smartphone-based microscopy a powerful analytical tool (Banik et al., 2021).

Microfluidic chips are miniaturized devices that facilitate the controlled movement of tiny liquid samples for analysis, often integrated with smartphone microscopy for detecting foodborne pathogens, toxins, and chemical contaminants. Their integration enables real-time, on-site food analysis with minimal sample volume, reducing the reliance on complex laboratory equipment. In food safety applications, microfluidic chips assist in bacterial detection, identifying contaminants like *E. coli* in food and water samples (Gao et al., 2020). They also enable rapid screening of pesticide residues in fruits and vegetables and aid in milk freshness testing by detecting spoilage markers in dairy products (Cao & Chen, 2025). This technology enhances food safety monitoring by providing a portable, cost-effective, and efficient analytical tool. A comparison between smartphone-based microscopic techniques and traditional laboratory microscopes is shown in Table 6.1. While smartphone-based microscopy cannot fully replace high-end laboratory microscopes, it offers an accessible and practical solution for rapid, field-based food analysis.

Smartphone-assisted microfluidic sensors are particularly well-suited for point-of-care (POC) testing due to their high data processing capabilities and seamless connectivity. With its portability, affordability, and ease of use, smartphone-based microscopy presents a practical and accessible solution for real-time food analysis, bridging the gap between laboratory-based research and field applications.

TABLE 6.1
Comparison between Smartphone-Based Microscopic Techniques and Traditional Laboratory Microscopes

Feature	Smartphone-Based Microscopy	Traditional Laboratory Microscopes
Portability	Highly portable, field-deployable	Requires a fixed lab setup
Cost	Affordable, low-cost solution	Expensive, requires maintenance
Ease of Use	User-friendly, minimal training	Requires specialized training
Resolution	Moderate (up to 1,000× with lenses)	High-resolution (up to 100,000×)
Applications	On-site food safety, quick tests	Advanced research, detailed analysis

Source: Jahan-Tigh et al. (2016)

6.2.4 Electrochemical Technique

Smartphone-based electrochemical techniques operate by detecting and measuring electrochemical reactions at the electrode surface, which occur due to interactions between the analytes (target substances) and the sensor. This interaction generates electrical signals that provide qualitative and quantitative data based on amperometry, potentiometry, or conductometry measurements. The connection between the electrode and the smartphone is typically established via Bluetooth or USB, enabling the control of electrochemical measurements and real-time results display. By integrating miniaturized electrochemical components with intuitive smartphone applications, these sensors can rapidly detect various food contaminants, including pathogens, pesticides, heavy metals, and chemical reactions. Users simply connect the sensor to their smartphone, apply a small sample of the food or beverage being tested, and receive instant results and data analysis through the app (Zhang et al., 2022; Ji et al., 2022; Yumnam et al., 2024)

The concepts underlying electrochemical techniques for analysis are that they operate based on Ohm's law ($V = IR$), which defines the relationship between voltage (V), current (I), and resistance (R). The sensor circuit typically consists of two resistors in series: one representing the sensor resistance (RS) and the other a known resistor, connected with a constant bias. Electrochemical sensors are categorized into conductometric, potentiometric, and amperometric types, measuring conductance, voltage, and current, respectively (Bui et al., 2023).

Smartphones play a crucial role in electrochemical sensing by serving as a power source, data processor, and display interface. They provide the necessary power to operate the sensor, while their analog-to-digital converter (ADC) captures electrical signals from the sensor. The smartphone's processor then analyzes the data using dedicated software, calculates the target concentration, and presents the results in a user-friendly format on the screen, making the system portable and efficient (Ji et al., 2022).

6.2.5 Electro-Chemiluminescence Technique

Electro-chemiluminescence (ECL) is an analytical technique that combines electrochemical reactions with luminescence, converting electrical energy into light. It is widely used for the detection, separation, and quantification of biomolecules. The process starts at the electrode surface, where reactive species are electrochemically generated from stable precursors. These species then undergo an oxidation-reduction reaction, exciting the luminophores to a high-energy state. As the luminophores return to their ground state, they release energy in the form of light (photons), which can be detected and analyzed (Premnath & Zubair, 2023).

Luminophores are light-emitting substances that play a key role in electrochemiluminescence. During this process, electron transfer at the electrode surface triggers an oxidation-reduction reaction, exciting the luminophores to a high-energy state. As they return to their ground state, they release photons, producing detectable light. The phone's camera captures the emitted light, while specialized apps process and analyze the data. The built-in processor and software display results in a quantitative and visual format, enhancing accessibility for non-experts. This property makes luminophores valuable as labels for biomolecules, allowing for their detection and quantification by measuring the emitted light. This technique is widely used in biosensing and medical diagnostics due to its high sensitivity and precision (Premnath & Zubair, 2023).

ECL integrates well with separation techniques and benefits from nanomaterials, immunoassays, apta sensors, and microfluidics. Li et al. (2025) demonstrate the efficiency and accuracy of ECL by developing a detection system for choline. This smartphone-based ECL system offers a highly portable and efficient alternative to traditional laboratory methods, making on-site food safety monitoring more accessible and practical.

6.3 INTEGRATION OF COMPUTER VISION, AI, AND SMARTPHONES IN FOOD ANALYSIS

6.3.1 Computer Vision

Computer vision plays an important role in automating and improving the assessment of many aspects of food quality, identifying physical abuse and microbial spoilage, and analyzing the nutritional value of foods. For instance, in estimating the shelf life of food, computer vision plays a significant role in evaluating the visual characteristics of food to estimate its shelf life. Computer vision is useful in this context for visual changes (Zhu et al., 2021a), image processing techniques, machine learning models (Meenu et al., 2021), multimodal data fusion (Bahador et al., 2021), real-time monitoring systems (da Costa et al., 2022), and detection of food quality thresholds detection (ElMasry & Nakauchi, 2016). The visual changes in the analyzed images of food taken over a period of time to monitor changes in visual characteristics such as surface morphology, texture, color, and microbial growth indicate the degradation of food during storage and can be used to estimate the shelf life of foods as illustrated in Figure 6.2. However, image processing techniques are used to sort useful

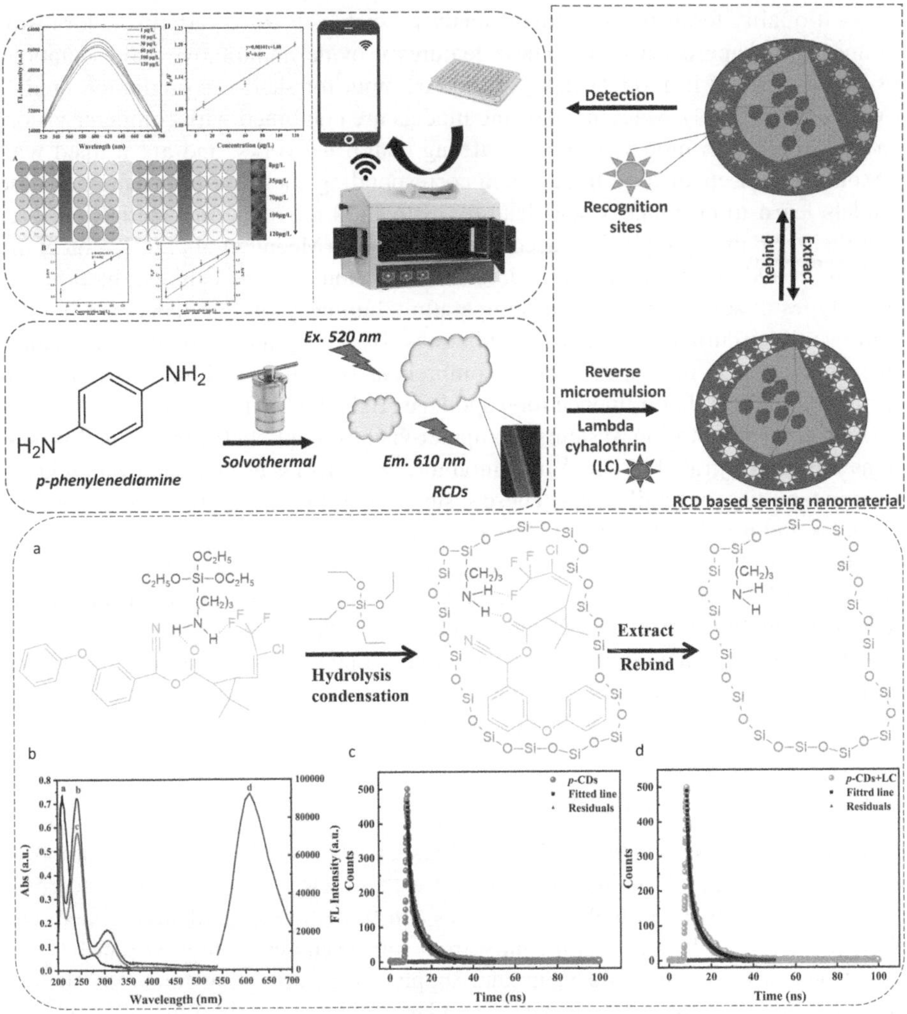

FIGURE 6.2 Smartphone-integrated design of the stage containing red emission carbon-dots-based optosensing nanomaterials for the detection of LC; (1) RCDs-based optosensing nanomaterials, developed by p-phenylenediamine precursor, in addition, SiO_2 coating, efficiently quench red fluorescence in LC. (2) This allows quantitative detection by means of fluorescence spectroscopy (incorporated in smartphone application; the smartphone was used as a signal output terminal) and represented as fluorescence responses of the *p*-CDs-based porous optosensing materials in the presence of LC at various concentrations. (3)(a) preparation method of the RCDs-based optosensing nanomaterials, (b) UV–vis spectra of LC, (c) fluorescence decay spectra and fitting curves of *p*-CDs, (d) fluorescence decline spectra and fitting curves of RCDs-based optosensing nanomaterials.

Source: Adapted from Zhu et al. (2021b)

or good-quality foods through image analysis, such as texture patterns, color histograms, and shape descriptors. These features provide quantitative representations of visual changes in food, which are used as input for shelf-life prediction models (Meenu et al., 2021). Machine learning models are combined with computer vision, such as regression models or deep learning neural networks, and are trained with annotated datasets of food images and corresponding shelf-life information. These models learn to correlate visual features with food degradation kinetics and predict their remaining shelf life based on new images (Meenu et al., 2021; Zhu et al., 2021b). In addition to multimodal data fusion, vision systems can also incorporate other types of sensor data such as temperature, humidity, and gas concentrations to improve the accuracy of shelf-life predictions. By fusing multimodal data streams, these systems can recognize a more comprehensive understanding of the environmental conditions that influence food spoilage (Bahador et al., 2021).

Real-time monitoring systems, computer-vision-based shelf-life prediction systems, can be integrated into real-time monitoring platforms used in food warehouses, storage facilities, or retail stores. These systems continuously capture images of food products and provide dynamic estimates of their remaining shelf life so that timely decisions can be made about inventory management, product rotation, and quality control (da Costa et al., 2022). Image processing algorithms can use visual cues to recognize quality thresholds beyond which food is classified as unfit for consumption. By comparing actual images with reference images of fresh products, these systems can trigger warnings or automatic actions when quality deterioration exceeds acceptable limits (ElMasry & Nakauchi, 2016).

6.3.2 Artificial Intelligence

AI is increasingly being integrated into various aspects of the food industry to improve efficiency, safety, and quality. By using AI technologies such as machine learning and data analytics, food analysis processes can be optimized and streamlined, and food industry can benefit significantly from improved sensitivity. AI can be used at various stages of the food value and supply chain, providing functions such as food monitoring, improved transparency and traceability, risk analysis and intelligent decision-making (Dora et al., 2022; Lei et al., 2022; Bačiulienė et al., 2023). AI can also help improve food quality and safety by using modeling techniques to assess and optimize food quality (Sahni et al., 2021). AI detects food spoilage through innovative methods that combine image classification with machine learning. The integration of AI in food analysis not only improves spoilage detection but also contributes to overall food safety measures (Sonwani et al., 2022).

Innovations in food production, particularly the development of nontraditional weaning foods for children, emphasize the role of antioxidants in grains, fruits, and vegetables in enhancing food quality, safety, and nutritional value. Smartphone applications such as 'Make Better Choices 2' support healthier eating habits by encouraging greater consumption of fruits and vegetables (Pellegrini et al., 2015; El-Sisy et al., 2016). By utilizing smartphone technology to track dietary intake, individuals can monitor their fruit and vegetable consumption, leading to improved dietary choices and overall health outcomes. Studies have shown that consumer perceptions of food quality, particularly in the purchase of fresh produce, are influenced by factors like food safety (Chamhuri & Batt, 2015).

The viability and development of chicken embryos during artificial incubation may be measured in a noninvasive way by monitoring their heart rates using smartphones' video recording capabilities and an external red-light source for photoplethy smography (Phuphanin et al., 2019). During processing, the quality of the raw poultry meat, e.g., concerning the pale poultry syndrome, can be assessed by analyzing its color based on the analysis of pictures captured with the smartphone camera (Barbin et al., 2016).

6.4 APPLICATION OF SMARTPHONES IN FOOD QUALITY AND SAFETY

6.4.1 Food and Beverages

Companies increasingly utilize AI tools not only for production optimization but also to influence consumer behavior through targeted marketing (Brooks et al., 2022). AI applications span across marketing, supply chain, logistics, and production, ensuring accurate and reproducible results throughout the food supply chain (Bačiulienė et al., 2023).

Smartphone-based sensor applications have been widely adopted across various stages of food and beverage manufacturing, enabling real-time monitoring and quality control from detecting ethanol levels in alcoholic beverages to assessing nutritional content (Pendergast et al., 2017; Choe et al., 2017). These technologies have facilitated the quantification of nutrients and the detection of key health-related compounds, including vitamin B12 (Kansay et al., 2024), food allergens (Ross et al., 2018), and biomarkers in saliva and sweat using colorimetric techniques (Oncescu et al., 2013). Apps have also been developed to support healthier dietary habits, such as identifying low-salt foods (Eyles et al., 2023) and reducing consumption of sugar-sweetened beverages (Septiyana et al., 2022), while offering personalized feedback and guidance. Advanced techniques like SERS have enabled precise identification and monitoring of beverage components, including ethanol content in real-time (Li et al., 2022). The integration of electrochemical sensing with nanomaterials such as copper nanoparticles has further expanded capabilities in measuring antioxidant capacity and nutrient content in products like fruit juices (Curulli, 2020). Moreover, smartphone-based spectrometric systems have been employed for the detection of key compounds such as ascorbic acid (vitamin C) in juices (Aguirre et al., 2019). Techniques like lab-on-a-disc systems and microwave-assisted processing have also been used to evaluate antioxidant activity and total phenolic content efficiently (Sattar et al., 2019).

The convergence of smartphones with advanced analytical technologies offers a powerful platform for enhancing food quality, safety, and personalized nutrition. These innovations hold immense potential for transforming the food and beverage industry, particularly through real-time, user-friendly, and cost-effective solutions.

6.4.2 Meat, Fish, and Poultry

The integration of deep learning algorithms in identifying images of deceased chickens has been implemented to enhance automation and efficiency in poultry farming (Liu et al., 2021). Similarly, in aquaculture, the utilization of intelligent equipment, IoT, edge computing, 5G, and AI algorithms has been reviewed to address challenges and explore future development prospects in fish farming (Wang et al., 2020). The practice of AI in the poultry industry extends to systems like intelligent poultry feed dispensing, which not only reduces labor but also enhances economic efficiency (Yang et al., 2022). In aquaculture and fisheries, applications such as deep learning, machine vision, big data, and IoT have been employed for tasks like fish disease detection using image-based machine learning approaches (Lim, 2024). Yumnam et al. (2024) developed a smartphone-based colorimetric sensor equipped with polyaniline to detect fish spoilage that provides a qualitative and quantitative analysis through a mobile application, which is a move toward a sustainable, rapid, and cost-effective alternative to traditional fish spoilage detection. These advancements highlight the role of smartphone-based sensors with AI technology efficiency, improving productivity and decision-making processes within the meat, fish, and poultry sectors.

Nondestructive methods based on artificial vision have proven successful in assessing beef meat quality, showcasing their potential in quality evaluation (El Barbri et al., 2014). Various technologies such as computer vision systems, e-noses, and e-tongues have been used for quality assessment applications in the food industry (Madhubhashini et al, 2024). These technologies play a crucial role in monitoring and measuring food quality, ensuring that standards of food safety are continuously improved through advanced technology applications (Eissa & Shehata, 2024). Deep learning techniques are also applied in the meat supply chain management to prevent cross-contamination and ensure the segregation of spoiled meat from fresh products, thereby enhancing food safety (Amani & Sarkodıe, 2022). Furthermore, intelligent packaging, which uses smartphone sensors integrating AI technologies, has emerged as a significant innovation in ensuring food safety and quality. Active and intelligent packaging systems interact with the product and its environment to extend shelf life, improve safety, and maintain the quality of packed food. Such packaging technologies have enormous potential to enhance the safety, quality, and traceability of food products, offering convenience for consumers. The integration of AI in the meat industry has resulted in various benefits, including reduced downtime, improved consumer experiences, and increased operational efficiency (Sahni et al., 2021). Such approaches, such as computer vision techniques, enable real-time assessment of chicken meat freshness, underscoring its role in maintaining meat quality (Taheri-Garavand et al., 2019). These technologies not only aid in quality evaluation but also enhance work processes and worker efficiency in the food sector. The meat processing industry, especially concerning meat quality and safety, has transformed monitoring, assessment, and management practices. From nondestructive evaluation methods to intelligent packaging solutions, AI technologies play a crucial role in upholding food safety standards and enhancing the overall quality of meat products.

Smartphone-based technologies are increasingly being adopted in the fisheries and poultry industries to enhance food quality, safety, and operational efficiency. In the fisheries sector, smartphone-integrated tools such as Near-Infrared Spectroscopy (NIRS) are used for on-site food safety evaluation of fish and fishery products (Qu et al., 2015). Smartphone-based sensors, including battery-free RFID systems, enable real-time monitoring of fish freshness and detection of bacterial contamination (Potyrailo et al., 2012). Like smart freezing systems operated or monitored via smartphones, AI technology has also been developed to improve the quality and efficiency of frozen fish products (Fu & Cheng, 2024). Moreover, smartphones play a vital role in smart fish farming, assisting in species classification, real-time identification, and water quality assessment through connected imaging and sensing systems (Yang et al., 2022).

In the poultry industry, smartphone-based imaging systems help monitor flock health and detect early signs of illness, improving product quality and survival rates (Neethirajan, 2022). Hyperspectral and spectral imaging adapted to smartphone platforms are used to detect contaminants in poultry products and ensure compliance with safety standards (Falkovskaya & Gowen, 2020). Mobile applications also support automated tracking and management of poultry, facilitating better farm practices. Additionally, smartphones are employed in disease monitoring and prevention, such as detecting risk factors for avian influenza (Wang et al., 2014). Smartphone-based imaging and sensing systems are further used in quality control processes, offering accessible and efficient tools for both diagnostic and operational applications.

The portability, connectivity, and affordability of smartphone-based AI technologies make them especially valuable in enhancing efficiency and safety throughout the food production chain, particularly in resource-limited environments, providing real-time solution for improving food quality, safety, disease control, and farm management in the fisheries and poultry sectors.

6.4.3 Bakery and Confectionery

Neural networks and neuro-fuzzy systems adapted for use with smartphone-connected imaging devices can aid in the real-time detection of foreign bodies in baked goods, significantly enhancing consumer safety. Smartphones equipped with high-resolution cameras and AI-based image processing are also being used to evaluate quality characteristics such as color, texture, and volume of bread and confectionery items (Olakanmi et al., 2023). In small-scale bakeries, smartphone-supported AI tools can assist in identifying Halal Control Points, streamlining certification processes, and supporting compliance with religious and quality standards (Sucipto et al., 2022). Additionally, mobile apps powered by AI and data analytics are being used in bakeries and food outlets to monitor and optimize production processes, helping improve product consistency, reduce waste, and boost overall efficiency.

Intelligent packaging systems, which combine sensor technology with smartphone interfaces, play a vital role in ensuring food safety. These systems can track spoilage indicators, indicators, shelf life, freshness, pH levels, and even the presence of pathogens, providing real-time data that can be easily accessed via smartphones (Shen et al., 2024). Consumers and producers alike benefit from this interactive and accessible approach to quality monitoring. The integration of smartphone-based technologies with AI in the bakery and confectionery industries offers a robust approach to modern food safety and quality control. From contaminant detection and process optimization to intelligent packaging and real-time monitoring, these innovations empower producers to deliver safer, higher-quality products while enhancing transparency and efficiency across the supply chain.

6.5 RECENT DEVELOPMENT OF MOBILE APPLICATIONS FOR SMARTPHONES IN FOOD QUALITY AND SAFETY

The commercialization of smartphone-integrated analysis empowers consumers to participate directly in food testing, promoting proactive monitoring, increasing public awareness, and strengthening food security. However, despite its potential, only a few commercial products utilizing smartphone-based technologies for food safety have been introduced (Tsagkaris et al., 2023). One such product is the MyDx® handheld analyzer, which uses an AChE chip to detect organophosphate and carbamate pesticides in fruits and vegetables (OrganaDX, 2014) and heavy metals in water (AquaDX, 2017). The device pairs with a proprietary smartphone app to deliver one-click results. Another example is Test4all, developed by Zeulab, which combines a portable analyzer with a smartphone app for on-site antibiotic detection in milk (ZeuLab, 2018).

Recent developments in smartphone-integrated sensing technologies have significantly advanced food safety and quality assessment by enabling real-time, portable, and cost-effective analysis. Electrochemical sensors such as modified screen-printed electrodes and impedance systems offer precise biomolecule detection (Shen et al., 2020; Ji et al., 2022), while electro-chemiluminescence (ECL) platforms provide highly sensitive tools for detecting glucose and aflatoxin B1 (Calabria et al., 2023; Chen et al., 2025). Smartphone-based imaging and microfluidic systems further expand capabilities by enabling colorimetric analysis, calorie estimation, and rapid diagnostics (Wang et al., 2023; Tsagkaris et al., 2023, Adam et al., 2024). These innovations collectively demonstrate the potential of smartphones as versatile tools for on-site, point-of-care testing in food safety and health monitoring.

Khumngern et al. (2024) developed a Digital Image Colorimetry (DIC) smartphone app to monitor histamine (HA) levels in fish. Their sensor, using 2,3-naphthalenedicarboxaldehyde (NDA) embedded in a polyvinyl alcohol (PVA) sol-gel, produced a blue color in the presence of HA. A microtube containing the PVA@NDA sol-gel enabled semiquantitative detection, and the Color Lab® app analyzed the color intensity via the red channel. The sensor showed high selectivity, stability, and reproducibility, aligning well with traditional ELISA results, supporting reliable on-site fish freshness detection. In another study, Park et al. (2021) introduced a smartphone-based approach to evaluate pH-responsive inverse opal photonic gel (IOPG) sensors. By analyzing the hue (h) parameter in the HSV color space, they ensured consistent performance across various smartphone models and lighting conditions (LED, fluorescent, or flash). In poultry, Kilic et al. (2022) demonstrated that color changes in FG-UV-CD100 films correlate with microbial growth and total volatile basic nitrogen (TVB-N) in chicken breast, allowing freshness evaluation through colorimetric analysis.

Paper-based sensors have also proven effective in colorimetric applications, with image analysis conducted in RGB, HSV, and L*ab** spaces (Lopez-Ruiz et al., 2014; Jia et al., 2015; Morsy et al., 2016). For example, Jung et al. (2015) used a smartphone camera and paper-based test converting RGB to HSV to detect alcohol levels in saliva. Nontraditional smartphone methods have been used for fruit maturity estimation and chlorine detection in water using vial-based colorimetric approaches (Intaravanne et al., 2015). More recently, Su et al. (2023) developed a smartphone-based method using the Color Analyzer app to quantify the blue oxidized product (TMBox), indirectly detecting glucose via H_2O_2 oxidation, showcasing its promise as a low-cost, accurate tool for food analysis.

Among the early pioneers, R-Biopharm developed the RIDA®SMART APP Mycotoxin, a smartphone-based colorimetric tool for detecting multiple mycotoxins (e.g., zearalenone, aflatoxins, fumonisins, deoxynivalenol, T2, and HT-2) in food. Designed for industrial and primary producers, this system provides semiquantitative results within 10–15 minutes, using a color reference chart and QR code to calibrate lighting and classify samples. Results can be instantly shared via email. However, its compatibility is limited to specific Android smartphones, revealing a key barrier in cross-platform functionality.

Some companies are shifting away from smartphone cameras to dedicated hardware. For example, Neogen's Raptor® system and Abingdon Health's AppDx offer portable devices for lateral flow immunoassays (LFIA), with the latter supporting a broader range of smartphones, including iOS. However, these are proprietary systems, often limited to the manufacturers' assays. More universal tools have emerged. The Cube-Reader by ChemBio is a versatile, low-cost camera system that works with any LFIA and is open for a third-party use. Similarly, Scienion AG's SciReader LF1 supports real-time assay tracking and can communicate results to any smartphone, independent of the phone's camera (ChemBio C.-R., 2019).

On a different front, the SCAN4CHEM app offers a barcode-based public information tool for detecting substances of very high concern (SVHC) in consumer products (including food packaging, cosmetics, and household chemicals). While informative and widely accessible, it lacks bioanalytical capabilities, limiting direct consumer engagement in actual testing. To advance smartphone-based food analysis, EU-funded initiatives like FoodSmartphone and Phasma FOOD aim to develop compact, spectroscopic, and user-friendly biosensors for detecting allergens, pesticides, and food contaminants. These platforms integrate bioanalytical assays and spectroscopic sensors (UV/IR), paving the way for on-site testing and nutritional analysis, especially in detecting food adulteration. While these devices offer key advantages such as portability, ease of use, and real-time data connectivity, information regarding critical performance parameters like bio-affinity, detection limits, and test duration remains limited. Broader adoption and confidence in smartphone-based food safety tools will depend on transparent performance validation and regulatory approval.

6.6 CHALLENGES AND LIMITATIONS

They are frequently used as easy data gathering and processing tools, as well as for providing a graphic, user-friendly interface in place of personal computers; however, peripherals are still required for real analysis. In some cases, the analytical technique required to provide a relevant result is comparatively complex and requires the use of instruments generally available only in laboratories. Potential end users may be put off by multi-stage sampling and time-consuming preparation methods. Furthermore, such complexity drastically limits the application of the proposed methods, particularly in field situations, and contradicts the primary benefits of adopting smartphone-based procedures in the first place.

Smartphones are not meant for optical measurements as their image sensors record polychromatic light, reducing measurement accuracy and precision (Capitán-Vallvey et al., 2015). While advanced camera sensors in smartphones with more than 40-megapixel resolution (in current models up to 100-megapixel) appear to be more than adequate for photometry, their performance is limited by small pixel size and 8-bit digitization, which reduce S/N ratio and precision. The latter, however, will be substantially improved, as smartphone models using 10-bit encoding are being released at the time of writing, thus extending the color palette from 2,563 to 10,243 colors (Tonelli et al., 2019), though at the increase in expense of higher file size and computing work.

A key future enhancement in smartphone-based sensing for food analysis is expanding the spectral response range of smartphone cameras. Currently limited by bandpass filters that restrict detection to the visible spectrum, this range could be extended to approximately 310–900 nm by removing or modifying these filters (Wilkes et al., 2016). Although this may reduce the phone's conventional photography capabilities, the increasing use of multi-camera systems (often five or more sensors) makes this a feasible upgrade. Such spectral expansion would significantly benefit noninvasive food content analysis, especially in the near-infrared (NIR) range, which is already a standard in food quality control (Porep et al., 2015). Additionally, incorporating external UV light sources can enhance the sensor's ability to capture fluorescence signals, widening its analytical range (Fang et al., 2016). Another promising development involves the integration of metal-oxide-semiconductor field-effect transistor (MOSFET) gas detectors into smartphone modules to act as electronic noses. These sensors offer low energy consumption and compact design, making them ideal for portable gas sensing. However, challenges such as signal stability and reliability have limited their widespread adoption, even years after their initial proposal (Wojnowski et al., 2019). Moreover, the growing reliance on computational photography by smartphone manufacturers used to enhance image quality creates barriers to sensor readout consistency, as companies are reluctant to provide access to proprietary image-processing protocols. To overcome this, researchers are increasingly employing machine learning techniques to standardize and interpret image-based sensor data (Abdalla et al., 2019). Beyond cameras, other smartphone sensors also hold potential for food-related applications. For instance, the microphone has been used to assess honeybee activity by correlating sound intensity with hive microclimate, which in turn affects honey production and crop pollination (Lima et al., 2021). Similarly, the microphone and speaker together can enable ultrasonic sensing, which may be useful for monitoring and optimizing food processing operations like emulsification, drying, crystallization, and fermentation (Wang et al., 2020). Given the limitations of smartphone cameras as spectrometers, a growing consensus points to combining smartphones with low-cost optical attachments as a practical solution for advancing food inspection technologies.

Enhancing spectral sensitivity, incorporating diverse sensors (e.g., gas detectors, microphones), and using external optical attachments are key strategies for advancing smartphone-based food sensing. While technical and proprietary barriers remain, innovations in hardware and machine learning continue to push the boundaries of what is possible in portable food analysis.

6.7 FUTURE SCOPE OF SMARTPHONE-BASED SENSORS IN FOOD

The future of smartphone-based sensors in food analysis is prepared for significant advancements, driven by innovations in sensor technology, AI, and real-time data connectivity. Miniaturized spectroscopy techniques such as NIR, Raman, and fluorescence, along with electrochemical sensors, will enable more precise and portable food quality assessments (Rateni et al., 2017; Azizi et al., 2024). The integration of AI and machine learning will enhance image processing and analytical detection. Additionally, smart labels and IoT-connected sensors will provide real-time food freshness status, reduce food waste and improve supply chain transparency. The adoption of smartphone-based food testing kits will empower consumers to assess food safety at home, while cloud-based data sharing and block chain integration will enhance traceability and regulatory compliance. Industries and regulatory bodies are likely to incorporate these tools for automated quality control and on-field inspections. As the food industry moves toward sustainability, smartphone sensors will play a crucial role in minimizing food waste by providing real-time freshness assessments beyond conventional techniques. With continued progress in nanotechnology, AI, and the Internet of Things (IoT), smartphone-based fluorescent techniques are expected to evolve further. Future research should focus on enhancing sensitivity, miniaturizing devices, and developing standardized protocols to ensure regulatory compliance. Similarly, advancements in smartphone technology, such as improved camera sensors, AI-driven image analysis, and cloud-based data processing, are continually enhancing the accuracy and reliability of smartphone-based colorimetric sensors (Bahador et al., 2021; Adam et al., 2024). Emerging designs, including 3D-printed optical attachments, advanced light sources, and AI-based calibration techniques, are expanding their potential as cost-effective and user-friendly tools for food quality monitoring and safety assurance. However, challenges such as background fluorescence interference, variations in smartphone camera specifications, and the need for standardization as variations in smartphone camera specifications such as sensor quality and automatic image processing can lead to inconsistent results must be addressed to ensure accuracy and reliability. Addressing this requires optimized optical filters and advanced software correction techniques to enhance signal clarity. Ensuring reproducibility across different devices necessitates calibration protocols and standardized imaging conditions. Additionally, integrating smartphone-based techniques with robust analytical methods is essential, as they must be validated against established laboratory procedures to ensure accuracy and reliability in real-world applications. Overcoming these limitations will strengthen the credibility and adoption of smartphone-based colorimetric sensors in both scientific and industrial settings. By improving food safety, reducing waste, and ensuring quality control, smartphone-based sensors are set to revolutionize the food industry. They will benefit not only scientists and industry professionals but also everyday consumers, making food testing more convenient and reliable for all.

REFERENCES

Abdalla, A., Cen, H., Abdel-Rahman, E., Wan, L., & He, Y. (2019). Color calibration of proximal sensing RGB images of oilseed rape canopy via deep learning combined with K-means algorithm. *Remote Sensing*, 11(24), 3001.

Adam, A., Asikin, H., Ipa, A., Sahrir, S., & Imran, A. (2024). Development of a food detector using image processing through camera sensor innovation. *Healthcare in Low-Resource Settings*, 12(3).

Aguirre, M. Á., Long, K. D., Canals, A., & Cunningham, B. T. (2019). Point-of-use detection of ascorbic acid using a spectrometric smartphone-based system. *Food Chemistry*, 272, 141–147.

Amani, M. A., & Sarkodie, S. A. (2022). Mitigating spread of contamination in meat supply chain management using deep learning. *Scientific Reports*, 12(1), 5037.

AquaDX. (2017). *MyDX, AquaDX*. https://www.mydxlife.com/aqua-sensor/

Azizi, N., Hallaj, T., & Samadi, N. (2024). A dual-mode fluorometric and smartphone-based colorimetric sensor for cyanide detection using tungsten disulfide quantum dots and silver nanoparticles. *Journal of Food Composition and Analysis*, 129, 106081.

Bačiulienė, V., Bilan, Y., Navickas, V., & Civín, L. (2023). The aspects of artificial intelligence in different phases of the food value and supply chain. *Foods*, 12(8), 1654.

Bahador, N., Ferreira, D., Tamminen, S., & Kortelainen, J. (2021). Deep learning–based multimodal data fusion: Case study in food intake episodes detection using wearable sensors. *JMIR mHealth and uHealth*, 9(1), e21926.

Banik, S., Melanthota, S. K., Arbaaz Vaz, J. M., Kadambalithaya, V. M., Hussain, I., . . . & Mazumder, N. (2021). Recent trends in smartphone-based detection for biomedical applications: A review. *Analytical and Bioanalytical Chemistry*, 413, 2389–2406.

Barbin, D. F., Mastelini, S. M., Barbon Jr, S., Campos, G. F., Barbon, A. P. A., & Shimokomaki, M. (2016). Digital image analyses as an alternative tool for chicken quality assessment. *Biosystems Engineering*, 144, 85–93.

Brooks, R., Nguyen, D., Bhatti, A., Allender, S., Johnstone, M., Lim, C. P., & Backholer, K. (2022). Use of artificial intelligence to enable dark nudges by transnational food and beverage companies: Analysis of company documents. *Public Health Nutrition*, 25(5), 1291–1299.

Bui, T. H., Thangavel, B., Sharipov, M., Chen, K., & Shin, J. H. (2023). Smartphone-based portable bio-chemical sensors: Exploring recent advancements. *Chemosensors*, 11(9), 468.

Calabria, D., Lazzarini, E., Pace, A., Trozzi, I., Zangheri, M., Cinti, S., . . . & Mirasoli, M. (2023). Smartphone-based 3D-printed electrochemiluminescence enzyme biosensor for reagentless glucose quantification in real matrices. *Biosensors and Bioelectronics*, 227, 115146.

Cao, Q., & Chen, X. (2025). Microfluidics detection technologies and applications. In Oliveira, M., Holló, B. B., Zafar, M. & De Aguiar Andrade, E. H. (eds.), *Advances in Analytical and Coordination Chemistry–Applications and Innovations*. Radanovic MM IntechOpen.

Capitán-Vallvey, L. F., Lopez-Ruiz, N., Martinez-Olmos, A., Erenas, M. M., & Palma, A. J. (2015). Recent developments in computer vision-based analytical chemistry: A tutorial review. *Analytica Chimica Acta*, 899, 23–56.

Chamhuri, N., & Batt, P. J. (2015). Consumer perceptions of food quality in Malaysia. *British Food Journal*, 117(3), 1168–1187.

ChemBio, C.-R. (2019). Accessed 6th May 2020. ChemBio, Cube-Reader. https://www.optricon.de/cube-reader.php

Chen, M. M., Zhang, M. L., Song, X., Jiang, J., Tang, X., Zhang, Q., . . . & Li, P. (2025). Smartphone-assisted electrochemiluminescence imaging test strips towards dual-signal visualized and sensitive monitoring of aflatoxin B1 in corn samples. *Chinese Chemical Letters*, 36(1), 109785.

Choe, E. K., Abdullah, S., Rabbi, M., Thomaz, E., Epstein, D. A., Cordeiro, F., . . . & Kientz, J. A. (2017). Semi-automated tracking: A balanced approach for self-monitoring applications. *IEEE Pervasive Computing*, 16(1), 74–84.

Curulli, A. (2020). Nanomaterials in electrochemical sensing area: Applications and challenges in food analysis. *Molecules*, 25(23), 5759.

da Costa, T. P., Gillespie, J., Cama-Moncunill, X., Ward, S., Condell, J., Ramanathan, R., & Murphy, F. (2022). A systematic review of real-time monitoring technologies and its potential application to reduce food loss and waste: Key elements of food supply chains and IoT technologies. *Sustainability*, 15(1), 614.

Deshpande, K., & Kanungo, L. (2023). Chemiluminescence and fluorescence biosensors for food application: A review. *Sensors and Actuators Reports*, 5, 100137.

Dora, M., Kumar, A., Mangla, S. K., Pant, A., & Kamal, M. M. (2022). Critical success factors influencing artificial intelligence adoption in food supply chains. *International Journal of Production Research*, 60(14), 4621–4640.

Eissa, F. I., & Shehata, A. M. (2024). Eggs and egg products contamination: Analysis of the EU RASFF notifications from 2000 to 2022. *Food Control*, 158, 110249.

El Barbri, N., Halimi, A., & Rhofir, K. (2014). A nondestructive method based on an artificial vision for beef meat quality assessment. *International Journal of Innovative Research in Electrical, Electronics, Instrumentation and Control Engineering*, 2, 2060–2063.

El Masry, G. M., & Nakauchi, S. (2016). Image analysis operations applied to hyperspectral images for non-invasive sensing of food quality–a comprehensive review. *Biosystems Engineering*, 142, 53–82.

El-Sisy, T. T., El-Gendy, M. H., & Ali, S. A. (2016). Production of nontraditional food for children after weaning. *Journal of Food and Dairy Sciences*, 7(11), 463–470.

Eyles, H., Grey, J., Jiang, Y., Umali, E., McLean, R., Te Morenga, L., . . . & Mhurchu, C. N. (2023). Effectiveness of a sodium-reduction smartphone app and reduced-sodium salt to lower sodium intake in adults with hypertension: Findings from the salt alternatives randomized controlled trial. *JMIR mHealth and uHealth*, 11(1), e43675.

Falkovskaya, A., & Gowen, A. (2020). Literature review: Spectral imaging applied to poultry products. *Poultry science*, 99(7), 3709–3722.

Fang, J., Qiu, X., Wan, Z., Zou, Q., Su, K., Hu, N., & Wang, P. (2016). A sensing smartphone and its portable accessory for on-site rapid biochemical detection of marine toxins. *Analytical Methods*, 8(38), 6895–6902.

Fu, J., He, Y., & Cheng, F. (2024). Intelligent cutting in fish processing: Efficient, high-quality, and safe production of fish products. *Food and Bioprocess Technology*, 17(4), 828–849.

Gao, H., Yan, C., Wu, W., & Li, J. (2020). Application of microfluidic chip technology in food safety sensing. *Sensors*, 20(6), 1792.

Gao, X., & Wu, N. (2016). Smartphone-based sensors. *The Electrochemical Society Interface*, 25(4), 79.

Geng, Z., Miao, Y., Zhang, G., & Liang, X. (2023). Colorimetric biosensor based on smartphone: State-of-art. *Sensors and Actuators A: Physical*, 349, 114056.

Intaravanne, Y., & Sumriddetchkajorn, S. (2015). Android-based rice leaf color analyzer for estimating the needed amount of nitrogen fertilizer. *Computers and Electronics in Agriculture*, 116, 228–233.

Jahan-Tigh, R. R., Chinn, G. M., & Rapini, R. P. (2016). A comparative study between smartphone-based microscopy and conventional light microscopy in 1021 dermatopathology specimens. *Archives of Pathology & Laboratory Medicine*, 140(1), 86–90.

Ji, D., Low, S. S., Zhang, D., Liu, L., Lu, Y., & Liu, Q. (2022). Smartphone-based electrochemical system for biosensors and biodetection. *Biomedical Engineering Technologies*, 1, 493–514.

Jia, M.-Y., Wu, Q.-S., Li, H., Zhang, Y., Guan, Y.-F., & Feng, L. (2015). The calibration of cellphone camera-based colorimetric sensor array and its application in the determination of glucose in urine. *Biosensors and Bioelectronics*, 74, 1029–1037.

Jiang, W., Zhao, Y., Zhang, D., Zhu, X., Liu, H., & Sun, B. (2021). Efficient and robust dual modes of fluorescence sensing and smartphone readout for the detection of pyrethroids using artificial receptors bound inside a covalent organic framework. *Biosensors and Bioelectronics*, 194, 113582.

Jung, Y., Kim, J., Awofeso, O., Kim, H., Regnier, F., & Bae, E. (2015). Smartphone-based colorimetric analysis for detection of saliva alcohol concentration. *Applied Optics*, 54(31), 9183–9189.

Kansay, V., Sharma, V. D., Chandan, G., Srivastava, V., Batra, N., Mittal, A., Chakrabarti, S., & Bera, M. K. (2024). Smartphone platform for low-cost point-of-care quantitative detection of vitamin B12 utilizing a biodegradable bioplastic nanocomposite-based fluorescence Nanosensor. *ACS Applied Electronic Materials*, 6(3), 1971–1981.

Khumngern, S., Nontipichet, N., Thavarungkul, P., Kanatharana, P., & Numnuam, A. (2024). A simple colorimetric histamine sensor based on smartphone digital image processing for fish quality assessment. *Journal of Food Composition and Analysis*, 126, 105934.

Kilic, B., Dogan, V., Kilic, V., & Kahyaoglu, L. N. (2022). Colorimetric food spoilage monitoring with carbon dot and UV light reinforced fish gelatin films using a smartphone application. *International Journal of Biological Macromolecules*, 209, 1562–1572.

Kılıç, V., Horzum, N., & Solmaz, M. E. (2018). From sophisticated analysis to colorimetric determination: Smartphone spectrometers and colorimetry. In *Color Detection*. IntechOpen.

Lei, M., Xu, L., Liu, T., Liu, S., & Sun, C. (2022). Integration of privacy protection and blockchain-based food safety traceability: Potential and challenges. *Foods*, 11(15), 2262.

Li, J., Tian, Y., & Zheng, T. (2022). A multifunctional nanoprobe for real-time SERS monitoring of invasion ability affected by photodynamic therapy. *Chemical Communications*, 58(45), 6542–6545.

Li, X., Li, G., Wu, S., Cao, X., Sun, D., Li, S., . . . & Wang, N. (2025). Sensitive and rapid electrochemiluminescent detection of choline based on luminol derivate-doped silica nanoparticles. *Journal of Analysis and Testing*, 1–7.

Lim, L. W. K. (2024). Implementation of artificial intelligence in aquaculture and fisheries: Deep learning, machine vision, big data, internet of things, robots and beyond. *Journal of Computational and Cognitive Engineering*, 3(2), 112–118.

Lima, M., De Queiroz, J., Pascoal, L., Saraiva, E., Soares, K., & Evangelista-Rodrigues, A. (2021). Smartphone-based sound level meter application for monitoring thermal comfort of honeybees Apis mellifera L. *Biological Rhythm Research*, 52(7), 1073–1086.

Liu, H.-W., Chen, C.-H., Tsai, Y.-C., Hsieh, K.-W., Lin, H.-T. (2021). Identifying images of dead chickens with a chicken removal system integrated with a deep learning algorithm. *Sensors*, 21(11), 3579. https://doi.org/10.3390/s21113579

Lopez-Ruiz, N., et al. (2014). Smartphone based simultaneous pH and nitrite colorimetric determination for paper microfluidic devices. *Analytical Chemistry*, 86(19), 9554–9562.

Lv, Y., Yang, Y., Wu, R., Xu, Y., Li, J., Li, N., . . . & Li, L. S. (2022). A CdSe/ZnS core/shell competitive quantum dot-based fluorescence-linked immunosorbent assay for the sensitive and accurate detection of aflatoxin B1 in corn sample. *Journal of Food Measurement and Characterization*, 1–10.

Madhubhashini, M. N., Liyanage, C. P., Alahakoon, A. U., & Liyanage, R. P. (2024). Current applications and future trends of artificial senses in fish freshness determination: A review. *Journal of Food Science*, 89(1), 33–50.

McGonigle, A. J., Wilkes, T. C., Pering, T. D., Willmott, J. R., Cook, J. M., Mims III, F. M., & Parisi, A. V. (2018). Smartphone spectrometers. *Sensors*, 18(1), 223.

Meenu, M., Kurade, C., Neelapu, B. C., Kalra, S., Ramaswamy, H. S., & Yu, Y. (2021). A concise review on food quality assessment using digital image processing. *Trends in Food Science & Technology*, 118, 106–124.

Mishra, M. (2022). Spectroscopic techniques for the analysis of food quality, chemistry, and function. In Food Chemistry, Function and Analysis, *Advanced Spectroscopic Techniques for Food Quality* (pp. 246). Royal Society of Chemistry.

Morsy, M. K., Zór, K., Kostesha, N., Alstrøm, T. S., Heiskanen, A., El-Tanahi, H., Sharoba, A., Papkovsky, D., Larsen, J., Khalaf, H., & Jakobsen, M. H. (2016). Development and validation of a colorimetric sensor array for fish spoilage monitoring. *Food control*, 60, 346–352.

Neethirajan, S. (2022). Automated tracking systems for the assessment of farmed poultry. *Animals*. 12(3), 232. https://doi.org/10.3390/ani12030232

Olakanmi, S. J., Jayas, D. S., & Paliwal, J. (2023). Applications of imaging systems for the assessment of quality characteristics of bread and other baked goods: A review. *Comprehensive Reviews in Food Science and Food Safety*, 22(3), 1817–1838.

OrganaDX. (2014). *MyDX, OrganaDX*. https://www.mydxlife.com/organadx/

Ozcan, A. (2014). Mobile phones democratize and cultivate next-generation imaging, diagnostics and measurement tools. *Lab on a Chip*, 14, 3187–3194. [CrossRef] [PubMed]

Oncescu, V., O'Dell, D., & Erickson, D. (2013). Smartphone based health accessory for colorimetric detection of biomarkers in sweat and saliva. *Lab on a Chip*, 13(16), 3232–3238.

Qu, J. H., Liu, D., Cheng, J. H., Sun, D. W., Ma, J., Pu, H., & Zeng, X. A. (2015). Applications of near-infrared spectroscopy in food safety evaluation and control: A review of recent research advances. *Critical Reviews in Food Science and Nutrition*, *55*(13), 1939–1954.

Park, H., Koh, Y. G., & Lee, W. (2021). Smartphone-based colorimetric analysis of structural colors from pH-responsive photonic gel. *Sensors and Actuators B: Chemical*, 345, 130359.

Potyrailo, R. A., Nagraj, N., Tang, Z., Mondello, F. J., Surman, C., & Morris, W. (2012). Battery-free radio frequency identification (RFID) sensors for food quality and safety. *Journal of Agricultural and Food Chemistry*, 60(35), 8535–8543.

Pellegrini, C. A., Steglitz, J., Johnston, W., Warnick, J., Adams, T., McFadden, H. G., Siddique, J., Hedeker, D., & Spring, B. (2015). Design and protocol of a randomized multiple behavior change trial: Make Better Choices 2 (MBC2). *Contemporary Clinical Trials*, 41, 85–92.

Pendergast, F. J., Ridgers, N. D., Worsley, A., & McNaughton, S. A. (2017). Evaluation of a smartphone food diary application using objectively measured energy expenditure. *International Journal of Behavioral Nutrition and Physical Activity*, 14, 1–10.

Phuphanin, A., Sampanporn, L., & Sutapun, B. (2019). Smartphone-based device for non-invasive heart-rate measurement of chicken embryos. *Sensors*, 19(22), 4843.

Porep, J. U., Kammerer, D. R., & Carle, R. (2015). On-line application of near infrared (NIR) spectroscopy in food production. *Trends in Food Science &Technology*, 46(2), 211–230.

Radotić, K., Stanković, M., Bartolić, D., & Natić, M. (2023). Intrinsic fluorescence markers for food characteristics, shelf life, and safety estimation: Advanced analytical approach. *Foods*, 12(16), 3023.

Rateni, G., Dario, P., & Cavallo, F. (2017). Smartphone-based food diagnostic technologies: A review. *Sensors*, 17(6), 1453.

Ross, G. M., Bremer, M. G., & Nielen, M. W. (2018). Consumer-friendly food allergen detection: Moving towards smartphone-based immunoassays. *Analytical and Bioanalytical Chemistry*, 410, 5353–5371.

Sahni, V., Srivastava, S., & Khan, R. (2021). Modelling techniques to improve the quality of food using artificial intelligence. *Journal of Food Quality*, 2021, 1–10.

Sattar, A., Naveed, M., Ali, M., Zahir, Z. A., Nadeem, S. M., Yaseen, M., Meena, V.S., Farooq, M., Singh, R., Rahman, M., & Meena, H. N. (2019). Perspectives of potassium solubilizing microbes in sustainable food production system: A review. *Applied Soil Ecology*, 133, 146–159.

Septiyana, S. M., Aji, A. S., Nuryandini, E., Penggalih, M. H. S. T., & Nurwanti, E. (2022). The smartphone app (MyFitnessPal) reduce sugar-sweetened beverages intake among overweight and obese college students. *Jurnal Gizi dan Dietetik Indonesia (Indonesian Journal of Nutrition and Dietetics)*, 9(3), 130–138.

Shen, X., Ju, F., Li, G., & Ma, L. (2020). Smartphone-based electrochemical potentiostat detection system using pedot: Pss/chitosan/graphene modified screen-printed electrodes for dopamine detection. *Sensors*, 20(10), 2781.

Shen, D., Zhang, M., Mujumdar, A. S., & Ma, Y. (2024). Consumer-oriented smart dynamic detection of fresh food quality: Recent advances and future prospects. *Critical Reviews in Food Science and Nutrition*, 64(30), 11281–11301.

Sikorska, E., & Khmelinskii, I. (2016). Glowing colours of foods: Application of fluorescence and chemometrics in food studies. *Spectroscopy Europe*, 28(6), 10–13.

Sonwani, E., Bansal, U., Alroobaea, R., Baqasah, A. M., & Hedabou, M. (2022). An artificial intelligence approach toward food spoilage detection and analysis. *Frontiers in Public Health*, 9, 816226.

Su, K., Xiang, G., Cui, C., Jiang, X., Sun, Y., Zhao, W., & He, L. (2023). Smartphone-based colorimetric determination of glucose in food samples based on the intrinsic peroxidase-like activity of nitrogen-doped carbon dots obtained from locusts. *Arabian Journal of Chemistry*, 16(3), 104538.

Sucipto, S., Damayanti, R. W., Perdani, C. G., Kamal, M. A., Astuti, R., & Hasanah, N. (2022). Decision tree of materials: A model of Halal Control Point (HCP) Identification in Small-Scale Bakery to Support Halal Certification. *International Journal of Food Science, 2022*(1), 5244586.

Taheri-Garavand, A., Fatahi, S., Omid, M., & Makino, Y. (2019). Meat quality evaluation based on computer vision technique: A review. *Meat Science*, 156, 183–195.

Tonelli, A., Candiani, A., Sozzi, M., Zucchelli, A., Foresti, R., Dall'Asta, C., Selleri, S., & Cucinotta, A. (2019). The geek and the chemist: Antioxidant capacity measurements by DPPH assay in beverages using open source tools, consumer electronics and 3D printing. *Sensors and Actuators B: Chemical*, 282, 559–566.

Tsagkaris, A. S., Nelis, J. L., Campbell, K., Elliott, C. T., Pulkrabova, J., & Hajslova, J. (2023). Smartphone and microfluidic systems in medical and food analysis. In *Microfluidic Biosensors* (pp. 233–257). Academic Press.

Wang, A., Zhu, Y., Zou, L., Zhu, H., Cao, R., & Zhao, G. (2022). Combination of machine learning and intelligent sensors in real-time quality control of alcoholic beverages. *Food Science and Technology*, 42, e54622.

Wang, B., Li, Y., Zhou, M., Han, Y., Zhang, M., Gao, Z., . . . & Liu, B. F. (2023). Smartphone-based platforms implementing microfluidic detection with image-based artificial intelligence. *Nature Communications*, 14(1), 1341.

Wang, T., Mei, Q., Tao, Z., Wu, H., Zhao, M., Wang, S., & Liu, Y. (2020). A smartphone-integrated ratiometric fluorescence sensing platform for visual and quantitative point-of-care testing of tetracycline. *Biosensors and Bioelectronics*, 148, 111791

Wang, Y., Li, P., Wu, Y., Sun, X., Yu, K., Yu, C., & Qin, A. (2014). The risk factors for avian influenza on poultry farms: A meta-analysis. *Preventive Veterinary Medicine*, 117(1), 1–6.

Wilkes, T. C., McGonigle, A. J., Pering, T. D., Taggart, A. J., White, B. S., Bryant, R. G., & Willmott, J. R. (2016). Ultraviolet imaging with low cost smartphone sensors: Development and application of a Raspberry Pi-based UV camera. *Sensors*, 16(10), 1649.

Wojnowski, W., Kalinowska, K., Majchrzak, T., Płotka-Wasylka, J., & Namieśnik, J. (2019). Prediction of the biogenic amines index of poultry meat using an electronic nose. *Sensors*, 19(7), 1580.

Yang, X., Chai, L., Bist, R. B., Subedi, S., & Wu, Z. (2022). A deep learning model for detecting cage-free hens on the litter floor. *Animals*, 12(15), 1983.

You, M., Liu, J., Zhang, J., Xv, M., & He, D. (2020). A novel chicken meat quality evaluation method based on color card localization and color correction. *IEEE Access,* 8, 170093–170100.

Yu, J., Huang, M., Tian, H., & Xu, X. (2022). Colorimetric sensor based on Ag-Fe NTs for H_2S sensing. *ACS Omega*, 7(48), 44215–44222.

Yumnam, M., Gopalakrishnan, K., Dhua, S., Srivastava, Y., & Mishra, P. (2024). A Comprehensive review on smartphone-based sensor for fish spoilage analysis: Applications and limitations. *Food and Bioprocess Technology*, 17(12), 4575–4597.

Yumnam, M., Hatiboruah, D., Mishra, R., Sathyaseelan, K., Nath, P., & Mishra, P. (2023). A smartphone-based optical sensor with polyaniline label for quantitative determination of freshness of freshwater fish fillets. *Sensors and Actuators A: Physical*, 361, 114557.

ZeuLab, T. A. (2018) ZeuLab, Test4all. https://www.zeulab.com/products.html/veterinary-residues/202-test4all-antibiotic-control-milk.html

Zhang, M., Cui, X., & Li, N. (2022). Smartphone-based mobile biosensors for the point-of-care testing of human metabolites. *Materials Today Bio*, 14, 100254.

Zhu, L., Spachos, P., Pensini, E., & Plataniotis, K. N. (2021a). Deep learning and machine vision for food processing: A survey. *Current Research in Food Science*, 4, 233

Zhu, X., Yuan, X., Han, L., Liu, H., & Sun, B. (2021b). A smartphone-integrated optosensing platform based on red-emission carbon dots for real-time detection of pyrethroids. *Biosensors and Bioelectronics*, 191, 113460.

7 Aerogel
An Emerging Functional Material for Food Quality and Safety Analysis

Subhamoy Dhua, Keshram Dulait, Arnab Bayan, and Poonam Mishra

7.1 INTRODUCTION

Aerogel, the name itself defines it as the gel composed of a solid colloidal network dispersed in the air. According to the International Union of Pure and Applied Chemistry (IUPAC), "Aerogels are non-fluid networks composed of interconnected colloidal particles as a dispersed phase in a gas (typically air)". In other words, aerogel can be defined as "a coherent highly porous unique material (solid-state) of low density and high specific surface area, fabricated by the sol-gel method followed by a drying process, comprising three-dimensional networks of interconnected polymers with numerous air-filled pores" (Zheng et al., 2020). It is a light weight porous solid material having low density and thermal conductivity, high porosity and specific surface area, and excellent mass transfer properties. Generally, the diameter of the pores in aerogel ranges from 2 to 50 nm. However, the aerogel can be classified on the basis of microstructure into three groups, namely macroporous (diameter of pore >50 nm), mesoporous (2–50 nm), and microporous (<2 nm) (Abdullah et al., 2023). Based on the nature of precursor material, aerogel is also classified into three categories: (i) organic, (ii) inorganic, and (iii) hybrid (Yang et al., 2019).

Aerogel was first introduced by Steven Kistler in 1931. Kistler used silica gel as precursor material, and the pressure and temperature of jellies increased beyond its critical point to replace the liquid with gas. Since aerogel's first development, mostly inorganic polymers like silica, stannic oxide, and graphene have been used to make aerogel (Zhao et al., 2018). The rising concern toward biodegradability, sustainability, environmental impact, etc., of inorganic polymer-based precursors of aerogel necessitates the need for alternatives like bio-based precursors (polysaccharides, proteins, etc.). However, the concept of biopolymer-based aerogel (cellulose, gelation, egg albumin, etc.) is not new, as it was also introduced by Kistler for the first time in 1931. Based on the desired application, all of the three aerogels, as categorized previously, are being produced by the researchers.

DOI: 10.1201/9781003482611-7

Production of aerogel requires four steps: gelation, aging, solvent exchange, and drying. Processing conditions related to those four steps are very crucial for the end properties and formation of aerogel. The development process regulates the porosity (macro-, micro-, and mesoporous) of aerogel. Moreover, the drying of wet gel plays an important role in regulating the physicochemical properties of aerogel (Dhua et al., 2022). On the basis of drying technique, the dried gel can be categorized as follows: (i) aerogel (supercritical, vacuum, etc.), (ii) cryogel (freeze-dried), and (iii) xerogel (hot air oven). The different drying technique leads to differ in morphological characteristics of the dried gel. The interesting characteristics (light weight, low conductivity, high porosity, etc.) of aerogel pave the way of its application in various fields. Aerogel finds its application in the field of moisture absorption, oil-water separation, biomolecules immobilization, air filtration, thermal insulation, food and drug delivery systems, sugar and salt reduction, biosensing, contaminant detection, food packaging, etc. In recent times, application of aerogel in food systems is increasing. Nowadays, aerogel is often used as a biosensor in the determination of food quality and safety. Therefore, in this chapter, we have summarized the development, application, challenges, limitations, and future scope of functional aerogel in food quality and safety analysis.

7.2 DIFFERENT PRECURSOR MATERIALS OF FUNCTIONAL AEROGEL

The term 'functional' is related to specific quality attribute of an aerogel which decides its application. However, precursor materials required for the synthesis of functional aerogel is broadly categorized as (i) inorganic and (ii) organic.

7.2.1 Inorganic Materials

Inorganic materials are very common precursor of aerogel. Inorganic-materials-based aerogel possesses superior qualities like high transparency and thermal insulation. Based on the nature of material, inorganic materials can be divided into subcategories like fluoride, silica, metal, oxides, and chalcogens. TiO_2, silica, and its oxides are commonly used to develop inorganic aerogel framework. This inorganic aerogel possesses some disadvantages like low mechanical strength (collapse under small stress), less environment friendliness, and low biodegradability which restrict their broad range of application especially in the food systems where direct contact may cause deterioration.

7.2.2 Organic Materials

The disadvantages of inorganic aerogel necessitate the development of another aerogel (i.e., organic aerogel). Organic aerogels can be classified into two categories: (i) polymers and (ii) carbon. Polymers can be categorized into two subcategories, namely, (i) synthetic (resorcinol, polyvinyl alcohol (PVA), melamine, etc.) and (ii) biopolymers (polysaccharides, protein, mucilage, etc.). Organic aerogel exhibits several advantages over inorganic aerogel like non-hazardous, high biodegradability and mechanical strength, environment friendly, etc. These organic aerogels have gained application potentiality in the food systems, more specifically in food quality and safety analysis. For example, gelatin/gum arabic aerogel is used in on-site determination of seafood freshness (Li et al., 2024), chitosan nanofiber, and anthocyanin, and starch-nanoparticles-based aerogel is used in spoilage detection of protein-based food products (Sohrabi et al., 2024). However, some organic aerogels (natural polymer-based aerogels) possess several disadvantages like high stiffness, structural collapse, and solubility in water.

Nowadays, the concept of hybrid material (inorganic and organic) based functional aerogel is emerging as it exhibits good characteristics and finds a broad range of applications. These aerogels are mostly used in the area of sensing. For example, zeolitic imidazolate framework-8, PVA, and phenol-red-based colorimetric aerogel is used in the sensing of formic acid gas that is injurious to health (Xia et al., 2023).

7.3 FABRICATION OF FUNCTIONAL AEROGEL

Fabrication of aerogel is a time-consuming process that involves several steps such as gel preparation, solvent exchange, and drying of wet gel. These steps are almost common in all aerogels (inorganic, organic, and hybrid materials based). All the steps related to the fabrication of functional aerogel have been discussed briefly in the following sections.

7.3.1 Gel Preparation

Gel preparation is the foremost step of aerogel preparation. This step includes three important processes, namely (i) aqueous solution preparation, (ii) gel formation, and (iii) aging. Briefly, the precursor material is first added into water or other organic solvents to form a homogeneous solution. Then the solution is converted into gel via different gelation methods like sol-gel, emulsion-gelation, one pot reaction, and cross-linking. The gelation method is selected on the basis of gel-forming capability of precursor material. Sol-gel and emulsion-gelation methods are very common in the development of organic-material-based aerogel. To develop bio-polymer-based aerogel, these two gelation method are usually followed (Dhua & Mishra, 2023). One pot reaction method is used for hybrid-material-based aerogel development. Resorcinol-formaldehyde (RF) and metal oxide (MOx) based hybrid aerogel is developed using one pot reaction method (Leventis, 2011).

7.3.2 Solvent Exchange

In the preparation of aerogel, solvent exchange plays an important role. During solvent exchange, the aged gel is subjected to solvent substitution process to form alcogel. Normally, for solvent exchange, methanol, ethanol, acetone, etc., are used as the solvents. This substitution process can be done directly with the concentrated solvent as well as solution of solvents having progressively increased concentration. The aged gel is soaked directly into the solutions for a stipulated period of time to exchange the solvents effectively. These two different ways of substitution show different impacts in the final properties of aerogel as maximum shrinkage takes place during solvent exchange. The need of solvent exchange is related to higher affinity and solubility of alcohol, acetone, etc., toward supercritical CO_2 than water during supercritical drying of alcogel.

7.3.3 Drying of Wet Gel

This is a very critical step in the aerogel fabrication process in which removal of pore fluid takes place. During the removal of pore fluid, the microstructure should be maintained that results in superior end properties (porosity, specific surface area, etc.) of aerogel. The end properties (density, porosity, specific surface area, etc.) of aerogel are dependent on the drying technique (Zhao et al., 2018). Some common drying techniques used for the development of aerogel are atmospheric, freeze-, supercritical drying, etc.

7.3.3.1 Atmospheric Drying

Atmospheric drying possesses several disadvantages like time consuming (> 48 h), significant shrinkage, structural collapse, high density, and less porosity. (Zuo et al., 2015). The adhesive force between solid and liquid and liquid-gas surface tension in gel pores results in a strong capillary force that restricts the development of internal matrix porosity by the means of shrinkage and structural collapse of the gel pores (Thomas et al., 2018; Zheng et al., 2020). These negative impacts of air drying restrict their use in aerogel drying. However, minimizing the effect of shrinkage can increase its acceptance in aerogel drying through the following strategies: (i) if the gel contains an organic solvent that is having low vapor pressure and surface tension (Ganesan et al., 2016), (ii) modification of internal matrix through replacing the hydrophilic groups with hydrophobic groups, which reduces the capillary pressure gradient due to changes in contact angle (Pour et al., 2015), (iii) development of a strong network by using monomers that can resist the capillary force and impede structural collapse during drying (Xu et al., 2016).

7.3.3.2 Freeze Drying

In this kind of drying, solvent exchange step is not needed as water is directly evaporated from the aged wet gel through sublimation (bypassing the triple point) process. No structural collapse is observed in aerogel as there is no existence of capillary force due to the nonavailability of liquid-gas interface during the entire drying process. During sublimation, the tiny crystal developed in the gel matrix is removed gently which creates pores in the aerogel matrix. The overall drying process consists of two parts: (i) primary drying and (ii) secondary drying.

7.3.3.3 Supercritical Drying

Supercritical drying is the most used, effective, and widely accepted technique in the field of pore fluid elimination from the wet gel. The supercritical fluid behaves like liquid (density, dissolving capability, etc.) as well as gas (low viscosity, high diffusion coefficient, etc.) beyond critical point. Therefore, the liquid-gas interface and surface tension are avoided in this technique that results in negligible microstructural damages to the network of aerogel. Liquid carbon dioxide is widely used as supercritical fluid over other fluids as it possesses superior qualities like low critical point (temperature 31.1°C and pressure 7.4 MPa), chemically inert, nontoxic, ease of availability, low cost, and noncombustible (Ubeyitogullari & Ciftci, 2017). Solvent exchange is a prerequisite as water is having low solubility in supercritical fluid.

7.3.4 Incorporation of Functional Material to Develop Functional Aerogel

Functional-material-loaded aerogel or functional aerogel is the most utilized form of aerogel. Functional aerogels are made to perform special functions like bioactive or drug delivery, wound healing, and sensing. To develop functional aerogels, functional materials can be incorporated into the aerogel matrix at various stages of aerogel production depending upon the solubility of functional materials in the solvents (organic solvents, water, supercritical CO_2, etc.): direct incorporation into the solution dispersed with precursor material, at the time of hydrogel preparation, during solvent exchange, during supercritical drying, post supercritical drying (Selvasekaran & Chidambaram, 2021). Wet impregnation and supercritical impregnation are the two well-accepted and effective methods of functional material incorporation. In the wet impregnation method, the alcogel is dipped in functional material containing organic solvent for a fixed time period. Thereafter, the functional material is incorporated through the extraction of organic solvent and precipitation of solute by anti-solvent mechanism during supercritical drying (Viganó et al., 2019). In the supercritical impregnation, a saturated solution of supercritical fluid and functional material is diffused into the pores of aerogel matrix, and the impregnation takes place through chemical adsorption of functional materials into the surface of aerogel matrix or by precipitation upon depressurization (Lopes et al., 2017).

7.4 APPLICATION OF FUNCTIONAL AEROGEL FOR FOOD QUALITY AND SAFETY

Rodríguez et al. (2022) suggest that functional aerogels can be used for a variety of applications, including intelligent components that govern nutrient release, active compound delivery systems, fat alternatives, and biodegradable food packaging (Figure 7.1). For example, aerogels made from starch-based polymers have been found useful in food component delivery, packaging, and heat isolation (Rahayu & Septiani, 2021). Aerogels can operate as structural elements in food matrices, producing better food product stability and quality (Rege et al., 2018). Advanced food applications include smart nutrition-release ingredients, active component delivery systems, fat alternatives, and smart packaging materials (Rodríguez et al., 2022). Aerogels with antioxidant capabilities are especially valuable in bioactive encapsulation, controlled release, food conservation, and packaging, demonstrating their effectiveness as moisture regulators in food systems (Bakhori et al., 2023). Bioactive aerogels derived from natural sources such as seaweed have shown promise in controlled-release systems for bioactive substances in food packaging (Oliveira et al., 2019).

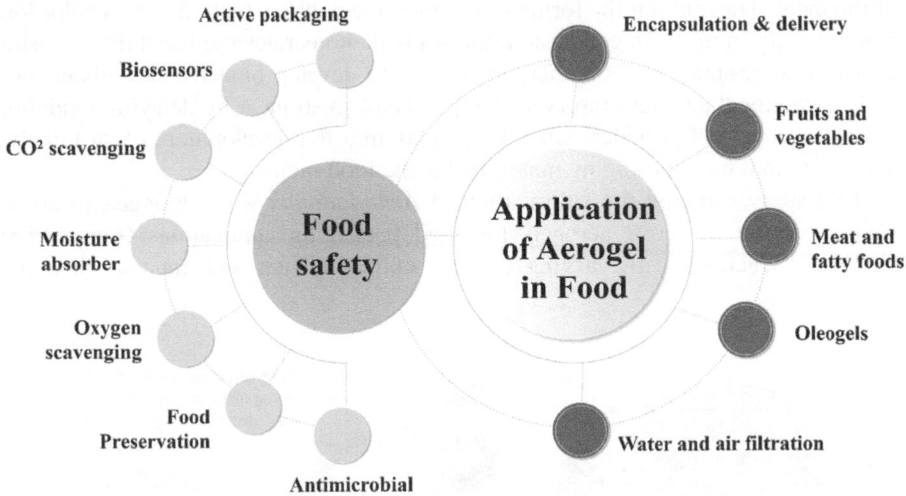

FIGURE 7.1 Applications of aerogels in food safety.

Aerogels have received much attention for their ability to act as barriers against gases such as oxygen and carbon dioxide, which can extend the shelf life of many items by preventing oxidation and spoiling. Silica aerogels have been acknowledged for their intriguing qualities, including thermal protection, comfort, and gas-sensing ability (Shaid et al., 2015; Wang et al., 2013). Aerogels in food packaging have showed potential in prolonging product shelf life by combining non-active barrier qualities with active activities to reduce pathogen risk and preserve product freshness (Hron & Macak, 2015).

7.4.1 Controlling Moisture in Food Products

Aerogel have a high capacity to absorb liquids, giving them control over moisture absorption, which is helpful in various applications, including food product shelf life (Mashabela et al., 2022). Aerogels have shown potential in food quality preservation by reducing moisture-induced deterioration.

The research study examines the use of agricultural residues in biorefinery processes to create bioactive absorbent food pads (Figure 7.2). It focuses on developing chitosan aerogels reinforced with lignocellulose and cellulose micro/nanofibers from bay tree pruning waste (Rincón et al., 2023). The study investigates the impact of residual lignin on the material's properties and incorporates bay leaf extract to make the aerogels bioactive. The release profiles of the extract and its antioxidant power were evaluated, with a focus on their use as absorbent pads for fresh meat to delay food oxidation during storage. Specifically, in meat packaging, absorbent pads are crucial for moisture control, and there is growing interest in developing materials with antioxidant capacity to inhibit the oxidation process during storage and maintain meat quality. The study discovers that bioactive aerogels, especially those with 2% or more extract, delay meat oxidation during storage and preserve the freshness of the meat. They inhibit the formation of metmyoglobin, which indicates color loss and rancidity in meat. In conclusion, the study demonstrates the feasibility of using agricultural residues in biorefinery processes to develop bioactive absorbent food pads. Functional aerogel preserved the quality of fresh meat by delaying oxidation during storage and provides valuable insights into the development of sustainable and functional food packaging materials for the food industry.

This study explored the valorization of *Arundodonax* waste biomass to create superabsorbent bioactive aerogels for food packaging applications (Figure 7.3). Cellulosic fractions with varying degrees of purification and aqueous bioactive

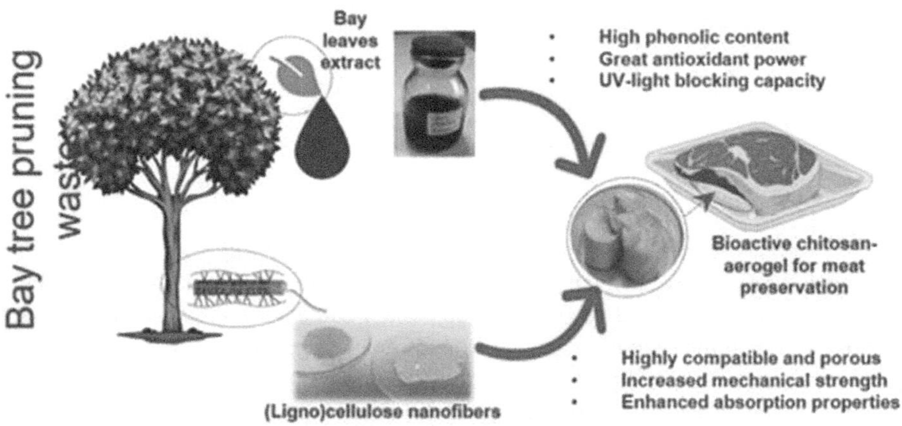

FIGURE 7.2 Bioactive absorbent chitosan aerogels reinforced with bay tree pruning waste nanocellulose with antioxidant properties for burger meat preservation.

Source: Rincón et al. (2023)

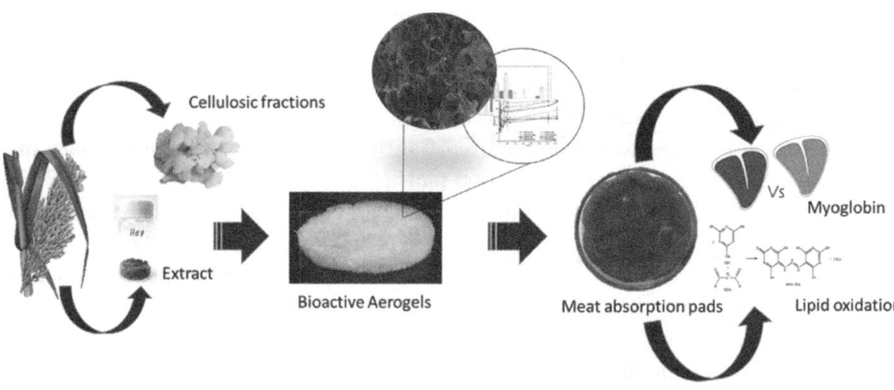

FIGURE 7.3 Superabsorbent food packaging bioactive cellulose-based aerogels from *Arundo donax* waste biomass.

Source: Fontes-Candia et al. (2019)

extracts were combined to develop these aerogels. All aerogels demonstrated excellent water and oil absorption capacities, with hemicelluloses contributing to increased porosity and hydrophilicity (Fontes-Candia et al., 2019). The hot water extract (S-HW) showed the highest antioxidant capacity and was incorporated into aerogels made from less purified stem fractions (F2A and F3A). These hybrid aerogels exhibited potential as bioactive food packaging pads, with the F2A+S-HW aerogel being the most promising. This aerogel demonstrated complete extract release in hydrophilic media and effectively reduced color loss and lipid oxidation in refrigerated red meat (Fontes-Candia et al., 2019).

7.4.2 Antibacterial Properties of Aerogel

López-Iglesias et al. (2020) found that chitosan aerogels containing vancomycin microparticles had a strong antibacterial action, completely inhibiting the growth of bacteria within 6 hours. Alginate-based aerogels with silver nanoparticles demonstrated antibacterial properties, especially against multidrug-resistant bacteria (Martínez-García, 2024). These studies highlight the effectiveness of modified aerogels in controlling microbial pathogens. Similarly, combining diverse materials in aerogels has resulted in improved antibacterial characteristics. According to Zhou et al. (2015), hybrid aerogels containing polypyrrole, microfibrillated cellulose, and silver nanoparticles demonstrated enhanced antibacterial characteristics due to the components' synergistic effects. Integrating several elements into aerogels could be the primary route to developing more effective antibacterial treatments.

Aerogels' high surface area and porous structure have also led to research on their potential use in pharmaceutical, medical, nutraceutical, and food safety and preservation due to their antibacterial activity (Figure 7.4) (Ubeyitogullari et al., 2018). This

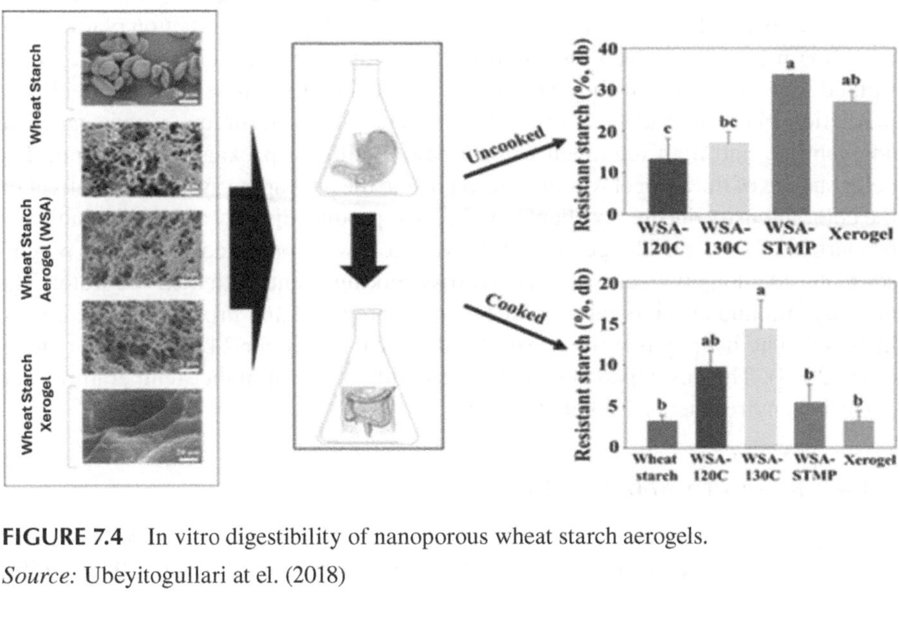

FIGURE 7.4 In vitro digestibility of nanoporous wheat starch aerogels.

Source: Ubeyitogullari at el. (2018)

work presents the starch aerogels' in vitro digestibility for the first time. The wheat starch aerogels (WSAs) with the addition of sodium metaphosphate (STMP) (WSA-STMP), xerogel, and gelatinization temperatures of 120°C (WSA-120C), 130°C (WSA-130C), and 120°C, all had comparable relative crystallinities. Nevertheless, the amylose-lipid complex concentration was highest in WSA-120C. Some bound starch with a 0.023% phosphorus concentration was produced by the addition of STMP. When left uncooked before digestion, the resistant starch (RS) levels of xerogel (26.9%) and WSA-STMP (33.5%) were greater than those of the other samples. However, following cooking, the RS contents of xerogel and WSA-STMP dropped significantly. The RS contents of WSA-120C and WSA-130C increased by 4.5 and 3.0 times, respectively, and remained steady when cooked. Even after heating, WSA retains its high RS level, making it a promising functional food ingredient.

7.4.3 Real-Time Monitoring of Food Product Freshness

Real-time monitoring of temperature, pH, and freshness in food product storage environments is made possible by integrating sensors with aerogels. Freshness sensors with RFID tags are used commercially to monitor temperature, humidity, light exposure, pressure, and pH, among other environmental parameters that impact food quality (Alam et al., 2021).

Zheng et al. (2024) developed anthocyanin complex aerogels for determining the freshness of fish (Figure 7.5). The study integrated machine learning models, with radial basis function neural networks and genetically optimized backpropagation networks, showing excellent performance in predicting remaining shelf life. A dual-channel dataset provided more comprehensive information, enhancing the system's descriptive capabilities. The proposed three-layer monitoring platform, combining smart aerogels and machine learning, offers a new paradigm for intelligent food packaging.

The developed sodium alginate aerogels were incorporated with anthocyanin complexes using the freeze-drying method. For developing a prediction platform for real-time monitoring based on optimized machine learning algorithms, they have trained four machine learning models with the color characteristics of aerogel. The prediction platform had high response efficiency, high accuracy, a wide range of adaptability, and high intelligence for a new intelligent packaging paradigm. The color change of the aerogel was monitored inside the packaging system. The effect on the color intensity due to any change in the food product directly affects the indicator properties of aerogel. The gelatin/PVA hydrogels were prepared by mixing precursor material, along with citric acid as a cross-linking agent, with distilled water. The anthocyanin and co-pigmented solutions were prepared and mixed with the cooled hydrogel. The hydrogels were freeze-dried after freezing for 24 h at −80°C (Zheng et al., 2024). The developed aerogels were highly stable at higher temperatures and corrosive environments (Sun et al., 2021).

7.4.4 Regulated Additives' Release

This regulated release mechanism can ensure the safe and efficient delivery of nutrients or bioactive chemicals in food products (Singh & Acevedo-Fani, 2022). Edible

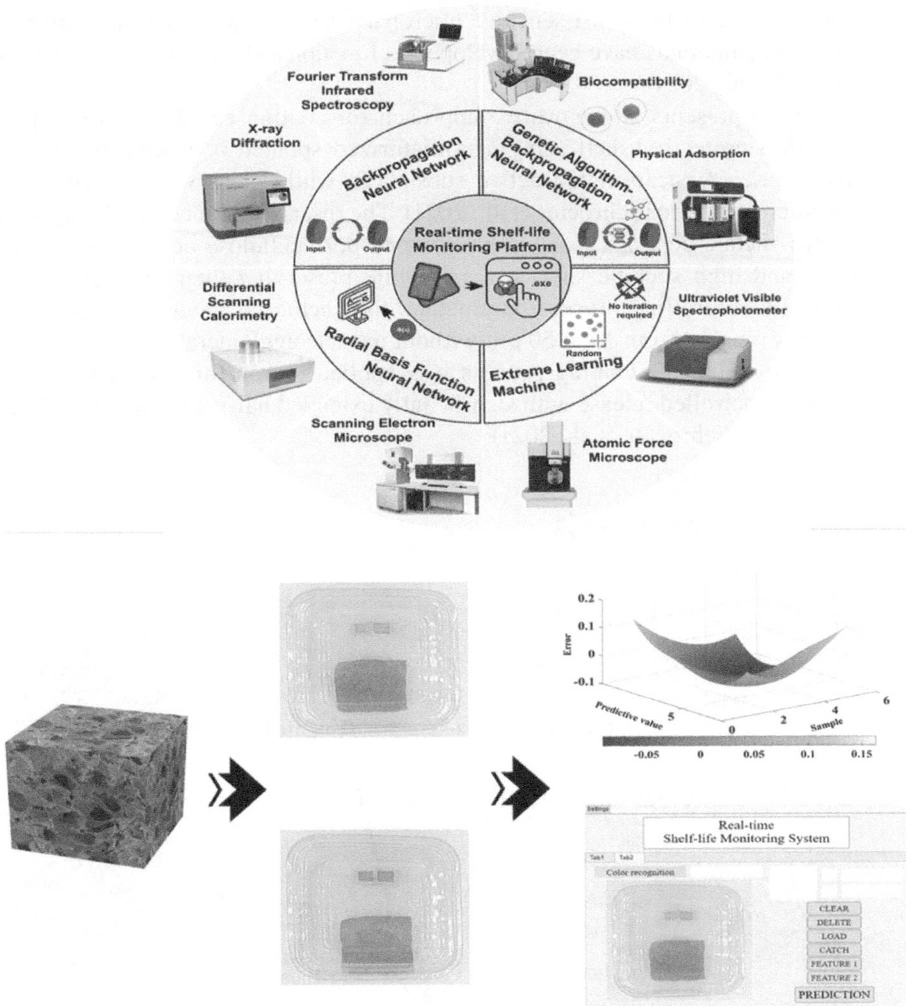

FIGURE 7.5 A new paradigm for smart packaging: A dual-channel freshness monitoring platform based on aerogels of sodium alginate-anthocyanin complex with high colorimetric sensitivity and stability.

Source: Zheng et al. (2024)

aerogels can improve food safety by imitating nature's nutrient delivery method and packaging by encasing active ingredients in structures that offer protection and distribute nutrients at the right places (Onwulata, 2012). It has been shown that encasing nutrients in mesoporous particles like aerogel shields them from outside influences like oxidation, extending their stability and bioavailability (Kamakura et al., 2021; Ruiz-Rico et al., 2017). Furthermore, the regulated release of phytonutrients in food products has been made possible by developing innovative food-grade coatings and dispersants. As an illustration of the possibility of improving food safety through regulated release systems, pH-sensitive microparticles containing antibacterial and antioxidant compounds have been developed to function as food-preserving materials (Biswal et al., 2021).

Figure 7.6 presents an innovative approach for coating cellulose aerogel particles with a protective shell using a miniaturized spouted bed setup for a controlled release of encapsulated active substances while preserving the aerogel's porous microstructure (Schroeter et al., 2021). The researchers successfully applied protective shellac coatings to hydrophilic open-pore cellulose aerogels with low densities and high specific surface areas while preserving their microstructure. By carefully controlling process parameters, they achieved precise coating layer thicknesses ranging from 10 to 50 μm without particle agglomeration or pore clogging. Additionally, the coating process was applied to vanillin-loaded particles, resulting in controlled release with significantly extended half-life times from 20 to 1,600 minutes (Schroeter et al., 2021).

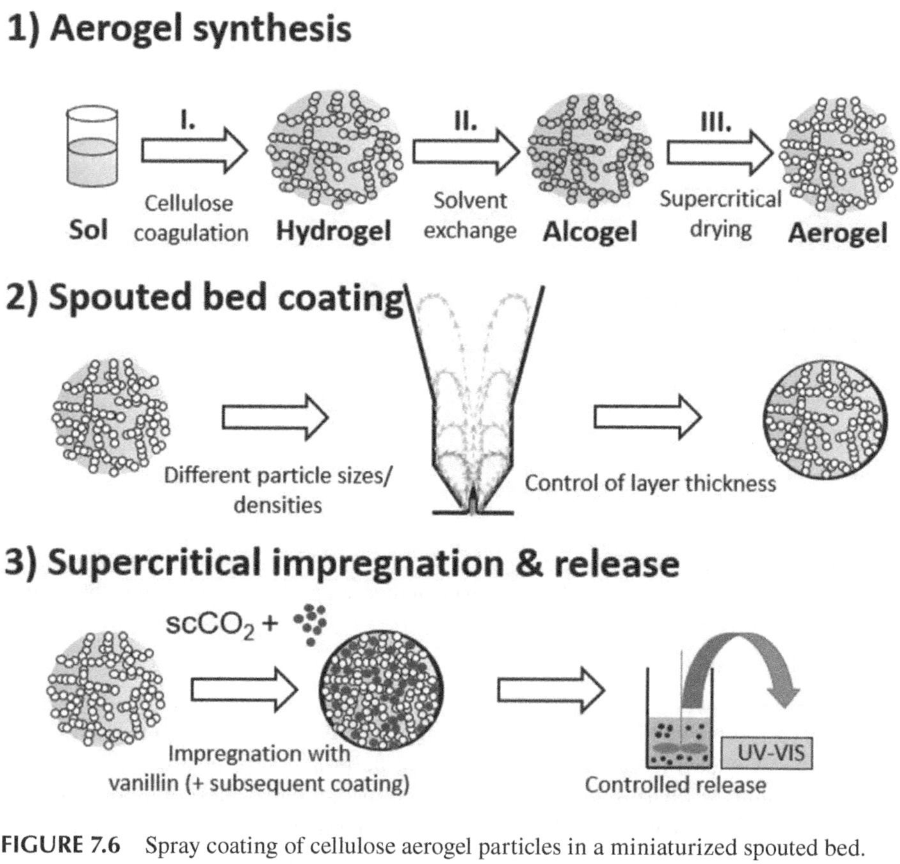

FIGURE 7.6 Spray coating of cellulose aerogel particles in a miniaturized spouted bed.
Source: Schroeter at el. (2021)

7.4.5 Aerogels in Food Packaging

Studies have indicated that food preservation could be enhanced by aerogels derived from diverse waste materials. For instance, cardboard aerogels have been produced as efficient oil absorbents (Shi et al., 2021), and aerogels supplemented with nanocellulose for chitosan have demonstrated antioxidant characteristics for preserving meat (Rincón et al., 2023). Furthermore, this emphasizes the importance of food packaging, especially in meeting the food industry's demand for packaging that not only protects food but also enhances its nutritional value by incorporating beneficial compounds like antioxidants (Peydayesh, 2024).

Additionally, research has shown that fresh-cut produce, such as cucumbers, can preserve their quality, safety, and shelf life by using chitosan coating and modified environment packaging (Olawuyi et al., 2019). Likewise, it has been discovered that the combination of chitosan and changed environment packaging extends the shelf life of cheese (Gammariello et al., 2011). A sustainable method of waste reduction is provided by using biopolymeric nano building blocks from agri-food waste to make sophisticated materials such as films, hydrogels, and aerogels (Peydayesh, 2024).

Bakhori et al. (2023) have integrated the phenolic-rich natural extracts into aerogels as a packed for the regulated release of bioactive substances, such as antioxidants, in the food. This strategy improves the nutritional value and quality of food products in addition to providing an innovative way to increase their shelf life. Furthermore, the creation of materials with special oil absorption and bioactive delivery properties is made possible by the development of food-grade hydrophobic bio-aerogels, which could completely change the way that food is packaged and preserved (Manzocco et al., 2021a). There is a rising interest in investigating these materials for many applications, including in food matrices for high-performance composites, as a result of continuing research concentrating on the synthesis of aerogels from biomass and industrial wastes such as lignin (Asim et al., 2019).

7.4.6 Mechanical and Insulation Properties of Aerogel

Aerogels have been modified to improve their hydrophobicity and mechanical qualities in food packaging, making them suitable for applications like adsorbent pads in packaging or oil spill cleanup (Benito-González et al., 2020). Polysaccharide-based aerogels have been developed with thermal insulation and antimicrobial capabilities, making them suitable for packaging of temperature-sensitive foods (Wang et al., 2019). Aerogels improved the loading capacity of beneficial substances, including sea buckthorn pomace extract, for food applications (Švermickaitė et al., 2022).

Silica aerogels are highly effective thermal insulators due to their nanoporous structure, with porosities surpassing 90% (Venkataraman, 2017; Venkataraman et al., 2018). Their thermal conductivity is as low as 0.013 $Wm^{-1}K^{-1}$. Polyimide aerogels have been developed to have low thermal conductivity, excellent thermal stability, and flame retardancy, making them suitable for thermal insulation applications (Feng et al., 2016; Liu et al., 2021). Composite aerogels, such as graphene/

MXene aerogels, have been designed to exhibit superior thermal insulation capabilities at extremely low densities compared to other insulating materials (Yang et al., 2022).

Liu et al. (2021) developed a silica–polyimide aerogel blanket for self-floating artificial islands (Figure 7.7), with key properties of low density (0.036 to 0.196 g/cm³), excellent thermal insulation (0.018 W/mK), flame resistance (up to 1,300°C), mechanical durability through 50 compression cycles, and waterproof characteristics (contact angle of 153°). The aerogel blanket was found to be suitable for lightweight and safe-floating platforms. Silica aerogels are used in packaging primarily for their excellent thermal insulation properties, making them ideal for temperature-sensitive products. Silica–polyimide aerogel blankets have potential applications in environmental protection as floating platforms for oil spill cleanup, providing thermal insulation for sensitive equipment in harsh conditions, and serving as lightweight barriers to prevent contamination in water bodies. Their flame-retardant properties also make them suitable for use in fire prevention and control in vulnerable areas.

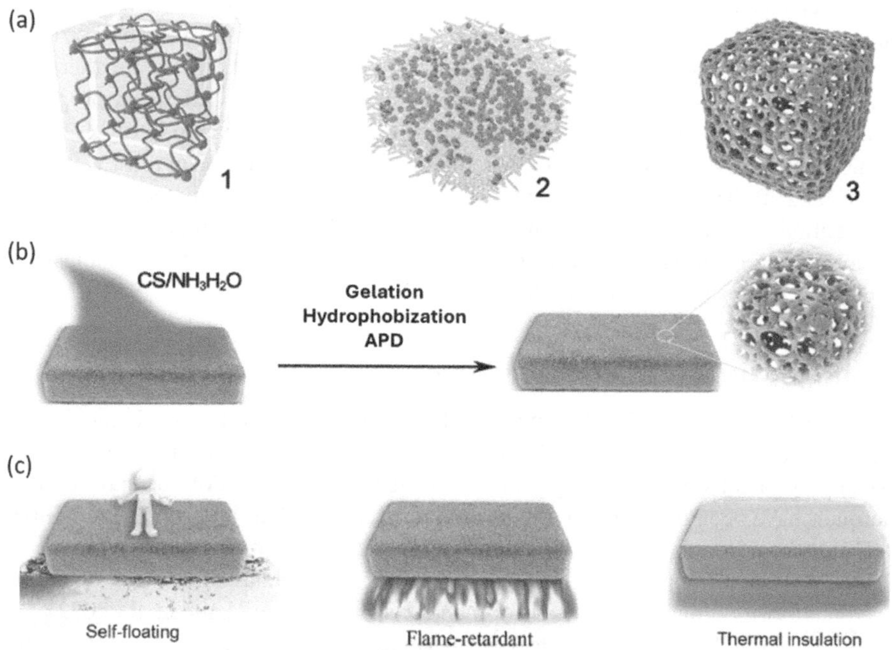

FIGURE 7.7 (a) Mechanical enhancement of the silica aerogel by organic groups or polymers (1), silica aerogel blanket (2), and proposed structure of silica foam blanket (3); (b) schematic description of the in-situ synthesis of silica-PI aerogel blanket; and (c) properties of the aerogel blanket and potential applications.

Source: Liu at al. (2021)

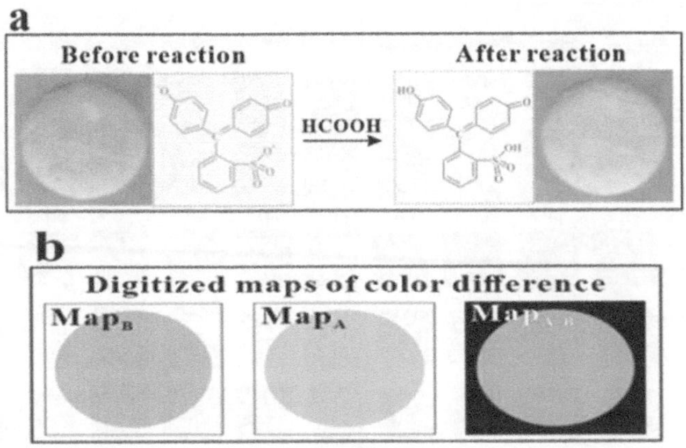

FIGURE 7.8 (a) Sensing mechanism for detecting formic acid based on colorimetric changes. (b) A schematic diagram is presented to illustrate the production process of color difference maps. (Colorimetric aerogel gas sensor with high sensitivity and stability)

Source: Xia et al. (2023)

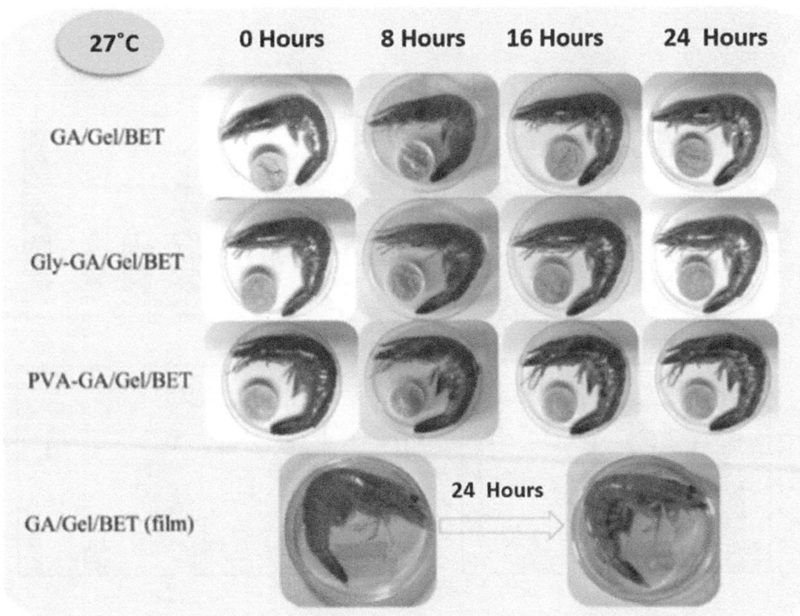

FIGURE 7.9 Different forms of based smart labels to monitor the freshness of shrimp under 27°C storage conditions (Effects of glycerol and PVA on gelatin/gum arabic aerogels: Microstructure, mechanical properties, and applications for food monitoring).

Source: Li et al. (2024)

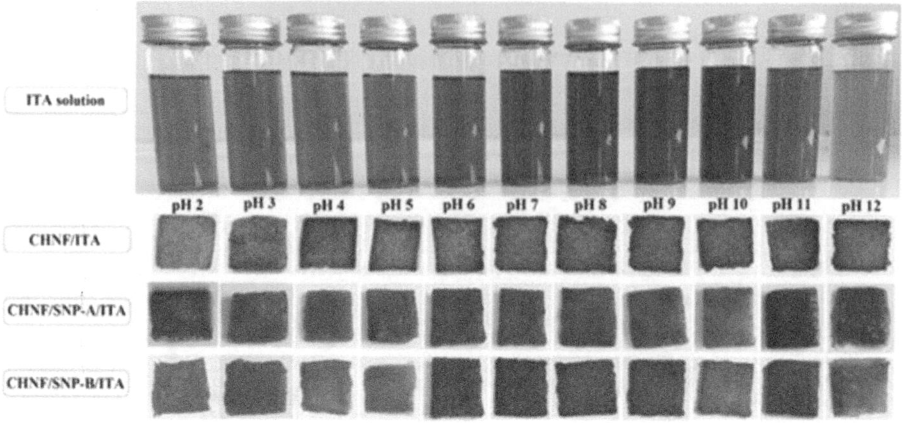

FIGURE 7.10 Color change response of ITA solution and CHNF-based aerogel indicators at pH 2–12 (controlled porosity intelligent aerogel based on chitosan nanofiber, anthocyanins, and starch nanoparticles: Preparation, characterization and application in monitoring food spoilage).

Source: Sohrabi et al. (2024)

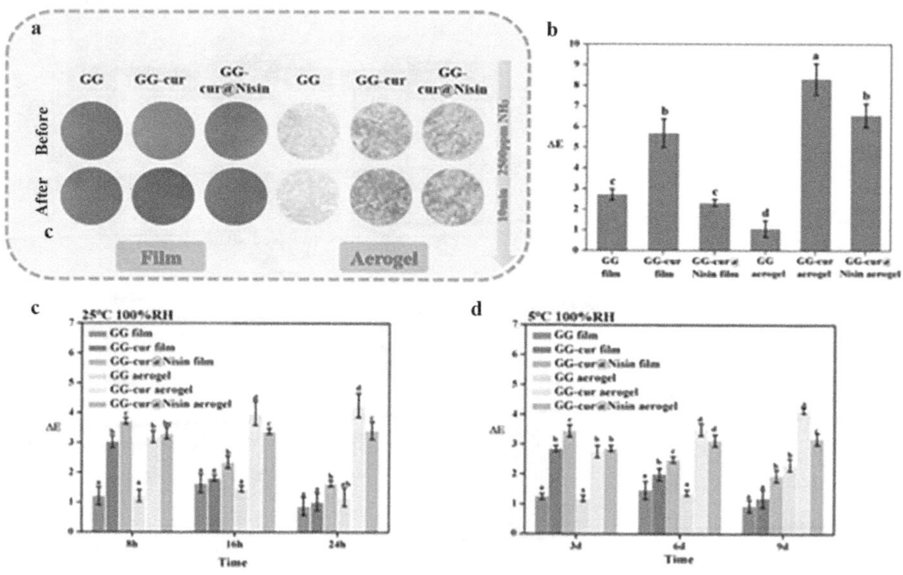

FIGURE 7.11 (a) Photographs and (b) ΔE values of aerogels and films before and after immersion in 2,500 ppm ammonia solution for 10 minutes. ΔE values of aerogels and films at (c) 25°C + 100% RH for 8–24 hours and (d) 5°C + 100% RH for 3–9 d (an aerogel-based intelligent active packaging (IAP) with the dual functions of spoilage detection and freshness preservation).

Source: Wu et al. (2024)

7.5 CHALLENGES AND LIMITATION OF AEROGEL-BASED TECHNOLOGIES IN FOOD SAFETY AND ANALYSIS

For food-grade aerogels, ensuring product stability during storage and preventing the leakage of encapsulated particles are crucial, while maintaining their free-flowing characteristics. For instance, a minor dust formation was observed in whey protein and sodium caseinate aerogels over a 12-week storage period. Additionally, a thin cemented layer developed on the top of egg white aerogels stored for extended periods, which also exhibited increased dust accumulation attributed to their smaller particle size (Selmer et al., 2019). Consequently, a thorough investigation into the shelf life of both loaded and unloaded food-grade aerogels is essential.

Biocompatibility is a crucial property for any biomaterial intended for biomedical applications. Recent studies have begun to evaluate the biocompatibility of various types of aerogels. Among these, silica particles have been extensively investigated and have demonstrated excellent biocompatibility with living tissues (Bharali et al., 2005). A paramount challenge in the realm of functional compound-loaded aerogels lies in the meticulous delivery of these compounds to their designated locales. The intricate process of nutrient absorption, particularly that of lipids, transpires predominantly through the intestinal mucosa, contingent upon their degradation by pancreatic enzymes (Kleemann et al., 2020; Selvasekaran & Chidambaram, 2021). The incorporation of STMP as a cross-linking agent diminishes the digestibility of WSAs when the digestibility test is performed without the cooking pretreatment (Ubeyitogullari et al., 2018). To ameliorate this predicament, it is imperative to delve into the nuances of digestibility and swelling kinetics, especially when modifications like pre-gelatinization of the native wheat starch are employed. According to the insights of Plazzotta et al. (2020), aerogels endowed with a more malleable structure exhibit a propensity to discharge their encapsulted compounds with alacrity, while those with a more robust constitution demonstrate a commendable restraint in release. This underscores the exigency of meticulous scrutiny into the physical, structural, and constructional properties of these loaded aerogels. Additionally, it is worth noting that the deployment of egg white protein and whey protein-based aerogels may be circumscribed owing to their allergenic proclivities.

One key challenge in applying polysaccharide, protein, and seed mucilage aerogels in food systems is ensuring that they do not compromise the rheological and sensory properties of the foods. Comin et al. (2015) reported observing brown discoloration and uneven pigmentation on lignan-loaded flax seed mucilage aerogels, noting that such color changes could negatively impact their potential use in food products. Moreover, given the distinct properties of aerogels, compared to their bulk counterparts, there is limited understanding of how they may interact in environments involving human contact (Kaur & Singh, 2023). Therefore, it would be prudent to quantify aerogel particle concentrations to better define exposure limits and determine appropriate safety measures.

In real-world applications, the weight-bearing capacity of aerogels is a crucial factor, as their robust load-bearing ability can effectively safeguard packaged food from mechanical stresses encountered during transportation, loading, and unloading, thus preventing damage to any smart labels due to external pressure. Furthermore, Li et al. (2024) reported that an increase in the entanglement of the aerogel's skeletal molecules resulted in a more uniform distribution of pressure within the aerogel structure.

While aerogels demonstrate promising capabilities in laboratory environments, translating this potential to meet the industrial demands of food analysis remains a formidable challenge. Ensuring consistent quality at larger production scales will require further technological innovations. The intricate and diverse composition of food matrices poses challenges for the sensitivity and precision of aerogel-based sensors in analytical applications. Addressing these matrix effects is imperative to ensure the dependability of food analysis methodologies. As a novel technology in food applications, aerogels require thorough regulatory compliance and safety approvals across different areas of food analysis. Negotiating the regulatory terrain proves intricate and time-intensive.

7.6 RECENT INNOVATION IN AEROGEL-BASED TECHNIQUES FOR FOOD ANALYSIS

Aerogels have been proven to be the ideal candidates for high-performance sensors, exhibiting low detection limits, high sensitivity, rapid response and recovery rates, as well as other exceptional properties. The highly porous construction, significant specific surface area, and unique interconnected 3D network of aerogels contribute to these capabilities. Additionally, integrating active building blocks with low-dimensional and 3D porous configurations can endow aerogel-based sensors with multifunctional attributes, enhancing their applicability in various sensor technologies.

Xia et al. (2023) reported the fabrication of colorimetric aerogel gas sensor (CAS) assembled from metal-organic framework (MOF) particles with chemo-responsive dye phenol red as an indicator into a 3D porous structure via ice-template-assisted method (Figure 7.8). Compared to conventional film sensors (CFSs) for the detection of highly mobile target molecules of formic acid, aerogel sensors exhibit an 8-fold lower limit of detection, 15-fold better sensitivity at low concentrations, 34-fold faster response time, and higher mechanical stability, when a target vapor concentration of 100 ppm is transferred through the gas sensors by an airflow of 200 mL/min. The high mobility of airflow presents significant challenges in the intake volume and reaction probability of gaseous molecules on solid sensors, complicating the detection of trace gases such as formic acid and volatile organic compounds (VOCs) etc. When gas flows through the sensor, an amplified airflow rate leads to a diminished adsorption efficiency of gaseous molecules. Conventional gas sensor films (CFS) exhibit low gas-volume intake in the sensitive area under high airflow conditions, resulting in a decreased contact probability for target molecule adsorption and reaction. However, the design of hierarchical pore structures has been found to improve mass transfer of target molecules and increase reaction rates of active sites owing to minimizing diffusion barriers. Aerogel sensor with staggered lamellae structures significantly provides a high air-volume intake of flowing gas, generating a sufficient probability of contact reactions. The detection mechanism primarily relies on the intermolecular interactions between phenol red and ZIF-8, leading to the formation of a quinone phenol salt structure. This interaction is visually indicated by the pink color observed in the colorimetric aerogel sensors prior to any reaction. When exposed to HCOOH (formic acid) molecules, phenol red undergoes a transformation into an acidic structure, resulting in a noticeable shift toward a yellow hue. This color change is quantitatively confirmed by the evaluated numerical values of Euclidean distance (ED) and color difference maps.

Intelligent response colorimetric labels have emerged as a significant area of research in food technology, incorporating sensitive dyes such as curcumin (Cai et al., 2022), anthocyanin (Tang et al., 2023), and betanin (Yang et al., 2024). These dyes exhibit visually perceptible color changes, enabling consumers to assess food freshness by observing these changes. In the meat processing industry, the core principle behind intelligent response labels involves the interaction with volatile amines, produced during the enzymatic and microbial decomposition of proteins in meat and seafood, leading to a change in the pH of the surrounding environment. This pH shift subsequently influences the sensitive dye encapsulated within the smart label, providing an indicator of food deterioration. Li et al. (2024), in their study, employed the complex coacervation of gelatin and gum arabic under acidic conditions to fabricate an intelligent colorimetric aerogel, to which they added betanin (BET). They then used ammonia (2,500 ppm) as a stimulant to investigate the responsiveness of the produced aerogels to biogenic amines, and response sensitivity was compared against smart film labels. The findings indicated that spoilage gases diffused swiftly through the porous channels of the aerogel, enhancing the interaction between the BET encapsulated in the gelatin/gum arabic aerogel and the spoilage gases. This interaction led to more rapid and pronounced color changes that outperformed the traditional colorimetric film sensors. The color changes (ΔE) of the smart labels were recorded and the accuracy was validated measuring total volatile basic nitrogen (TVB-N) levels in shrimp meat. For the smart label aerogels, pink denotes fresh, yellow indicates spoilage, and orange signifies severe spoilage. During the storage of shrimp samples at 27°C, the smart aerogel labels transitioned from lavender to light yellow after 8 hours, signifying a decline in shrimp freshness (Figure 7.9). After 24 hours of storage, the labels further changed from light yellow to orange, indicating that the shrimp had undergone significant spoilage, attributed to the pH sensitivity of BET causing the degradation of betacyanins resulting in the fading of red–purple color.

Sohrabi et al. (2024) designed food freshness colorimetric aerogel (Figure 7.10) with controlled porosity based on chitosan nanofibers (CHNF), *Ixioliriontataricum* anthocyanins (ITA), and starch nanoparticles (SNPs) via nanoprecipitation (SNP-A) and H_2SO_4 hydrolysis methods (SNP-B). The findings from this study demonstrate that the developed smart aerogel is capable of differentiating minced meat into three distinct stages of freshness when stored at 4°C: fresh, medium fresh, and spoiled. The CHNF/SNP-B/ITA aerogel exhibited an initial bright red color when fresh. By the third day of storage, this color transitioned to violet, indicating a medium fresh state. On the sixth day, the aerogel turned dark blue, and by the ninth day, it had shifted to green, signaling the spoilage of minced beef. This colorimetric progression is of considerable practical importance, enabling consumers to visually monitor the freshness and spoilage stages of meat over time. The exposure of the aerogels with a 0.8 M ammonia solution was employed to simulate their responsiveness to volatile nitrogen compounds generated during meat spoilage. During meat storage, spoilage microorganisms produce alkaline substances, such as ammonia and amines, leading to an increase in pH. This pH rise results in a color change in the aerogel indicator, which can be correlated with the levels of Total Volatile Basic Nitrogen (TVB-N) and Total Microbial Count (TMC)-established measures for detecting meat spoilage. The enhanced and precise responsiveness of the CHNF/SNP-B/ITA indicator is attributed to the aerogel's porous network (91% to 91.86%) that increases the diffusion rate of ammonia vapors and enhances the reactivity of anthocyanin molecules with volatile compounds, thereby improving sensitivity.

Intelligent active packaging (IAP) systems utilizing pH indicators are designed to exhibit a color change, providing a visual cue when food begins to lose its freshness or starts to spoil (Figure 7.11). Wu et al. (2024) developed an IAP system from gelatin, gum arabic, curcumin, and nisin (GG-cur@Nisin) by cross-linking glutaraldehyde to achieve the dual functions of spoilage detection and freshness preservation, utilizing it as an absorbent substrate to assess the quality of shrimp and pork in acidic environments (pH 4.2). The aerogels produced in this study exhibited open, porous structures with average pore sizes ranging from 63.5 to 90.3 μm, as determined through SEM imaging, suggesting a microporous nature. Following the addition of curcumin and nisin, the average pore size increased, and the pore walls became thinner. This alteration is likely attributable to the enhanced dispersion and stability imparted by curcumin and nisin to the solution-gel matrix. Consequently, this stabilization allowed the aerogels to preserve their pore structures during the cooling process and prevented contraction following the sublimation of ice crystals from the pores.

The porous structure of aerogel containing curcumin effectively absorbs volatile biogenic amines produced during shrimp spoilage, leading to alterations in curcumin's molecular structure. In the presence of GG-cur@Nisin aerogel, the pH increase was significantly lower compared to other groups, demonstrating the material's efficacy in retarding shrimp spoilage. This is attributed to its low water vapor absorption capacity (WVAC) and the antimicrobial properties of nisin. The porous structure of the aerogel, along with the physical encapsulation of nisin, facilitates the gradual release of nisin into the storage environment, inhibiting microbial growth and maintaining shrimp freshness, when tested against *E. coli* and *S. aureus*. Consequently, GG-cur@Nisin aerogel exhibited effective preservation, successfully delaying shrimp deterioration to a notable extent.

Electrochemical biosensors have garnered significant interest for glucose monitoring due to their ability to integrate with portable electronic devices like smartphones, enabling real-time sensing. To fully realize this potential, the performance of these glucose probes needs to be further improved in terms of response time, sensitivity, and selectivity. Glucose sensors are developed and designed using aerogels as the fundamental materials due to their advantageous characteristics, such as high and tunable porosity, substantial pore volumes, and surface area (Mahmoudpour et al., 2021). Jeong et al. (2017) developed an electrochemical approach to detect glucose using a graphene and 2DMoS2 nanocomposite that formed a 3D aerogel. This platform was employed for enzymatic recognition of glucose. The large surface area of the aerogel enabled the capture of a high amount of glucose oxidase within its porous structure. The resulting flow-injection-based electrochemical sensor could measure glucose concentrations ranging from 2 to 20 mM (in 4s), with a low detection limit of approximately 0.29 mM and a sensitivity of 3.36 μA/mM. The majority of glucose detection methods reported to date have relied on enzymatic reactions. Nevertheless, due to certain drawbacks of glucose oxidase-based nanoprobes, such as complex enzyme anchoring, environmental sensitivity, and high costs, enzyme-free glucose sensing technologies have garnered significant academic and commercial interest.

Fluorescent aerogels are hybrid materials with fluorescent agents incorporated into their highly porous structure, enabling various applications. Study by Wu et al. (2019) introduced fluorescent aerogels composed of carbon dots (CDs) and carbon nanofibers (CNFs), which could function as gas biosensors for nitric oxide detection. The fluorescence of these aerogels can be rapidly and completely quenched by exposure to nitrogen oxides (NOx). The CDs bonded to the surface of the CNFs act as a stabilizer, providing the fluorescent aerogel with shape recovery properties in water. The fluorescence quenching is attributed to the inhibition of radiative recombination of electrons due to the interaction between NOx and the CDs. The porous structure of the aerogel facilitates the trapping and adsorption of gas molecules, allowing this system to be successfully utilized for the sensitive detection of glutaraldehyde.

Aerogel-based techniques have potential for its application in food quality and safety monitoring specially for freshness monitoring and preservation of highly perishable foods, offering high sensitivity, selectivity and absorption of substances (e.g., volatile biogenic amines), etc. The inexorable march of research and development in the field of aerogel portends even more groundbreaking applications, ensuring the zenith of food safety and quality in the future.

7.7 FUTURE SCOPE OF FUNCTIONAL AEROGEL FOR FOOD ANALYSIS

Aerogels possess rapid response, high selectivity, and exceptional sensitivity, enabling them to detect a diverse range of targets in biomedical environments. Consequently, this has galvanized the scientific community to develop an array of promising aerogel-based sensors, leading to numerous promising advancements in the field. Extensive research on food spoilage sensing aerogel is essential for its application in the field of food quality and safety. Aerogels have demonstrated the ability to respond to various factors including changing pH, temperature, and humidity, as well as detecting volatile chemicals. This responsiveness makes them valuable tools for ensuring the freshness and safety of food products.

From a safety perspective, creating aerogels from consumable materials poses significant challenges for researchers due to their unique properties, which differ from those of their bulk counterparts. Additionally, the development of cost-effective and time-efficient methods for producing aerogels remains a critical area of focus. Presently, aerogels are manufactured using processes such as supercritical CO_2 drying and freeze drying, both of which are expensive and time-consuming, limiting their usage in commercial application (Manzocco et al., 2021b). Reducing the use of fresh solvent coupled with employing a continuous CO_2 drying method could enable more efficient aerogel manufacturing techniques and lower costs compared to traditional batch processing.

Current research at the laboratory level is focused on exploring the structure of aerogels, particularly in relation to their molecular composition, processing techniques, and potential applications in the food industry. However, there is still a significant gap in understanding the fate of aerogel ingredients within food matrices that remains largely unexplored.

While exploring the application of aerogel in food packaging, researchers and developers must address the issue of transparency as the visual appearance of packaging materials is significant for consumer experience. Also, the challenge pertains to the end-of-life options for packaging materials as these can be recycled into new products. The composition of aerogels and their integration into composite materials will play a critical role in determining the feasibility of recycling and disposal options that remain an area for further research.

To mainstream the application of aerogel technology in the food industry, it is quintessential to establish clear guidelines for their safe use in human consumption and food contact. Evaluating the properties of nanostructured aerogels to clearly define nature and kinetics of their interaction, especially when incorporated into foods, is a complex task that requires the development of methods to thoroughly understand their interaction with biological systems. Consequently, comprehensive research on both unloaded and loaded aerogels is imperative, encompassing practical applications as well as assessments of stability, shelf life, digestibility, and sensory properties.

Researchers have exerted substantial effort to develop smart aerogel materials to be used in the realm of food safety and quality. Nanomaterials such as CDs and quantum dots are grafted into the aerogel structural framework, enabling them to exhibit photoluminescent properties with good quantum yield. Further research is necessary to explore the correlation between the microstructure of aerogels and their mechanical stability. According to Xia et al. (2023), it was found that under identical loading conditions, aerogels synthesized from precursor solutions with higher concentration factors exhibited greater internal stress and demonstrated recuperative elastic deformation. Conversely, aerogels with lower concentration factors experienced lower internal stress and significant deformation, resulting in irreversible plastic deformation.

Furthermore, exploratory studies like aerogel-based IAP with dual functions of spoilage detection and freshness preservation using curcumin as a colorimetric agent and nisin as an antimicrobial agent (Wu et al., 2024) are required to explore the multifunctional role of aerogel-based smart sensors, thereby offering a promising approach for freshness monitoring and preservation of highly perishable foods. It is noteworthy to highlight that food-grade edible aerogel has not been commercialized yet; nevertheless, the future prospects of these products are a significant area of concern.

REFERENCES

Abdullah, Zou, Y., Farooq, S., Walayat, N., Zhang, H., Faieta, M., . . . & Huang, Q. (2023). Bio-aerogels: Fabrication, properties and food applications. Critical Reviews in Food Science and Nutrition, 63(24), 6687–6709.

Alam, A., Rathi, P., Beshai, H., Sarabha, G., & Deen, M. (2021). Fruit quality monitoring with smart packaging. *Sensors* 21(4): 1509. https://doi.org/10.3390/s21041509

Asim, N., Badiei, M., Alghoul, M., Mohammad, M., Fudholi, A., Akhtaruzzaman, M., & Sopian, K. (2019). Biomass and industrial wastes as resource materials for aerogel preparation: Opportunities, challenges, and research directions. *Industrial & Engineering Chemistry Research* 58(38): 17621–17645. https://doi.org/10.1021/acs.iecr.9b02661

Bakhori, N., Ismail, Z., Hassan, M., & Dolah, R. (2023). Emerging trends in nanotechnology: Aerogel-based materials for biomedical applications. *Nanomaterials* 13(6): 1063. https://doi.org/10.3390/nano13061063

Benito-González, I., Martínez-Sanz, M., López-Rubio, A., & Gómez-Mascaraque, L. (2020). Confocal Raman imaging as a useful tool to understand the internal microstructure of multicomponent aerogels. *Journal of Raman Spectroscopy* 51(10): 2022–2035. https://doi.org/10.1002/jrs.5936

Bharali, D. J., Klejbor, I., Stachowiak, E. K., Dutta, P., Roy, I., Kaur, N., Bergey, E. J., Prasad, P. N., & Stachowiak, M. K. (2005). Organically modified silica nanoparticles: A non-viral vector for in vivo gene delivery and expression in the brain. *Proceedings of the National Academy of Sciences* 102(32): 11539–11544.

Biswal, A., Thodikayil, A., & Saha, S. (2021). Ph-sensitive acetalated dextran/plga-based double-layered microparticles and their application in food preservation. *ACS Applied Bio Materials* 4(3): 2429–2441. https://doi.org/10.1021/acsabm.0c01361

Cai, R., Pan, S., Li, R., Xu, X., Pan, S., & Liu, F. (2022). Curcumin loading and colon release of pectin gel beads: Effect of different de-esterification method. *Food Chemistry* 389: 133130.

Comin, L. M., Temelli, F., & Saldaña, M. D. (2015). Flax mucilage and barley beta-glucan aerogels obtained using supercritical carbon dioxide: Application as flax lignan carriers. *Innovative Food Science & Emerging Technologies* 28: 40–46.

Dhua, S., Gupta, A. K., & Mishra, P. (2022). Aerogel: Functional emerging material for potential application in food: A review. *Food and Bioprocess Technology* 15(11): 2396–2421.

Dhua, S., & Mishra, P. (2023). Development of highly reusable, mechanically stable corn starch-based aerogel using glycerol for potential application in the storage of fresh spinach leaves. *International Journal of Biological Macromolecules* 242: 125102.

Feng, J., Wang, X., Jiang, Y., & Du, D. (2016). Study on thermal conductivities of aromatic polyimide aerogels. *ACS Applied Materials & Interfaces* 8(20): 12992–12996. https://doi.org/10.1021/acsami.6b02183

Fontes-Candia, C., Erboz, E., Martínez-Abad, A., López-Rubio, A., & Martínez-Sanz, M. (2019). Superabsorbent food packaging bioactive cellulose-based aerogels from Arundo donax waste biomass. *Food Hydrocolloids* 96(2019): 151–160.

Gammariello, D., Conte, A., Attanasio, M., & Nobile, M. (2011). A study on the synergy of modified atmosphere packaging and chitosan on stracciatella shelf life. *Journal of Food Process Engineering* 34(5): 1394–1407. https://doi.org/10.1111/j.1745-4530.2009.00537.x

Ganesan, K., Dennstedt, A., Barowski, A., & Ratke, L. (2016). Design of aerogels, cryogels and xerogels of cellulose with hierarchical porous structures. *Materials & Design* 92: 345–355.

García-González, C. A., Uy, J. J., Alnaief, M., & Smirnova, I. (2012). Preparation of tailor-made starch-based aerogel microspheres by the emulsion-gelation method. *Carbohydrate polymers*, *88*(4), 1378–1386.

Hron, J., & Macak, T. (2015). Mixture design for food packaging in a modified atmosphere. *Agricultural Economics* 61(9): 393–399. https://doi.org/10.17221/182/2015-agricecon

Jeong, J. M., Yang, M., Kim, D. S., Lee, T. J., & Choi, B. G. (2017). High performance electrochemical glucose sensor based on three-dimensional MoS2/graphene aerogel. *Journal of Colloid and Interface Science* 506: 379–385.

Kamakura, R., Raza, G., Mäkilä, E., Riikonen, J., Kovalainen, M., Ueta, Y., & Herzig, K. (2021). Colonic delivery of α-linolenic acid by an advanced nutrient delivery system prolongs glucagon-like peptide-1 secretion and inhibits food intake in mice. *Molecular Nutrition & Food Research* 66(4). https://doi.org/10.1002/mnfr.202100978

Kaur, A., & Singh, H. (2023). Nanomaterials in environment: Sources, risk assessment, and safety aspect. In *Advanced Functional Nanoparticles "Boon or Bane" for Environment Remediation Applications: Combating Environmental Issues* (pp. 75–93). Cham: Springer International Publishing.

Kleemann, C., Schuster, R., Rosenecker, E., Selmer, I., Smirnova, I., & Kulozik, U. (2020). In-vitro-digestion and swelling kinetics of whey protein, egg white protein and sodium caseinate aerogels. *Food Hydrocolloids* 101: 105534.

Leventis, N. (2011). *Interpenetrating organic/inorganic networks of resorcinol-formaldehyde/metal oxide aerogels. In Aerogels handbook* (pp. 287–313). New York, NY: Springer New York.

Li, B., Chen, M., Wu, Y., Liu, B., & Huang, G. (2024). Effects of glycerol and PVA on gelatin/Gum Arabic aerogels: Microstructure, mechanical properties, and applications for food monitoring. *Food Hydrocolloids* 149: 109515.

Liu, R., Wang, J., Liao, J., & Zhang, X. (2021). Robust silica–polyimide aerogel blanket for water-proof and flame-retardant self-floating artificial Island. *Frontiers in Materials* 8: 659655. https://doi.org/10.3389/fmats.2021.659655

Lopes, J. M., Mustapa, A. N., Pantić, M., Bermejo, M. D., Martín, A., Novak, Z., Knez, Ž., & Cocero, M. J. (2017). Preparation of cellulose aerogels from ionic liquid solutions for supercritical impregnation of phytol. *The Journal of Supercritical Fluids* 130: 17–22.

López-Iglesias, C., Barros, J., Ardao, I., Gurikov, P., Monteiro, F., Смирнова, И., & García-González, C. (2020). Jet cutting technique for the production of chitosan aerogel microparticles loaded with vancomycin. *Polymers* 12(2): 273. https://doi.org/10.3390/polym12020273

Mahmoudpour, M., Dolatabadi, J. E. N., Hasanzadeh, M., & Soleymani, J. (2021). Carbon-based aerogels for biomedical sensing: Advances toward designing the ideal sensor. *Advances in Colloid and Interface Science* 298: 102550.

Manzocco, L., Basso, F., Plazzotta, S., & Calligaris, S. (2021a). Study on the possibility of developing food-grade hydrophobic bio-aerogels by using an oleogel template approach. *Current Research in Food Science* 4: 115–120. https://doi.org/10.1016/j.crfs.2021.02.005

Manzocco, L., Mikkonen, K. S., & García-González, C. A. (2021b). Aerogels as porous structures for food applications: Smart ingredients and novel packaging materials. *Food Structure* 28: 100188.

Martínez-García, K. (2024). Radical scavenging, hemocompatibility, and antibacterial activity against mdracinetobacterbaumannii in alginate-based aerogels containing lipoic acid-capped silver nanoparticles. *ACS Omega* 9(2): 2350–2361. https://doi.org/10.1021/acsomega.3c06114

Mashabela, L. T., Maboa, M. M., Miya, N. F., Ajayi, T. O., Chasara, R. S., Milne, M., Mokhele, S., Demana, P. H., Witika, B. A., Siwe-Noundou, X., & Poka, M. S. (2022). A comprehensive review of cross-linked gels as vehicles for drug delivery to treat central nervous system disorders. *Gels* 8(9): 563. https://doi.org/10.3390/gels8090563

Olawuyi, I., Park, J., Lee, J., & Lee, W. (2019). Combined effect of chitosan coating and modified atmosphere packaging on fresh-cut cucumber. *Food Science & Nutrition* 7(3): 1043–1052. https://doi.org/10.1002/fsn3.937

Oliveira, J., Bruni, G., Fabra, M., Zavareze, E., López-Rubio, A., & Martínez-Sanz, M. (2019). Development of food packaging bioactive aerogels through the valorization of gelidiumsesquipedale seaweed. *Food Hydrocolloids* 89: 337–350. https://doi.org/10.1016/j.foodhyd.2018.10.047

Onwulata, C. (2012). Microencapsulation and functional bioactive foods. *Journal of Food Processing and Preservation* 37(5): 510–532. https://doi.org/10.1111/j.1745-4549.2012.00680.x

Peydayesh, M. (2024). Sustainable materials via the assembly of biopolymericnanobuilding blocks valorized from agri-food waste. *Sustainability* 16(3): 1286. https://doi.org/10.3390/su16031286

Plazzotta, S., Calligaris, S., & Manzocco, L. (2020). Structural characterization of oleogels from whey protein aerogel particles. *Food Research International* 132: 109099.

Pour, G., Beauger, C., Rigacci, A., & Budtova, T. (2015). Xerocellulose: Lightweight, porous and hydrophobic cellulose prepared via ambient drying. *Journal of Materials Science* 50: 4526–4535.

Rahayu, W., & Septiani, I. (2021). The role of food science and technology for indonesian indigenous food product development to achieve food security. *Indonesian Food and Nutrition Progress* 17(1): 12. https://doi.org/10.22146/ifnp.53294

Rege, A., Preibisch, I., Schestakow, M., Ganesan, K., Gurikov, P., Milow, B., & Itskov, M. (2018). Correlating synthesis parameters to morphological entities: Predictive modeling of biopolymer aerogels. *Materials* 11(9): 1670. https://doi.org/10.3390/ma11091670

Rincón, E., Espinosa, E., Pinillos, M., & Cantador, L. (2023). Bioactive absorbent chitosan aerogels reinforced with bay tree pruning waste nanocellulose with antioxidant properties for burger meat preservation. *Polymers* 15(4): 866. https://doi.org/10.3390/polym15040866

Rodríguez, H., Salgado-Delgado, R., Paredes, A., Delgado, A., Uribe, J., & Pinedaa, E. (2022). Partial characterization of aerogels made from chayotextle and potato starch. *Emirates Journal of Food and Agriculture*. https://doi.org/10.9755/ejfa.2022.v34.i8.2837

Ruiz-Rico, M., Pérez-Esteve, É., Bernardos, A., Sancenón, F., Martínez-Máñez, R., Marcos, M. D., & Barat, J. M. (2017). Enhanced antimicrobial activity of essential oil components immobilized on silica particles. *Food Chemistry* 233: 228–236.

Schroeter, B., Yonkova, V. P., Goslinska, M., Orth, M., Pietsch, S., Gurikov, P., Smirnova, I., & Heinrich, S. (2021). Spray coating of cellulose aerogel particles in a miniaturized spouted bed. *Cellulose* 28(12): 7795–7812.

Selmer, I., Karnetzke, J., Kleemann, C., Lehtonen, M., Mikkonen, K. S., Kulozik, U., & Smirnova, I. (2019). Encapsulation of fish oil in protein aerogel micro-particles. *Journal of Food Engineering* 260: 1–11.

Selvasekaran, P., & Chidambaram, R. (2021). Food-grade aerogels obtained from polysaccharides, proteins, and seed mucilages: Role as a carrier matrix of functional food ingredients. *Trends in Food Science & Technology* 112: 455–470.

Shaid, A., Wang, L., & Padhye, R. (2015). The thermal protection and comfort properties of aerogel and pcm-coated fabric for firefighter garment. *Journal of Industrial Textiles* 45(4): 611–625. https://doi.org/10.1177/1528083715610296

Shi, G., Wu, M., Zhong, Q., Mu, P., & Li, J. (2021). Superhydrophobic waste cardboard aerogels as effective and reusable oil absorbents. *Langmuir* 37(25): 7843–7850. https://doi.org/10.1021/acs.langmuir.1c01216

Singh, H., & Acevedo-Fani, A. (2022). New perspectives on the delivery of nutrients and bioactive compounds through gastric restructuring of foods. *Journal of Nutritional Science and Vitaminology* 68(Supplement): S149–S151. https://doi.org/10.3177/jnsv.68.s149

Sohrabi, N., Almasi, H., & Moradi, M. (2024). Controlled porosity intelligent aerogel based on chitosan nanofiber, anthocyanins and starch nanoparticles: Preparation, characterization and application in monitoring food spoilage. *Food Hydrocolloids* 110261.

Sun, Y., Chu, Y., Wu, W., & Xiao, H. (2021). Nanocellulose-based lightweight porous materials: A review. *Carbohydrate Polymers* 255: 117489.

Švermickaitė, G., Eisinaitė, V., Vinauskienė, R., Jasutienė, I., & Leskauskaitė, D. (2022). Characterisation of hydrogels and aerogels as carriers for sea buckthorn pomace extract. *International Journal of Food Science & Technology* 57(7): 4441–4450. https://doi.org/10.1111/ijfs.15777

Tang, Q., Hu, J., Li, S., Lin, S., Tu, Y., Gui, X., & Dong, Y. (2023). Preparation of an aramid nanofiber-reinforced colorimetric hydrogel employing natural anthocyanin as an indicator for shrimp and fish spoilage monitoring. *European Polymer Journal* 187: 111889.

Thomas, S., Pothan, L. A., & Mavelil-Sam, R. (Eds.). (2018). *Biobased Aerogels: Polysaccharide and Protein-Based Materials* (Vol. 58). Royal Society of Chemistry.

Ubeyitogullari, A., Brahma, S., Rose, D. J., & Ciftci, O. N. (2018). In vitro digestibility of nanoporous wheat starch aerogels. *Journal of Agricultural and Food Chemistry* 66(36): 9490–9497. https://doi.org/10.1021/acs.jafc.8b03231

Ubeyitogullari, A., & Ciftci, O. N. (2017). Generating phytosterol nanoparticles in nanoporousbioaerogels via supercritical carbon dioxide impregnation: Effect of impregnation conditions. *Journal of Food Engineering* 207: 99–107.

Venkataraman, M. (2017). Thermal insulation properties of electrospunnanofibrous layers. *Journal of Fiber Bioengineering and Informatics* 10(4): 187–199. https://doi.org/10.3993/jfbim00276

Venkataraman, M., Mishra, R., Militký, J., Xiong, X., Marek, J., Yao, J., & Zhu, G. (2018). Electrospunnanofibrous membranes embedded with aerogel for advanced thermal and transport properties. *Polymers for Advanced Technologies* 29(10): 2583–2592. https://doi.org/10.1002/pat.4369

Viganó, J., Meirelles, A. A., Náthia-Neves, G., Baseggio, A. M., Cunha, R. L., Junior, M. R. M., Meireles, M. A. A., Gurikov, P., Smirnova, I., & Martínez, J. (2019). Impregnation of passion fruit bagasse extract in alginate aerogel microparticles. *International Journal of Biological Macromolecules* 155: 1060–1068.

Wang, R., Li, G., Dong, Y., Chi, Y., & Chen, G. (2013). Carbon quantum dot-functionalized aerogels for no2 gas sensing. *Analytical Chemistry* 85(17): 8065–8069. https://doi.org/10.1021/ac401880h

Wang, Y., Su, Y., Wang, W., Fang, Y., Riffat, S. B., & Jiang, F. (2019). The advances of polysaccharide-based aerogels: Preparation and potential application. *Carbohydrate Polymers* 226: 115242.

Wu, B., Zhu, G., Dufresne, A., & Lin, N. (2019). Fluorescent aerogels based on chemical crosslinking between nanocellulose and carbon dots for optical sensor. *ACS Applied Materials & Interfaces* 11(17): 16048–16058.

Wu, Y., Li, B. H., Chen, M. M., & Bing, L. (2024). An aerogel-based intelligent active packaging with the dual functions of spoilage detection and freshness preservation. *Food Hydrocolloids*: 110160.

Xia, X., Wu, R., Zhang, L., Chen, X., Yan, Y., Yin, J., Ren, J., Li, H., Yin, J., Xue, Z., & Wang, T. (2023). Colorimetric aerogel gas sensor with high sensitivity and stability. *Analytical Chemistry* 95(33): 12313–12320.

Xu, X., Zhang, Q., Yu, Y., Chen, W., Hu, H., & Li, H. (2016). Naturally dried graphene aerogels with superelasticity and tunable Poisson's ratio. *Advanced Materials (Deerfield Beach, Fla.)* 28(41): 9223–9230.

Yang, G., Yang, Y., Chen, T., Wang, J., Ma, L., & Yang, S. (2022). Graphene/mxene composite aerogels reinforced by polyimide for pressure sensing. *ACS Applied Nano Materials* 5(1): 1068–1077. https://doi.org/10.1021/acsanm.1c03722

Yang, L., Yuan, Q. Y., Lou, C. W., Lin, J. H., & Li, T. T. (2024). Recent advances of cellulose-based hydrogels combined with natural colorants in smart food packaging. *Gels* 10(12): 755.

Yang, J., Li, Y., Zheng, Y., Xu, Y., Zheng, Z., Chen, X., & Liu, W. (2019). Versatile aerogels for sensors. *Small* 15(41): 1902826.

Zhao, S., Malfait, W. J., Guerrero-Alburquerque, N., Koebel, M. M., & Nyström, G. (2018). Biopolymer aerogels and foams: Chemistry, properties, and applications. *AngewandteChemie International Edition* 57(26): 7580–7608.

Zheng, Q., Tian, Y., Ye, F., Zhou, Y., & Zhao, G. (2020). Fabrication and application of starch-based aerogel: Technical strategies. *Trends in Food Science & Technology* 99: 608–620.

Zheng, S., Wang, D., Ren, L., Wang, T., Meng, Y., Ma, R., Wang, S., Cui, F., Li, T., & Li, J. (2024). A new paradigm for smart packaging: A dual-channel freshness monitoring platform based on aerogels of sodium alginate-anthocyanin complex with high colorimetric sensitivity and stability. *International Journal of Biological Macromolecules* 267: 131485.

Zhou, S., Wang, M., Xiong, C., & Xu, F. (2015). Facile template synthesis of microfibrillated cellulose/polypyrrole/silver nanoparticles hybrid aerogels with electrical conductive and pressure responsive properties. *ACS Sustainable Chemistry & Engineering* 3(12): 3346–3354. https://doi.org/10.1021/acssuschemeng.5b01020

Zuo, L., Zhang, Y., Zhang, L., Miao, Y. E., Fan, W., & Liu, T. (2015). Polymer/carbon-based hybrid aerogels: Preparation, properties and applications. *Materials* 8(10): 6806–6848.

8 Rapid Microbial Techniques

Methods of Detection of Microbial in Food Processing

Neetu Gupta and Yashi Srivastava

8.1 INTRODUCTION

Food safety is vital for maintaining life and promoting good health. It is a key concern in every country, alongside economic and social development. According to the World Health Organization (WHO), over 3 million people die each year from foodborne illnesses, which account for 1.5 billion cases annually. The most common cause of foodborne illnesses is eating or drinking water or food tainted with bacteria or their poisons. Numerous microbes, such as bacteria, viruses, fungi, and parasites, can cause foodborne diseases. The majority of foodborne illness outbreaks worldwide are caused by pathogenic bacteria and viruses such as *Listeria monocytogenes, Salmonella species, Campylobacter species, Vibrio species, Staphylococcus aureus, Norovirus,* Hepatitis E virus, and pathogenic *Escherichia coli* (Yordanova and Andonova 2024). Numerous microbes may produce toxins that lead to foodborne illnesses. These include some species of *E. coli, Bacillus cereus, Vibrio cholerae, Clostridium species, S. aureus, and the genus Fusarium* (Fusco et al. 2011). Some fungal species create toxic substances called mycotoxins, which are commonly found in commonly consumed cereals including rice, maize, and wheat. These substances often contaminate staple foods and feed (Marroquín-Cardona et al. 2014). The popularity of street food and desire for minimally processed ready-to-eat goods have raised concerns among public health organizations regarding food safety (Lee et al. 2014). Foodborne microrganisms can be found in fruits, vegetables, and ready-to-eat goods which are ingested untreated (Chung et al. 2010, Lee et al. 2014). Failure to address food safety concerns may lead to foodborne illnesses. Also, foodborne illnesses are frequently linked to raw or undercooked seafood, beef, and poultry eating (Wingstrand et al. 2006). Analyzing food for foodborne pathogens is crucial for ensuring a safe supply and preventing foodborne infections.

Conventional approaches for detecting foodborne microorganism typically involve the preparation of culture medium. These methods are inexpensive, simple, but they are very time-consuming methods. Preliminary identification takes two to three days, whereas the confirmation of the particular infection takes over a week (Zhao et al. 2014). Furthermore, the conventional method has some drawbacks such

DOI: 10.1201/9781003482611-8

as it has limited sensitivity, and sometimes it gives false negative result due to viable but non-culturable (VBNC) pathogens (Lee et al. 2014). The risk of disease transmission would rise if foodborne pathogens are not identified. To overcome the limitations of the conventional method, researchers developed different rapid detection methods with the improvements of sensitivity, specificity, more labor-saving quality, time-efficiency, and the ability to minimize human mistakes. However, each rapid detection method has their own pros and cons. This book chapter focuses on the different types of rapid detection methods and their application along the challenges and limitations (Mandal et al. 2011).

8.2 PATHOGENIC/SPOILAGE-CAUSING MICROORGANISMS

Microbes are everywhere, and their multiplication on food under favorable conditions leads to food spoilage. Many sources of foodborne microorganisms are shown in Figure 8.1. Microorganisms cause spoilage defects such as putrefaction, foul smell, changes in color, off-odor, flavor, slime production, and gas production. The spoilage microorganisms are mainly bacteria, yeast, mold, prions, fungi, protozoa, and viruses in food. The prevalent spoiling defects caused by various pathogenic microorganisms in various foods are illustrated with examples in Table 8.1.

8.3 VARIOUS RAPID MICROBIAL TECHNIQUES FOR DETERMINING THE SAFETY AND QUALITY OF FOOD

8.3.1 Microscopic-Based Microbial Techniques

Microscopic-based approaches are mostly used for the foodborne pathogen detection, and they usually require staining and observation with microscope; however, numerous unstained observations and non-microscopic techniques exist as alternatives.

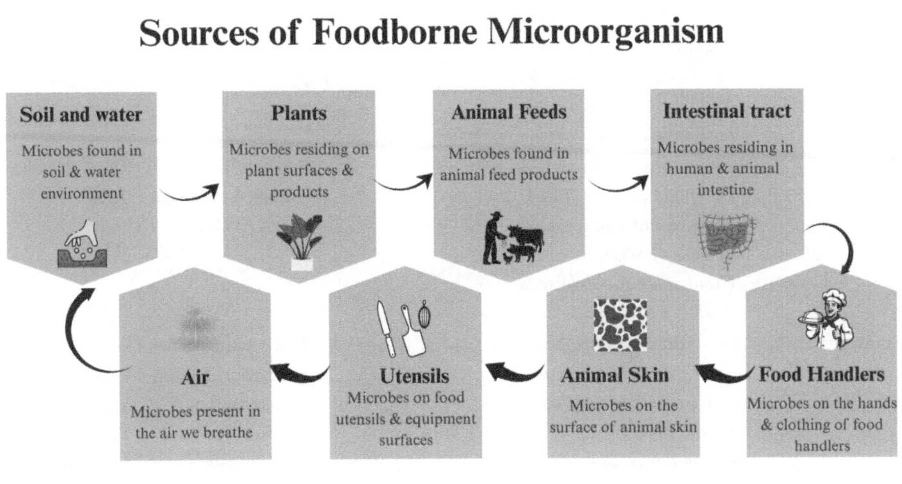

FIGURE 8.1 The schematic diagram illustrates various sources of foodborne microorganisms.

TABLE 8.1
Common Spoilage Defects Caused by Different Microorganisms

Types of Microorganism	Microorganism Involved	Common Spoilage Defects	Examples	References
Bacteria	*Bacillus subtilis, Alcaligenes, Clostridium, Proteus vulgaris, Pseudomonas fluorescens, Pseudomonas, Micrococcus, Lactobacillus, Leuconostoc*	Ropiness, sourness, slime, off-flavor, ropiness, discolorations, putrefaction, gas production	Fish, eggs, bread, fresh meats, concentrated juices, milk, and creams	(de Blackburn, 2006), (Bintsis, 2017)
Yeast	*Zygosaccharomyces spp, Saccharomyces spp., Candida spp, Dekkera spp*	Change in color, a change in texture, an unpleasant odor, or an undesirable taste	Specially grow on high salt or sugar content such as maple syrup, pickles, jams, soy sauce, and sauerkraut	(Deak, 2007), (de Blackburn, 2006)
Mold	*Mucor, Aspergillus* spp, *Rhizopus* spp, *Penicillium* spp, *Alternaria* spp, *Bothrytis, Byssochlamys, Fusarium* spp.	Off-flavors, mycotoxins contamination, discoloration, rotting	Specially grow in dried foods such as grains, beans, peanuts, and some spices	(de Blackburn, 2006)
Protozoa	*Giardia lamblia, Entamoeba histolytica, Cyclospora cayetanensis, Toxoplasma gondii, and Trichinella spiralis*	-	Water and food	(Bintsis, 2017)
Algae	*Gonyaulax catenella, Gonyaulax tamarensis, Gambierdiscus toxicus, Ptychodiscus brevis, Microcystis aeruginosa, blue-green algae*	-	Fish or other marine sea foods	(STEFAN et al.)
Prions	*Bovine spongiform encephalopathies (BSE), Scrapie, chronic wasting disease (CWD), Creutzfeldt-Jacob Disease (CJD)*	-	Consumption of infected meat and meat products	(Rao & Hsieh, 2016)
Viruses	*Norovirus, hepatitis A virus (HAV), hepatitis E virus (HEV), Astrovirus (AstV), rotavirus (RV), Coronavirus, Sapovirus (SaV)*	-	Water and food act as carriers for the transmission of viruses from one host to another	(Koopmans, von Bonsdorff, Vinjé, de Medici, & Monroe, 2002), (Bintsis, 2017)

8.3.1.1 Direct Epifluorescent Filter Technique (DEFT)

A microscopy-based technology called the direct epifluorescent filter technique, or DEFT, is intended for the direct counting of foodborne bacteria. Earlier, it was used for the detection of the bacteria loads in milk, and its application has been broadened to include microbial enumeration in meat and fish products (Logue and Nde 2017). The assay time for determining the microbial counts in raw milk is less than 10 minutes (Beck and Hehir 1982): significant advantage is the removal of the requirement for the overnight incubation. The DEFT process entails first homogenizing the sample in a fluid, then treating the homogenate with enzymes and detergents to extract protein, filtering the liquid through a polycarbonate membrane to collect cells, and finally using fluorescent labeling to allow for viewing. A typical stain for bacterial detection is acridine orange. DEFT detection limits fall between 10^4 and 10^6 CFU/mL (Logue and Nde 2017). After that, every microorganism is quickly identified and measured. In order to specifically count *Salmonella* using the DEFT method, as well as to detect the presence of *Listeria monocytogen* and other organisms in raw foods, the convoluted method of fluorescent-antibody staining was modified (Jan et al. 2018).

8.3.1.2 Fluorescent In Situ Hybridization (FISH)

In the early 1990s, FISH became a widely used method for detecting diseases and other microbes (Rohde et al. 2015). It has been used to identify microorganisms by focusing on the abundant RNAs found in bacterial cells. The principle of FISH encompasses several sequential steps, i.e., initially, a sample is subjected to fixation to preserve its cellular architecture. Subsequently, the sample undergoes permeabilization, facilitating the entry of fluorescent probes into the cells. Following this, the probes are hybridized to the target nucleic acids. A rinsing step is performed to eliminate any unbound excess stain. Finally, the target is visualized using a microscope (Logue and Nde 2017).

Certain adjustments to the procedures are essential for applying FISH in food microbiology, including enhancing the target to levels that are detectable and possibly isolating the target from pollutants or food-related items. These techniques necessitate optimization for the specific target, which includes protocols aimed at concentrating it for sufficient detection and eliminating contaminants that could interfere with the assay. FISH assays have been developed for a variety of pathogens, such as *E. coli* in meat, beverages, and milk; *Listeria spp.* in milk, meats, salmon, and cheese; *Salmonella spp.* found in meats, milk, cheese, and vegetables; and *Campylobacter* identified in poultry meat and associated products (Logue and Nde 2017).

8.3.1.3 Flow Cytometry

Flow cytometry is an optical technique that analyzes individual cells in a liquid sample, especially in food contexts. It consists of four primary components: a computer, an electronic network, fluid lines and controls (fluidics), and a light source (Givan 2013). The process entails suspending the target cells of interest in a liquid, which are then placed in a flow cytometer. Within the cytometer, the cells are examined on the basis of the patterns of light scatter that result when they are illuminated.

To facilitate detection, these cells are usually labeled with a fluorescent dye. The pattern of scattering that results from the light striking the cells provides the size, quantity, and shape of the cells in the sample (Logue and Nde 2017). This method is rapid, automatic, and can be highly specific and provides result at a rate of 1,000 cells per second. Applications of flow cytometry in food analysis can vary widely. They include determining the total viable count in a sample and assessing live and dead cell counts when an appropriate live/dead cell stain is used. Additionally, more specific assays can be performed if a specific antibody and fluorescent labels are employed (Givan 2013).

8.3.2 Nucleic Acid-Based Microbial Techniques

Nucleic acid-based methods, particularly DNA-based techniques are widely employed for the rapid detection of foodborne microorganisms. Nucleic acid-based approaches detect specific genes in target pathogens, avoiding unclear or incorrectly interpreted data. Recent methods for nucleic acid analysis include PCR, qPCR, Multiplex PCR, Nested PCR, Droplet Digital PCR, LAMP, NASBA.

8.3.2.1 Polymerase Chain Reaction (PCR)

PCR is among the most common molecular-based detection methods for foodborne bacterial pathogens. Approximately 30 years ago, PCR was invented, and it allows all the foodborne bacteria which are present in the food by targeting the specific DNA sequence (Velusamy et al. 2010). PCR uses a three-step cycle method to amplify a particular target DNA sequence (Mandal et al. 2011). First, target-specific double-stranded DNA is denatured into the single strands of DNA by applying heat. Through the process of polymerization, forward and reverse primers anneal to the DNA by using deoxyribo nucleotides and a thermostable DNA polymerase. The products of PCR amplification are visualized as bands on an electrophoresis gel after staining with ethidium bromide (Zhao et al. 2014). PCR has been employed to detect various microorganisms, including *Listeria monocytogenes, Escherichia coli O157:H7, Staphylococcus aureus, Campylobacter jejuni, Salmonella spp.*, and *Shigella spp* (Jan et al. 2018).

8.3.2.2 Real-Time or Quantitative PCR (qPCR)

Real-time PCR, sometimes called as qPCR, is a technique that can continuously track the PCR amplification during the reaction. It provides fast, concurrent amplification and targets genes specific to an amplification, and its application in food microbiology is growing (Jung et al. 2005). This technology enables the measurement of a single bacterium in food and the study of its behavior due to environmental factors (e.g., food composition, temperature, pH, oxygen) by the appropriate target gene's expression. Real-time monitoring of the process also removes the requirement for post-amplification treatments such as gel electrophoresis, thus reducing the analysis time. Grothaus et al. (2006) designed a qPCR to quantify the total aerobic bacteria and fungi on fresh produce, utilizing the centrifugation water produced during processing as a reference, instead of food matrix itself. Typically 35% of the native and 64% inoculated bacteria were found in the centrifugation water. The cell number count obtained through qPCR showed no significant differences compared to the plate assay, suggesting that it could serve as a viable replacement. This method may offer an alternative to plate assays for accurately measuring the aerobic bacterial load in fresh-cut products in less than 5 hours.

Fykse et al. (2007) developed both TaqMan and SYBR Green-based real-time PCR assays for the accurate detection and quantification of S. aureus strains carrying the enterotoxin gene cluster, irrespective of genetic variations among the strains. By utilizing optimized qPCR conditions, the assay successfully detected approximately 1×10^3 and 1×10^4 CFU of the pathogen per milliliter of raw (10 and 100 CFU equivalents of egc+ *S. aureus* for each reaction mixture) through the SYBR Green and TaqMan qPCR assays, respectively.

8.3.2.3 Multiplex PCR

It is the more advanced form of PCR technique in which multiple targets can be detected in a single reaction. This method uses multiple primer sets to detect several genes, gene variants, or genomic markers of different organisms and saves time and efforts by enabling the identification of multiple targets in a single test. Multiplex PCR was used to examine the presence of Salmonella species, *E. coli O157:H7 V parahaemolyticus* and *S. aureus* in milk and chicken meat matrices (Chiang et al. 2012). The detection limit for pathogens in food homogenates ranged from 10^2 to 10^4 CFU/mL without enrichment treatments. A multiplex PCR technique was used to detect pathogens such as *E. coli O157:H7, B. cereus, V. parahaemolyticus, Salmonella spp., L. monocytogenes,* and *S. aureus* in ready-to-eat (RTE) foods and evaluated food products such as chicken, steamed pork hocks, oysters, fresh tomato juice, sushi, and lettuce (Lee et al. 2014). These microorganisms were found in different types of food samples at a concentration of 10^4 CFU/mL or higher.

8.3.2.4 Nested PCR

Nested PCR is the advance form of PCR that aims to optimize the identification of a specific target and increase the specificity. This technology used initial set of primers to amplify a target DNA region, followed by a second set of primers that amplifies a smaller product within the first amplified region. It has two successive steps such that if during the first step, the wrong target is amplified then in the next step, it is less likely to be amplified due to the reason that the second set of primers only works within the first amplified target. *Shigella and enterotoxigenic Clostridium perfringens* microorganism are identified by nested PCR (Logue and Nde 2017).

8.3.2.5 Droplet Digital PCR

Droplet digital PCR is a type of digital PCR that differs slightly by using a tile-based method with small well for PCR reactions and readings. These techniques explore the formation of various droplets, each containing DNA as well as reagents, and each drop forms its own single PCR reaction whereas a droplet has a water-oil emulsion. A droplet generator produces droplets by oil-water emulsion, reagent, and DNA. After PCR amplification, a unique reader detects and counts the positive and negative PCR droplets. Recent studies showed droplet digital PCR technology used in the assay of *Cryptosporidium oocysts* in human fecal samples (Logue and Nde 2017), as well as to measure genetically modified organisms (GMOs) in intricate food and feed matrices (Morisset et al. 2013).

8.3.2.6 Loop-Mediated Isothermal Amplification (LAMP)

LAMP is the novel nucleic acid amplification technique that is affordable, fast, easy to operate, and highly sensitive toward the target with numerous applications across various fields (Li et al. 2012). This technique uses the Bst DNA polymerase for the autocycling strand displacement DNA synthesis. PCR requires two primers, whereas LAMP employs four to six primers to target six to eight specific regions of the gene. Amplification occurs at 60°C to 65°C, producing stem-loop and cauliflower-like structure. LAMP reaction can be speeded up with the addition of loop primers and is usually able to achieve around 1,000-fold or more amplification of stem-loop DNAs within 60 minutes, surpassing PCR. Instead of relying on gel electrophoresis, LAMP products may be detected visually using SYBR Green I dye, whose solution becomes green when amplicons are present and orange if there is no amplification. This technique was used to detect stxA2 in *Escherichia coli* O157:H7 cells (Maruyama et al. 2003). In situ LAMP technology induced fewer cell damages than in situ PCR due to its mild permeabilization conditions and low isothermal temperatures. Also, it generated images with higher contrast than in situ PCR. Chen et al. (2011) developed and evaluated for the detection and direct identification of acidophilic thermophilic bacteria contaminants in pure juices, which can identify 2.25×10^1 CFU/mL of ATB in juice samples within 2 hours.

In recent years, derivative LAMP assays, like the reverse transcription LAMP test (Chen et al. 2011), multiplex LAMP test (Jasson et al. 2010), in situ LAMP test (Zhao et al. 2014), and real-time reverse transcription LAMP test (Liu et al. 2009), were developed and employed to detect various food microorganisms, including *Bacillus anthracis* (Qiao et al. 2007), *Vibrio parahaemolyticus* (Zhao et al. 2014), *Staphylococcus aureus* (Zhao et al. 2011), *Salmonella* (Zhao et al. 2014), *Pseudomonas aeruginosa* (Zhao et al. 2011), *Escherichia coli O157* (Zhao et al. 2014), and *Listeria monocytogenes* (Wang et al. 2011).

8.3.2.7 Nucleic Acid Sequence-Based Amplification (NASBA)

NASBA is an isothermal amplification technique that is used for the identification of RNA or DNA (de Boer and Beumer 1999). The process typically involves three enzymes, which are T7 RNA polymerase, RNase H, and avian myeloblastosis virus (AMV) reverse transcriptase (RT), which work together to amplify sequences from a single-stranded RNA model. The reaction also involves buffering agents and two particular primers, and it takes place at around 41°C (Hahm and Bhunia 2006).

NASBA targets specific RNA or DNA sequences and is increasingly popular for microorganisms identification in clinical, environmental, and food items (Baeumner et al. 2003). Simpkins et al. (2000) demonstrated that NASBA can be specifically boost mRNA sequences from *Salmonella enterica* even when genomic DNA is present and indicated that NASBA may serve as an effective method for evaluating the vitality of cells. Min and Baeumner (2002) created a NASBA test to detect live *Escherichia coli.* Bolton et al. (2000) confirmed the assay's specificity for live *Escherichia coli* by demonstrating that heat-killed cells did not provide a signal above the instrument's background. Compton (1991) produced assays utilizing molecular beacons for the accurate time identification of *Campylobacter jejuni and Campylobacter coli* in chicken meat product samples.

NASBA enables for sensitive, specific tests to detect, quantify, and analyze RNA (and, in some circumstances, DNA) targets (Tsaloglou et al. 2011). Molecular beacons were utilized to produce fluorescence signals assays for the identification of *Vibrio cholera* (Gill and Ghaemi 2008). A detailed comprehensive list of examples demonstrating the use of nucleic-acid-based approaches in food is provided in Table 8.2.

8.3.3 Biosensor-Based Microbial Techniques

Biosensors are analytical devices that consist of recognition and transduction elements. The identification element relies on physiologically acquired chemicals, whereas the transduction element converts the physiochemical response into quantitative signals. Biosensors are categorized according to the kind of transduction elements. Biosensors that rely on affinity, utilizing antibodies, nucleic acids, and bacteriophages as recognition components, are favored over enzymatic biosensors for detecting microorganisms because of their improved specificity and selectivity (Jayan et al. 2020). Biosensors have been designed and used for microbial foodborne pathogen analysis, *like Escherichia coli O157:H7, S. aureus, Salmonella* species and *L. monocytogenes,* and other microbial toxins *like staphylococcal* enterotoxins and mycotoxins from genus *Fusarium* in peanuts, corn, and wheat (Asiello and Baeumner 2011). Some of the important biosensors are electronic nose sensors, mass-sensitive sensors, surface plasmon resonance (SPR), optic biosensors, and aptamer-based biosensors.

8.3.3.1 Electronic Nose Sensor

The use of electronic nasal sensors to identify microorganisms is based on the concept of identifying volatiles generated by living organisms in a sample. A population of specific organisms can be linked to a change in sensor resistance caused by the adsorption of volatiles onto a collection of conducting organic polymers. Balasubramanian et al. (2005) showed by effectively differentiating between contaminated and control samples that electronic nose technology was utilized to identify Salmonella in inoculation beef samples. Additionally, *E. coli* O157:H7 strains have been distinguished from non-O157 strains using similar techniques (Younts et al. 2002; Younts et al. 2003). Others have used commercial technology for the detection of *Bacillus* and *Pseudomonas* species in milk and milk products (Magan et al. 2001), and *B. cereus, V. parahaemolyticus,* and *Salmonella* could also be detected on the basis of their unique signals (Arshak et al. 2009). Siripatrawan and Harte (2015) utilized an electronic nose in combination with a Kohonen neural network to visualize and classify *Salmonella Typhimurium* contamination in packaged fresh alfalfa sprouts.

TABLE 8.2
Nucleic Acid-Based Method for the Foodborne Microorganism Identification

S. N.	Methods	Detection of Pathogen	Detection Limit	Detection Time	Detection Principle	Reference
1	Polymerase chain reaction	Salmonella spp., *Listeria monocytogenes*, and *Escherichia coli* O157:H7	1 cell per 25 g of infected pork sample or 10^3 CFU/ml for each pathogen by pure culture	30 h	With the help of DNA polymerase, two identical copies were produced on the basis of the principle of base complementary pairing	Kim et al. (2012)
		Salmonella spp., Listeria monocytogenes, and *Escherichia* coliO157:H	After 20 h of enrichment, 5 CFU/25 g of the injected sample	3 hour		Kawasaki et al. (2005)
2	Multiple PCR	*S. aureus E. coli* O157:H7 *V parahaemolyticus,* and *Salmonella*	10^2–10^4 CFU/mL of food homogenates.	-	In a single reaction tube, amplification was performed for several DNA templates or distinct areas of the same template.	Chiang et al. (2012)
		E. coli O157:H7, *B. cereus, V. parahaemolyticus, Salmonella* spp., *L. monocytogenes*, and *S. aureus*	Identified in various food samples at 10^4 CFU/mL or higher.			Lazcka et al. (2007)
3	LAMP	*Shigella* and *enteroinvasive*	Each reaction has eight CFUs.	2 hours	In order to replace the DNA polymerase with high activity and enable the continuous self-cycling synthesis of strand replacement DNA, four particular primers were employed at a temperature of 60–65°C, which is optimal for both DNA renaturation and extension.	Song et al. (2005)
		Escherichia coli *Streptococcus pneumonia*	The pure DNA of S. pneumonia in ten or more copies	1 hour		Seki et al. (2005)
		Salmonellae	Pure-cultured *Salmonella* cells range from 3.4 to 34, while spiked produce samples included 6.1 × 103 to 6.1 × 104 CFU/g.	3 hours		Chen (2008)
4	NASBA	*Chlamydophila pneunioniae*	C. pneumoniae's inclusion-forming unit (IFU) is 0.1	< 5 hours	In vitro, single-stranded RNA was used as a template for isothermal amplification of particular nucleotide sequences employing primers and three enzymes.	Loens et al. (2006)
		Aspergillus fumigatus	100 cells and 104 copies of RNA/ml			Yoo et al. (2008)
		Mycobacterium tuberculosis	1×10^2 CFU/ml			Gizeli and Lowe (1996)

8.3.3.2 Mass-Sensitive Biosensors

Mass-sensitive biosensors measure even the smallest changes in mass. Mass-based biosensors use piezoelectric crystals, which, when triggered by an electrical signal of a particular frequency, oscillate at that frequency. This crystal has the bioreceptors (like antigens) bonded to it. The vibrational frequency of the crystals will vary noticeably in proportion to the mass deposited on the crystal surface, once the target antigens have adhered to the immobilized antibodies on the crystal. Surface acoustic wave (SAW) resonators and bulk acoustic wave (BAW) resonaters, commonly referred to as quartz crystal microbalance (QCM), are the two primary types of mass-based biosensors (Law et al. 2015).

However, mass-based biosensors are frequently employed less in the field of foodborne pathogen detection than that of electrochemical and optical biosensors (Velusamy et al. 2010). *Salmonella Enteritidis* (Law et al. 2015) and *Escherichia coli* (Pohanka et al. 2007) are detected using piezoelectric immunosensors with a detection limit of 1×10^5 cells/mL and 10^6–10^9 CFU/mL, respectively. Berkenpas et al. (2006) described the application of a SAW biosensor to detect pathogenic *Escherichia coli O157:H7.* With a detection limit of 1×10^7 cells/mL, the QCM immunosensor was developed by Vaughan et al. (2001) for the identification of *Listeria monocytogenes.*

8.3.3.3 Surface Plasmon Resonance

SPR measures infections using reflectance spectroscopy and can detect minute changes in refractive index that occur when target cells connect to receptors on a transducer surface. Changes in medium density over time are quantified as a function of changes in reflected light. SPR uses mounted antibodies on a gold electrode surface to detect minuscule changes in resonance frequency brought on by antibody–antigen binding (Feng 2007).

Su and Li (2004) used a piezoelectric device in conjunction with SPR to detect *E. coli* O157:H7; a comparable method by Oh et al. (2004) could identify *Salmonella typhimurium*; and Leonard et al. (2005) reported the identification of *Listeria monocytogenes.* The creation of an SPR chip that can simultaneously identify multiple species, including *Salmonella typhimurium, Legionella pneumophila, Yersinia enterocolitica, and E. coli O157:H7,* was highlighted by Kim et al. (2004). SPR sensors have also been used to detect *Salmonella, E. coli O157:H7, Campylobacter*, and *Listeria monocytogenes* (Logue and Nde 2017)

The SPREETA and BIACORE 3000 biosensors are two commercially available SPRs that have been used to detect foodborne diseases like *Salmonella, Listeria*, and *E. coli O157* (Logue and Nde 2017)

8.3.3.4 Fiber Optic Biosensor

Fiber optic biosensors are used for the pathogen identification and function by using a fiber optic taper to route excitation laser light toward a detecting surface and then absorb the light that is emitted. This technology is used for detecting foodborne pathogens that evaluate variations in chemiluminescence, fluorescence energy transfer, and fluorescence emission refractive index. Geng et al. (2004) highlighted the use of a fiber optic-based detection equipment to detect *Listeria* in meat sample enrichments and mixed cultures, with a detection threshold as low as $10–10^3$ cells/mL. Keramas et al. (2004) developed an automated fiber-based biosensor with a minimum 50 CFU/g detection level for detecting Salmonella in the rinse water from sprouts using seeds that were infected. Fiber optics have found applications in detection of *Salmonella Listeria, E. coli O157:H7*, and *Clostridium botulinum* toxin (Logue and Nde 2017), and in order to identify *S. aureus* (Hsin Chang et al. 1996). Mendonca et al. (2012) used fiber optics and immune magnetic separation to detect *Listeria spp.*

8.3.3.5 Aptamer-Based Biosensors

Aptamers are single-stranded oligonucleotides or peptide compounds that exhibit excellent specificity and selectivity for their target targets. By using structural recognition, aptamers that are analogs of antibodies can attach to a variety of targets, including protein complexes and inorganic compounds. Aptamers are less expensive to produce than antibodies and are not immunogenic like antibodies. Furthermore, aptamers are tiny (10–100 times smaller than an antibody) and allow for a high ligand density on the sensor surface, increasing sensitivity. In contrast to antibodies, aptamers offer a wide range of dissociation constants, from micromolar (mM) to picomolar (pM). Because they are easily chemically modified, extremely thermostable, and able to enter tissues, aptamers are suitable ligands for identifying bacteria in food matrices (Röthlisberger and Hollenstein 2018). Further illustrations of biosensor-based methods in food and other samples are given in Table 8.3.

TABLE 8.3
Biosensor-Based Techniques for the Foodborne Microorganism Identification

S.N.	Methods of Detection	Detection of Pathogen	Detection Limit	Detection Time	Detection Principle	Reference
1	Electronic nose sensor	*Salmonella Pseudomonas*, and *Bacillus* *Salmonella, B. cereus*, and *V. parahaemolyticus*	-	-	Volatile adsorption on conducting polymers alters sensor resistance, indicating specific organism populations	Balasubramanian et al. (2005), Magan et al. (2001), Arshak et al. (2009)
2	Mass-sensitive biosensor	*Salmonella* Enteritidis *Escherichia coli* *Listeria monocytogenes*	1×10^5 cells/ mL $10^6–10^9$ CFU/mL 1×10^7 cells/mL	-	Based on the identification of small changes in mass	Law et al. (2015), Pohanka et al. (2007), Vaughan et al. (2001)

TABLE 8.3 (*Continued*)
Biosensor-Based Techniques for the Foodborne Microorganism Identification

S.N.	Methods of Detection	Detection of Pathogen	Detection Limit	Detection Time	Detection Principle	Reference
3	SPR biosensor	*Escherichia coli* O157:H7 *Escherichia coli* O157:H7 *Salmonella Typhimurium* *Salmonella* enteritidis and *Escherichia coli*	10^2 CFU/ml 8.7×10^6 CFU/ml 1×10^6 CFU/ml Salmonella has 23 CFU/ml while *Escherichia coli* has 25 CFU/ml.	2 min 35 min <1 h	Detects pathogens by measuring refractive index changes from target cell binding to surface receptors	(John Waswa, Irudayaraj, & DebRoy, 2007) (Meeusen, Alocilja, & Osburn, 2005) (Laura, Claudio, Cristina, & Gianfranco, 2011) (JW Waswa, Debroy, & Irudayaraj, 2006)
4	Fiber optic biosensor	*Listeria monocytogenes* *Salmonella*	$10–10^3$ cells/ml. 50 CFU/g in inoculated seeds	-	Detecting by measuring emitted light changes after excitation laser interaction with a detection surface	Geng et al. (2004) Keramas et al. (2004)

8.3.4 Luminescence-Based Microbial Techniques

8.3.4.1 Bioluminescence/ATP Detection

In the 1950s, NASA researchers developed ATP bioluminescence. The amount of ATP produced by microbes is measured using ATP bioluminescence, which is an indirect indicator of the amount of food residue on a surface that may support microbial development. Firefly light served as the foundation for ATP bioluminescence. Luciferin and Luciferase are two chemicals found in fireflies that react to the insect's ATP to produce bioluminescence light. Bioluminescence light is produced when the ATP reacts with the Luciferin/Luciferase compounds in the sample swab. The bioluminescence is measured by the luminometer; the result is shown in relative light units (RLU), which is correlated with the amount of ATP and, thus, the microbial biomass on the sampled surface. The total number of microorganisms in a sample can be determined using ATP-generated bioluminescence within 5 minutes, but only in cases where the bacterial concentration is high (more than 10^4CFU/g) (Scheu et al. 1998; Jasson et al. 2010).

8.3.5 Immunological-Based Microbial Techniques

Immunological tests for foodborne pathogen identification are commonly employed. Antibody–antigen interactions, in which a particular antibody binds to its particular antigen, constitute the basis of these techniques. The sensitivity and specificity of immunological techniques are determined by the strength of an antibody's affinity to its antigen. To identify microbial toxins and foodborne pathogens, a range of antibodies have been used in various assay types (Zhao et al. 2014). Although there are many other types of immunoassays, the two most modern immunological techniques for the identification of foodborne pathogens are lateral flow immunoassay (LFIA) and enzyme-linked immune sorbent assay (ELISA).

8.3.5.1 Enzyme-Linked Immunosorbent Assay

One of the most popular immunological tests for identifying foodborne microorganism is the enzyme-linked immunosorbent assay, which is a very sensitive and accurate method for identifying antigens (Vernozy-Rozand et al. 2004). Chromogenic reporters and substrates that produce a discernible color shift are typically used in traditional ELISA to indicate the presence of an antigen. The antigen from the enrichment cultures to be measured is positioned between two main antibodies, the capture antibody and the detection antibody, making the 'sandwich' test the most effective ELISA format. The sandwich format is utilized as it is robust and sensitive. Microtiter plate well walls are the most commonly used solid support; however, dipsticks, paddles, membranes, pipette tips, and other solid matrices have also been used to prepare ELISAs (Foley and Grant 2007). Although ELISA can employ a variety of enzymes, β-galactosidase, alkaline phosphatase, and horseradish peroxidase (HRP) are three of the most often used (Yeni et al. 2014).

The identification of many foodborne toxins is mostly dependent on the presence of immune reactions. *Staphylococcal enterotoxins* A, B, C, and E have been identified to have detection levels of less than 0.5 μg/100 g in ground beef, according to research using the commonly used ELISA method for toxin detection. Additionally, *E. coli*-produced enterotoxins and *botulinum* toxins have been identified using ELISA (Fusco et al. 2011).

8.3.5.2 Lateral Flow Immunoassay

One popular method for qualitatively determining target analytes is LFIA, which uses an immuno-chromatographic strip to detect microorganisms in the food matrix at the point of care. The paper strip made up of a simple pad, conjugated pad, absorbent pad, and nitrocellulose membrane works by allowing the liquid sample to flow through capillary action. The lines in LFA are coated with recognition elements such as antibodies or nucleic acids (Koczula and Gallotta 2016). Identifying microorganisms is most effectively done using the antibody sandwich assay, as it can recognize multiple epitopes on microbial surfaces, which enhance both sensitivity and specificity.

LFA can detect multiple diseases simultaneously with great sensitivity. For example, (Jayan, Pu et al. 2020) discovered two foodborne pathogens, *Shigella boydii* and *E. coli* O157:H7, with a limit of detection of 10^4 CFU/mL in bread, milk, and jelly.

8.4 APPLICATION

Rapid microbiological methods (RMMs) have emerged as crucial innovations within the contemporary food sector. As global demand continues to rise for food that is both safe and of high quality, and as supply chains grow more intricate and widespread, the need for efficient microbial detection has become increasingly critical. RMMs provide efficient, sensitive, and dependable tools for detecting and quantifying microbial threats. Their benefits are not limited to just one aspect of the food industry; rather, they are used across a wide range of functions from routine quality control procedures to urgent responses during contamination outbreaks, and even for continuous hygiene monitoring in real-time settings. By delivering speed, precision, and trustworthiness, RMMs support both regulatory compliance and consumer protection in a fast-moving, globalized food system (Shintani 2016).

8.4.1 Food Safety Assurance

RMMs are transforming food safety assurance by offering quicker, more sensitive, and reliable detection of microbial contaminants throughout food production and supply chains. Traditional culture-based techniques, while historically foundational, are often time-consuming, delaying critical responses in safety-critical environments. In contrast, advanced RMMs like PCR, enzyme-linked immunosorbent assay (ELISA), adenosine triphosphate (ATP) bioluminescence, and biosensor-based technologies allow near real-time analysis, improving outbreak prevention, production safety, and regulatory compliance.

Molecular techniques such as PCR, qPCR, and LAMP enable highly sensitive and specific detection of pathogens including *Listeria monocytogenes*, *Salmonella spp.*, *Escherichia coli* O157:H7, and *Campylobacter spp.* in various food matrices like meat, dairy, ready-to-eat meals, and produce. These approaches can detect pathogens at concentrations as low as 10^1–10^2 CFU/mL within 2–4 hours (Rossen et al. 1991; Law et al. 2015). Immunoassay-based methods such as ELISA and lateral flow devices (LFDs) detect microbial antigens and toxins, making them practical tools for screening for *shigella* toxins, *staphylococcal enterotoxins*, and *aflatoxins* in high-risk foods (Zhao et al. 2015). Additionally, biosensors are widely employed in industry settings because of their ability to provide real-time, point-of-care diagnostics with minimal sample preparation.

These rapid methods have also revolutionized foodborne outbreak response. During foodborne illness incidents, rapid detection tools, especially multiplex PCR assays, enable simultaneous identification of multiple pathogens in a single run. This significantly enhances the ability of epidemiologists and food safety authorities to identify contamination sources quickly and prevent further exposures. In quality control and routine monitoring of spoilage organisms and fermentation microbes, tools like flow cytometry allow quick enumeration of viable cells, distinguishing between live and dead microbes using fluorescence-based staining. This method is particularly useful in the dairy sector for assessing bacterial counts in raw milk and fermentation monitoring in yogurt and cheese production (Gunasekera et al. 2000). Similarly, the DEFT is utilized to assess the total microbial load in thermally processed foods, verifying the effectiveness of heat treatments in products such as milk, meat, and fish (Logue and Nde 2017), which are listed in Table 8.4.

TABLE 8.4
Summary of the Rapid Detection Method's Application, Advantages, and Disadvantages

S.N.	Techniques	Application	Advantage	Disadvantage	References
1	DEFT	Identification of bacterial counts in milk, meat, and fish	Removes the requirement for overnight incubation, low cost	Sample must be strongly illuminated, sometimes gives false results	Logue and Nde (2017)
2	FISH	Identification of *E. coli/E. coli* O157 in meat, beverages, and milk; *Listeria* spp. in milk, meats, salmon, and cheese; Salmonella spp. in meats, milk, cheese and vegetables	Specific to the targets, sensitivity, large number of sample can be done	Chances of contamination, low sensitivity, requires fluorescence microscopy and an image analysis system	Logue and Nde (2017), Gozzetti and Le Beau (2000)
3	PCR	Detection of *Listeria monocytogenes*, *Escherichia coli* O157:H7, *Staphylococcus aureus*, *Campylo bacterjejuni*, *Salmonella* spp., and *Shigella* spp	Highly sensitive, specificity, gives more reliable results	Problem in differentiating between live and dead cells, affected by PCR inhibitor enzyme, requires DNA purification	Jan et al. (2018), Law et al. (2015)
4	Real-time PCR	Detection of *S. aureus* strain in raw milk	Highly sensitive, specificity, no need for the post amplification products, reproducible	Costly, affected by PCR inhibitors, requires skilled person, chances of cross contamination, problem in differentiating between live and dead cells	Fykse et al. (2007), Law et al. (2015)
5	Multiple PCR	Detection of *E. coli* O157:H7, *B. cereus*, *V. parahaemolyticus*, *Salmonella* spp., L. monocytogenes, and S. aureus in RTE food products including chicken, steamed pork hocks, oysters, fresh tomato juice, sushi, and lettuce	Highly sensitive, able to detect multiple microorganisms, specific to targets, accurate result	Problem in differentiating between live and dead cells, affected by PCR inhibitors	Lee et al. (2014), Law et al. (2015)

TABLE 8.4 (*Continued*)
Summary of the Rapid Detection Method's Application, Advantages, and Disadvantages

S.N.	Techniques	Application	Advantage	Disadvantage	References
6	LAMP	Identification of stxA2 in *Escherichia coli* O157:H7 cells	Easy to use, budget friendly, highly sensitive, specific to target	Not able to detect unknown targets, complicated design of primers	Maruyama et al. (2003), Zhao et al. (2014)
7	NASBA	Detection of *Vibrio cholera*, *Campylobacter jejuni*, and *Campylobacter coli* in samples of chicken meat	Sensitive, low cost, detects live microorganism, specific to targets	Requirement of live microorganism, problem in handling RNA	Gill and Ghaemi (2008), Compton (1991), Law et al. (2015)
8	Electric nose biosensors	Detect *Salmonella* in inoculated beef samples, *Pseudomonas*, and *Bacillus* species in milk	Specific to the target, takes less time to give result, sensitivity toward target	High cost, require trained person	Magan et al. (2001), Law et al. (2015)
9	Mass sensitive biosensors	Detection of *Salmonella* Enteritidis,	Low cost, easy to handle, fast results	Low specificity, low sensitivity, require more for incubation of bacteria	Law et al. (2015)
10	Fibre optic biosensors	Detection of *Salmonella* in sprout rinse water	Highly sensitive, take less time to give result, more accurate	Cost-effective	Keramas et al. (2004), Law et al. (2015)
11	ELISA	Detection of botulinum toxins and enterotoxins produced by *E. coli*	Specific to the targets, operates large number of samples, labor saving, automated	Low sensitivity, sometimes gives false results, reacted with the closed antigens, requires trained personnel, need to label antigen	Fusco et al. (2011), Law et al. (2015)
12	Lateral flow immunoassay	Detection of *Shigella boydii* and *E. coli* O157:H7 in bread, milk, and jelly with limit of detection (LOD) of 4 CFU/mL	Budget friendly, easy to handle, specific to targets, sensitive, detects bacterial toxins	Need to label antigen	Jayan et al. (2020), Zhao et al. (2014)

8.4.2 Operational Applications and Strategic Value in Food Safety

In food production environments, sanitation lapses are a major factor in recurring microbial contamination. Rapid hygiene monitoring using ATP bioluminescence assays plays a critical role in verifying cleanliness by detecting biological residues like ATP within 10–60 seconds, guiding sanitation teams to re-clean surfaces as needed (Henriques et al. 2019). Additionally, immune chromatographic surface swab tests can detect pathogens like *Listeria spp.* and *Salmonella spp.* directly from food contact surfaces (Wang et al. 2016). Emerging technologies now integrate biosensors, digital imaging, and cloud-based data systems to automate environmental surveillance, identify trends, and recommend corrective actions more efficiently.

One of the most strategic advantages of rapid microbiological methods is their impact on decision-making within food production and distribution chains. By drastically reducing result turnaround times from several days to hours or even minutes, RMMs empower faster decisions on product release, reprocessing, or recalls. For example, a product can be cleared for shipment as soon as pathogen absence is confirmed or held back for reprocessing if microbial thresholds are exceeded. In cases of post-distribution contamination, rapid detection supports immediate recalls to prevent public health impacts (Lee et al. 2015).

The integration of RMMs not only reduces operational risks but also provides valuable economic and reputational benefits. By acting as an early warning system, these technologies reduce the likelihood of large-scale outbreaks, minimize product loss, and build consumer trust through improved food safety standards. As innovation continues, future developments aim to further refine in-situ, nondestructive detection techniques and minimize false positives, making RMMs even more embedded in proactive food safety systems worldwide.

8.4.3 Other Noteworthy Applications

Apart from their primary function in detecting pathogens and evaluating hygiene standards, rapid microbiological methods (RMMs) have increasingly been applied to other vital aspects of food safety and quality assurance. One prominent area is the monitoring of antibiotic resistance. Modern molecular techniques, including PCR and qPCR, are effectively used to identify resistance-related genes such as *blaCTX-M*, *tet(A)*, and *mecA* in bacteria found in food, helping to track and address the growing concern of antimicrobial resistance in the food chain (Zhao et al., 2020). RMMs are also critical in allergen detection, where quick-response immunological tools like ELISA and lateral flow assays are employed to pinpoint allergens such as gluten, peanut traces, and milk proteins on food contact surfaces and equipment. This application is essential for ensuring that allergen-free manufacturing standards are maintained and that sensitive consumers are protected from unintended exposure. Furthermore, RMMs play a key role in confirming product label accuracy. They are increasingly used to validate claims such as the presence of beneficial probiotics or the absence of harmful microorganisms, by quantifying microbial populations or confirming the nonexistence of targeted pathogens. These developments enhance both regulatory compliance and consumer trust. Overall, these expanded uses of RMMs illustrate their adaptability and growing significance in ensuring comprehensive food safety and quality control.

8.5 CHALLENGES AND FUTURE SCOPE

The effectiveness of microbial detection in food matrices is enhanced by significant developments in fast microbiological techniques. Even though many microbial detection techniques have been developed in recent decades, it is still challenge to create a single device that combines all analytical detection methods. For the most sensitive detection, the majority of microbial detection methods necessitate a pre-enrichment step, which eventually increases the assay's complexity and duration. Therefore, more investigation is required to guarantee the application and liability of quick methods for identifying foodborne microorganisms in food matrices. The development of the multiplex test, which uses a single analytical apparatus to detect several microbial species, is widely anticipated for the future.

The use of modern nucleic-based amplification techniques including PCR, real-time PCR, LAMP, and NASBA before detection is a current trend in the literature that significantly increases sensitivity and specificity. Additionally, the use of SPR-based in-line pathogen detection is expanded by the use of fiber optical biosensors as a transduction element. In the future, ELISA and immunological tests will be fully automated and widely employed for the detection of microorganisms in complicated food matrices, and biosensors will be implemented for Hazard Analysis Critical Control Point Programs.

8.6 CONCLUSION

Conventional methods used for detecting foodborne pathogens were mainly based on culturing methods and considered as the golden standard method due to their effectiveness. But these methods are of lower sensitivity, less precise, time consuming, and labor demanding. Consequently, to address these challenges, new detection methods are required. Advancement of rapid detection systems that can identify a foodborne microbes within the food matrices is vital for the ensuring the foodborne diseases. Such developments would not only offer significant commercial benefits to the food industry but also help in the prevention of disease outbreaks. Rapid detection methods providing accurate and real-time results would also ensure compliance with regulatory limits on pathogen level in specific food products, enhance the safety of imported food items, and protect public health from disease outbreaks.

REFERENCES

Arshak, K., V. Velusamy, O. Korostynska, K. Oliwa-Stasiak and C. Adley (2009). "Conducting polymers and their applications to biosensors: Emphasizing on foodborne pathogen detection." *IEEE Sensors Journal* **9**(12): 1942–1951.

Asiello, P. J. and A. J. Baeumner (2011). "Miniaturized isothermal nucleic acid amplification, a review." *Lab on a Chip* **11**(8): 1420–1430.

Baeumner, A. J., R. N. Cohen, V. Miksic and J. Min (2003). "RNA biosensor for the rapid detection of viable Escherichia coli in drinking water." *Biosensors and Bioelectronics* **18**(4): 405–413.

Balasubramanian, S., S. Panigrahi, C. Logue, M. Marchello and J. Sherwood (2005). "Identification of salmonella-inoculated beef using a portable electronic nose system." *Journal of Rapid Methods & Automation in Microbiology* **13**(2): 71–95.

Beck, C. and A. Hehir (1982). "The rapid fluorescence method for counting bacteria in milk." *Australian Journal of Dairy Technology* **37**(2): 66–67.

Berkenpas, E., P. Millard and M. P. Da Cunha (2006). "Detection of Escherichia coli O157: H7 with langasite pure shear horizontal surface acoustic wave sensors." *Biosensors and Bioelectronics* **21**(12): 2255–2262.

Bintsis, T. (2017). "Foodborne pathogens." *AIMS Microbiology* **3**(3): 529.

Bolton, F. J., E. Fritz, S. Poynton and T. Jensen (2000). "Rapid enzyme-linked immunoassay for detection of Salmonella in food and feed products: Performance testing program." *Journal of AOAC International* **83**(2): 299–304.

Chen, J., X. Ma, Y. Yuan and W. Zhang (2011). "Sensitive and rapid detection of Alicyclobacillus acidoterrestris using loop-mediated isothermal amplification." *Journal of the Science of Food and Agriculture* **91**(6): 1070–1074.

Chen, Z.-G. (2008). "Conductometric immunosensors for the detection of staphylococcal enterotoxin B based bio-electrocalytic reaction on micro-comb electrodes." *Bioprocess and Biosystems Engineering* **31**: 345–350.

Chiang, Y.-C., H.-Y. Tsen, H.-Y. Chen, Y.-H. Chang, C.-K. Lin, C.-Y. Chen and W.-Y. Pai (2012). "Multiplex PCR and a chromogenic DNA macroarray for the detection of Listeria monocytogens, Staphylococcus aureus, Streptococcus agalactiae, Enterobacter sakazakii, Escherichia coli O157: H7, Vibrio parahaemolyticus, Salmonella spp. and Pseudomonas fluorescens in milk and meat samples." *Journal of Microbiological Methods* **88**(1): 110–116.

Chung, M. S., C. M. Kim and S. D. Ha (2010). "Detection and enumeration of microorganisms in ready-to-eat foods, ready-to-cook foods and fresh-cut produce in Korea." *Journal of Food Safety* **30**(2): 480–489.

Compton, J. (1991). "Nucleic acid sequence-based amplification." *Nature* **350**(6313): 91–92.

de Blackburn, W. C. (2006). *Food spoilage microorganisms*. Woodhead Publishing.

de Boer, E. and R. R. Beumer (1999). "Methodology for detection and typing of foodborne microorganisms." *International Journal of Food Microbiology* **50**(1–2): 119–130.

Deak, T. (2007). *Handbook of food spoilage yeasts*. CRC press.

Feng, P. (2007). "Rapid methods for the detection of foodborne pathogens: Current and next-generation technologies." *Food Microbiology: Fundamentals and Frontiers* **2**: 911–934.

Foley, S. L. and K. Grant (2007). "Molecular techniques of detection and discrimination of foodborne pathogens and their toxins." In *Foodborne Diseases*, Springer: 485–510.

Fusco, V., G. M. Quero, M. Morea, G. Blaiotta and A. Visconti (2011). "Rapid and reliable identification of Staphylococcus aureus harbouring the enterotoxin gene cluster (EGC) and quantitative detection in raw milk by real time PCR." *International Journal of Food Microbiology* **144**(3): 528–537.

Fykse, E. M., G. Skogan, W. Davies, J. S. Olsen and J. M. Blatny (2007). "Detection of Vibrio cholerae by real-time nucleic acid sequence-based amplification." *Applied and Environmental Microbiology* **73**(5): 1457–1466.

Geng, T., M. T. Morgan and A. K. Bhunia (2004). "Detection of low levels of Listeria monocytogenes cells by using a fiber-optic immunosensor." *Applied and Environmental Microbiology* **70**(10): 6138–6146.

Gill, P. and A. Ghaemi (2008). "Nucleic acid isothermal amplification technologies—a review." *Nucleosides, Nucleotides and Nucleic Acids* **27**(3): 224–243.

Givan, A. L. (2013). *Flow Cytometry: First Principles*, John Wiley & Sons.

Gozzetti, A. and M. M. Le Beau (2000). *Fluorescence in situ hybridization: Uses and limitations*. Seminars in hematology, Elsevier.

Grothaus, G. D., M. Bandla, T. Currier, R. Giroux, G. R. Jenkins, M. Lipp, G. Shan, J. W. Stave and V. Pantella (2006). "Immunoassay as an analytical tool in agricultural biotechnology." *Journal of AOAC International* **89**(4): 913–928.

Gunasekera, T. S., P. V. Attfield and D. A. Veal (2000). "A flow cytometry method for rapid detection and enumeration of total bacteria in milk." *Applied and Environmental Microbiology* **66**(3): 1228–1232.

Hahm, B. K. and A. Bhunia (2006). "Effect of environmental stresses on antibody-based detection of Escherichia coli O157: H7, Salmonella enterica serotype Enteritidis and Listeria monocytogenes." *Journal of Applied Microbiology* **100**(5): 1017–1027.

Henriques, J., C. Cardoso and C. Vitorino (2019). "Rapid microbiological methods. They are rapid! Are they fast?" *MedDocs International-Research Trends of Microbiology* 1–12.

Hsin Chang, Y., T. C. Chang, E.-F. Kao and C. Chou (1996). "Detection of protein A produced by Staphylococcus aureus with a fiber-optic-based biosensor." *Bioscience, Biotechnology, and Biochemistry* **60**(10): 1571–1574.

Jan, T., A. Rouf and P. Sharma (2018). "Rapid detection techniques of microorganisms." *Journal of Pharmacognosy and Phytochemistry* **7**(5): 1606–1611.

Jasson, V., L. Jacxsens, P. Luning, A. Rajkovic and M. Uyttendaele (2010). "Alternative microbial methods: An overview and selection criteria." *Food Microbiology* **27**(6): 710–730.

Jayan, H., H. Pu and D.-W. Sun (2020). "Recent development in rapid detection techniques for microorganism activities in food matrices using bio-recognition: A review." *Trends in Food Science & Technology* **95**: 233–246.

Jung, B. Y., S. C. Jung and C. H. Kweon (2005). "Development of a rapid immunochromatographic strip for detection of Escherichia coli O157." *Journal of Food Protection* **68**(10): 2140–2143.

Kawasaki, S., N. Horikoshi, Y. Okada, K. Takeshita, T. Sameshima and S. Kawamoto (2005). "Multiplex PCR for simultaneous detection of Salmonella spp., Listeria monocytogenes, and Escherichia coli O157: H7 in meat samples." *Journal of Food Protection* **68**(3): 551–556.

Keramas, G., D. D. Bang, M. Lund, M. Madsen, H. Bunkenborg, P. Telleman and C. B. V. G. Christensen (2004). "Use of culture, PCR analysis, and DNA microarrays for detection of Campylobacter jejuni and Campylobacter coli from chicken feces." *Journal of Clinical Microbiology* **42**(9): 3985–3991.

Kim, H.-J., H.-J. Lee, K.-H. Lee and J.-C. Cho (2012). "Simultaneous detection of Pathogenic Vibrio species using multiplex real-time PCR." *Food Control* **23**(2): 491–498.

Koczula, K. M. and A. Gallotta (2016). "Lateral flow assays." *Essays in Biochemistry* **60**(1): 111–120.

Koopmans, M., C.-H. von Bonsdorff, J. Vinjé, D. de Medici and S. Monroe (2002). "Foodborne viruses." *FEMS Microbiology Reviews* **26**(2): 187–205.

Laura, A., B. Claudio, G. Cristina and G. Gianfranco (2011). "A lateral flow immunoassay for measuring ochratoxin A: Development of a single system for maize, wheat and durum wheat." *Food Control* **22**(12): 1965–1970.

Law, J. W.-F., N.-S. Ab Mutalib, K.-G. Chan and L.-H. Lee (2015). "Rapid methods for the detection of foodborne bacterial pathogens: Principles, applications, advantages and limitations." *Frontiers in Microbiology* **5**: 770.

Lazcka, O., F. J. Del Campo and F. X. Munoz (2007). "Pathogen detection: A perspective of traditional methods and biosensors." Biosensors and Bioelectronics **22**(7): 1205–1217.

Lee, N., K. Y. Kwon, S. K. Oh, H.-J. Chang, H. S. Chun and S.-W. Choi (2014). "A multiplex PCR assay for simultaneous detection of Escherichia coli O157: H7, Bacillus cereus, Vibrio parahaemolyticus, Salmonella spp., Listeria monocytogenes, and Staphylococcus aureus in Korean ready-to-eat food." *Foodborne Pathogens and Disease* **11**(7): 574–580.

Leonard, P., S. Hearty, G. Wyatt, J. Quinn and R. O'kennedy (2005). "Development of a surface plasmon resonance–based immunoassay for Listeria monocytogenes." *Journal of Food Protection* **68**(4): 728–735.

Li, Y., P. Cheng, J. Gong, L. Fang, J. Deng, W. Liang and J. Zheng (2012). "Amperometric immunosensor for the detection of Escherichia coli O157: H7 in food specimens." *Analytical Biochemistry* **421**(1): 227–233.

Liu, Y., C.-K. Chuang and W.-J. Chen (2009). "In situ reverse-transcription loop-mediated isothermal amplification (in situ RT-LAMP) for detection of Japanese encephalitis viral RNA in host cells." *Journal of Clinical Virology* **46**(1): 49–54.

Loens, K., T. Beck, H. Goossens, D. Ursi, M. Overdijk, P. Sillekens and M. Ieven (2006). "Development of conventional and real-time nucleic acid sequence-based amplification assays for detection of Chlamydophila pneumoniae in respiratory specimens." *Journal of Clinical Microbiology* **44**(4): 1241–1244.

Logue, C. M. and C. W. Nde (2017). "Rapid microbiological methods in food diagnostics." *Advances in Food Diagnostics*: 153–185.

Magan, N., A. Pavlou and I. Chrysanthakis (2001). "Milk-sense: A volatile sensing system recognises spoilage bacteria and yeasts in milk." *Sensors and Actuators B: Chemical* **72**(1): 28–34.

66, P. K., A. Biswas, K. Choi and U. Pal (2011). "Methods for rapid detection of foodborne pathogens: An overview." *American Journal of Food Technology* **6**(2): 87–102.

Marroquín-Cardona, A., N. Johnson, T. Phillips and A. Hayes (2014). "Mycotoxins in a changing global environment–a review." *Food and Chemical Toxicology* **69**: 220–230.

Maruyama, F., T. Kenzaka, N. Yamaguchi, K. Tani and M. Nasu (2003). "Detection of bacteria carrying the stx 2 gene by in situ loop-mediated isothermal amplification." *Applied and Environmental Microbiology* **69**(8): 5023–5028.

Meeusen, C. A., E. Alocilja and W. Osburn (2005). "Detection of E. coli O157: H7 using a miniaturized surface plasmon resonance biosensor." *Transactions of the ASAE* **48**(6): 2409–2416.

Mendonca, M., N. L. Conrad, F. R. Conceicao, A. N. Moreira, W. P. da Silva, J. A. Aleixo and A. K. Bhunia (2012). "Highly specific fiber optic immunosensor coupled with immunomagnetic separation for detection of low levels of Listeria monocytogenes and L. ivanovii." *BMC Microbiology* **12**: 1–15.

Min, J. and A. J. Baeumner (2002). "Highly sensitive and specific detection of viable Escherichia coli in drinking water." *Analytical Biochemistry* **303**(2): 186–193.

Morisset, D., D. Štebih, M. Milavec, K. Gruden and J. Žel (2013). "Quantitative analysis of food and feed samples with droplet digital PCR." *PLoS ONE* **8**(5): e62583.

Oh, B.-K., Y.-K. Kim, K. W. Park, W. H. Lee and J.-W. Choi (2004). "Surface plasmon resonance immunosensor for the detection of Salmonella typhimurium." *Biosensors and Bioelectronics* **19**(11): 1497–1504.

Pohanka, M., P. Skládal and O. Pavliš (2007). "Label-free piezoelectric immunosensor for rapid assay of Escherichia coli." *Journal of Immunoassay and Immunochemistry* **29**(1): 70–79.

Qiao, Y.-M., Y.-C. Guo, X.-E. Zhang, Y.-F. Zhou, Z.-P. Zhang, H.-P. Wei, R.-F. Yang and D.-B. Wang (2007). "Loop-mediated isothermal amplification for rapid detection of Bacillus anthracis spores." *Biotechnology Letters* **29**: 1939–1946.

Rao, Q. and Y.-H. P. Hsieh (2016). "Spread and control of prion diseases in the food and feed chains. In *Significance, prevention and control of food related diseases*. IntechOpen.

Rohde, A., J. A. Hammerl, B. Appel, R. Dieckmann and S. Al Dahouk (2015). "FISHing for bacteria in food–A promising tool for the reliable detection of pathogenic bacteria?" *Food Microbiology* **46**: 395–407.

Rossen, L., K. Holmstrøm, J. Olsen and O. Rasmussen (1991). "A rapid polymerase chain reaction (PCR)-based assay for the identification of Listeria monocytogenes in food samples." *International Journal of Food Microbiology* **14**(2): 145–151.

Röthlisberger, P. and M. Hollenstein (2018). "Aptamer chemistry." *Advanced Drug Delivery Reviews* **134**: 3–21.

Scheu, P. M., K. Berghof and U. Stahl (1998). "Detection of pathogenic and spoilage microorganisms in food with the polymerase chain reaction." *Food Microbiology* **15**(1): 13–31.

Seki, M., Y. Yamashita, H. Torigoe, H. Tsuda, S. Sato and M. Maeno (2005). "Loop-mediated isothermal amplification method targeting the lytA gene for detection of Streptococcus pneumoniae." *Journal of Clinical Microbiology* **43**(4): 1581–1586.

Shintani, H. (2016). "Methods of rapid microbiological assay and their application to pharmaceutical and medical device fabrication." *Biocontrol Science* **21**(4): 193–201.

Simpkins, S., A. Chan, J. Hays, B. Öpping and N. Cook (2000). "An RNA transcription-based amplification technique (NASBA) for the detection of viable Salmonella enterica." *Letters in Applied Microbiology* **30**(1): 75–79.

Siripatrawan, U. and B. R. Harte (2015). "Data visualization of Salmonella Typhimurium contamination in packaged fresh alfalfa sprouts using a Kohonen network." *Talanta* **136**: 128–135.

Song, T., C. Toma, N. Nakasone and M. Iwanaga (2005). "Sensitive and rapid detection of Shigella and enteroinvasive Escherichia coli by a loop-mediated isothermal amplification method." *FEMS Microbiology Letters* **243**(1): 259–263.

Su, X.-L. and Y. Li (2004). "A self-assembled monolayer-based piezoelectric immunosensor for rapid detection of Escherichia coli O157: H7." *Biosensors and Bioelectronics* **19**(6): 563–574.

Tsaloglou, M.-N., M. M. Bahi, E. M. Waugh, H. Morgan and M. Mowlem (2011). "On-chip real-time nucleic acid sequence-based amplification for RNA detection and amplification." *Analytical Methods* **3**(9): 2127–2133.

Vaughan, R., C. O'sullivan and G. Guilbault (2001). "Development of a quartz crystal microbalance (QCM) immunosensor for the detection of Listeria monocytogenes." *Enzyme and Microbial Technology* **29**(10): 635–638.

Velusamy, V., K. Arshak, O. Korostynska, K. Oliwa and C. Adley (2010). "An overview of foodborne pathogen detection: In the perspective of biosensors." *Biotechnology Advances* **28**(2): 232–254.

Vernozy-Rozand, C., C. Mazuy-Cruchaudet, C. Bavai and Y. Richard (2004). "Comparison of three immunological methods for detecting staphylococcal enterotoxins from food." *Letters in Applied Microbiology* **39**(6): 490–494.

Waswa, J., C. Debroy and J. Irudayaraj (2006). "Rapid detection of Salmonella enteritidis and Escherichia coli using surface plasmon resonance biosensor." *Journal of Food Process Engineering* **29**(4): 373–385.

Waswa, J., J. Irudayaraj and C. DebRoy (2007). "Direct detection of E. coli O157: H7 in selected food systems by a surface plasmon resonance biosensor." *LWT-Food Science and Technology* **40**(2): 187–192.

Wang DeGuo, W. D., W. Y. Wang YongZhen, W. J. Wang JiaHua, Z. X. Zhang XiaoWei and X. F. Xiao FuGang (2011). "Rapid detection of viable Listeria monocytogenes in raw milk using loop-mediated isothermal amplification with the aid of ethidium monoazide." *Milchwissenschaft* **66**(4): 426–429.

Wingstrand, A., J. Neimann, J. Engberg, E. M. Nielsen, P. Gerner-Smidt, H. C. Wegener and K. Mølbak (2006). "Fresh chicken as main risk factor for campylobacteriosis, Denmark." *Emerging Infectious Diseases* **12**(2): 280.

Xu, M., R. Wang and Y. Li (2016). "Rapid detection of Escherichia coli O157: H7 and Salmonella Typhimurium in foods using an electrochemical immunosensor based on screen-printed interdigitated microelectrode and immunomagnetic separation." *Talanta* **148**: 200–208.

Yeni, F., S. Acar, Ö. Polat, Y. Soyer and H. Alpas (2014). "Rapid and standardized methods for detection of foodborne pathogens and mycotoxins on fresh produce." *Food Control* **40**: 359–367.

Yoo, J.-H., S.-M. Choi, J.-H. Choi, E.-Y. Kwon, C. Park and W.-S. Shin (2008). "Construction of internal control for the quantitative assay of Aspergillus fumigatus using real-time nucleic acid sequence-based amplification." *Diagnostic Microbiology and Infectious Disease* **60**(1): 121–124.

Yordanova, R. and A. Andonova (2024). "Rapid methods for detection of foodborne pathogens and their toxins-a review of current trends and future perspectives." *Journal of IMAB* **30**(2): 5587–5592.

Younts, S., E. Alocilja, W. Osburn, S. Marquie, J. Gray and D. Grooms (2003). "Experimental use of a gas sensor–based instrument for differentiation of escherichia coli O157: H7 from non-O157: H7 escherichia coli field isolates." *Journal of Food Protection* **66**(8): 1455–1458.

Younts, S., E. Alocilja, W. Osburn, S. Marquie and D. Grooms (2002). "Differentiation of Escherichia coli 0157: H7 from Non–0157: H7 E. coli serotypes using a gas sensor–based, computer–controlled." *Transactions of the ASAE* **45**(5): 1681.

Zhao, X., C.-W. Lin, J. Wang and D. H. Oh (2014). "Advances in rapid detection methods for foodborne pathogens." *Journal of Microbiology and Biotechnology* **24**(3): 297–312.

Zhao, X., L. Wang, Y. Li, Z. Xu, L. Li, X. He, Y. Liu, J. Wang and L. Yang (2011). "Development and application of a loop-mediated isothermal amplification method on rapid detection of Pseudomonas aeruginosa strains." *World Journal of Microbiology and Biotechnology* **27**: 181–184.

9 NMR Techniques in Food Product Analysis

Avik Mazumder, Manpreet Kaur, and Yashi Srivastava

9.1 INTRODUCTION

The selection of analytical methods depends on the sample type, desired sensitivity, operational flexibility, analyte properties, and matrix composition. Multiple techniques are often preferred for the selective and specific detection of analytes. This requires adequate sample quantity and appropriate sample preparation. Sample preparation involves processes and operations that make the sample suitable for analysis by a specific analytical technique. It serves multiple purposes, enhancing the performance of analytical instruments by improving key metrics such as ease of experimentation, specific detector response, recovery, limits of detection and quantification, repeatability, and reproducibility, even with varying matrix compositions (Figure 9.1). (Mazumdar & Dubey, 2013).

The required sensitivity of the analysis will dictate the selection of appropriate sample preparation and analytical techniques, as well as the type of analysis tool to be used. This includes considerations of sample quantity, the nature and concentration of analytes, and the nature and concentration of matrix components, including any matrix interferents.

In some instances, extraction and/or enrichment are necessary to address the effects of diverse and often unknown matrices. In other cases, it may be required to chemically convert analytes into suitable derivatives under simple reaction conditions with high efficiency and yield, preferably without forming by-products. These derivatives often exhibit better stability during downstream sample processing and analysis, different volatility (high volatility for GC-based techniques and low volatility for NMR spectroscopy), and enhanced detectability such as through isotope tagging followed by NMR (Mazumder et al., (2013). High resolution 19F{1H} nuclear magnetic resonance spectroscopy and liquid chromatography–solid phase extraction–offline 1 H nuclear magnetic resonance spectroscopy is widely used for conclusive detection and identification of cyanide in water samples (Mazumder et al., 2012)).

Additionally, they can improve the availability of analytes for downstream processing and analysis, as seen in the de-masking of analytes complexed with certain metal ions. They can also enhance the extraction efficiency of specific compounds, such as using zirconia SPE for organophosphorus compounds or molecularly imprinted polymers for a particular compound, allowing for concentration and adequate measurement. Sample preparation steps also help to enhance the detectability of analytes

DOI: 10.1201/9781003482611-9

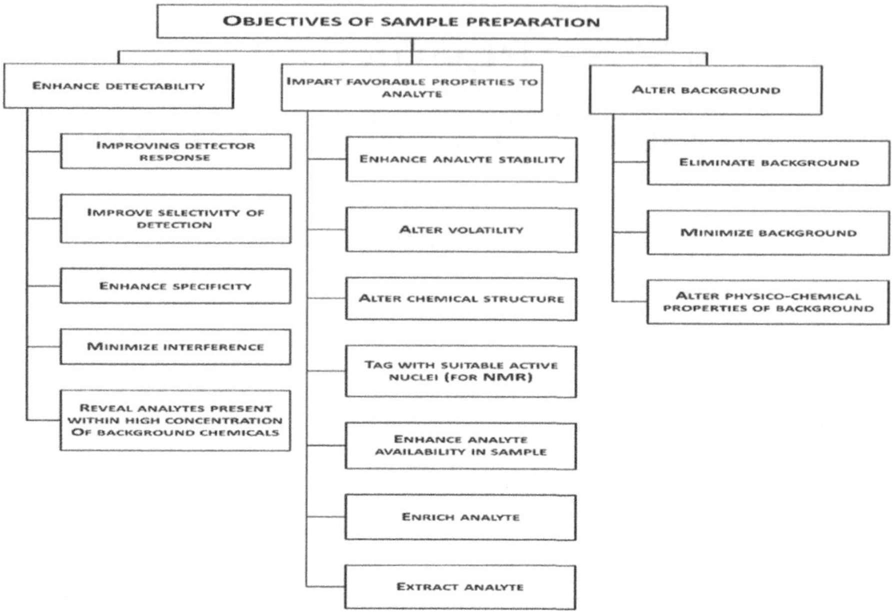

FIGURE 9.1 The different objectives that a sample preparation technique may need to achieve.

by improving detector response, selectivity, and specificity, and minimizing interference, as demonstrated by phosphorus-31 tagging of alcohols followed by NMR analysis (Figure 9.2).

9.2 ANALYSIS OF MIXTURES BY NMR SPECTROSCOPY: ADVANTAGES AND SHORTCOMINGS

Since Isidor Rabi recorded the first NMR spectrum in 1938, this analytical technique has undergone significant and rapid advancements. Chemists have employed NMR spectroscopy to gain structural and functional information about various compounds, complementing other analytical methods. NMR spectroscopy generates information such as resonance frequency, atom count, dipolar coupling, relaxation, diffusion properties, and the nuclear Overhauser effect (Figure 9.3).

In recent decades, there have been rapid advancements in high-field superconducting magnets, associated NMR hardware, pulse programs, data processing, and analysis software. These developments have enabled more accurate, precise, repeatable, and reproducible sample analysis within shorter time intervals. Consequently, NMR spectroscopy has found applications across various industries and research fields (Monakhova et al., 2014). Pure analytes (containing small and/or large molecules), complex mixtures, and complex systems (such as cells) can be categorized into different groups and their combinations (Figure 9.4).

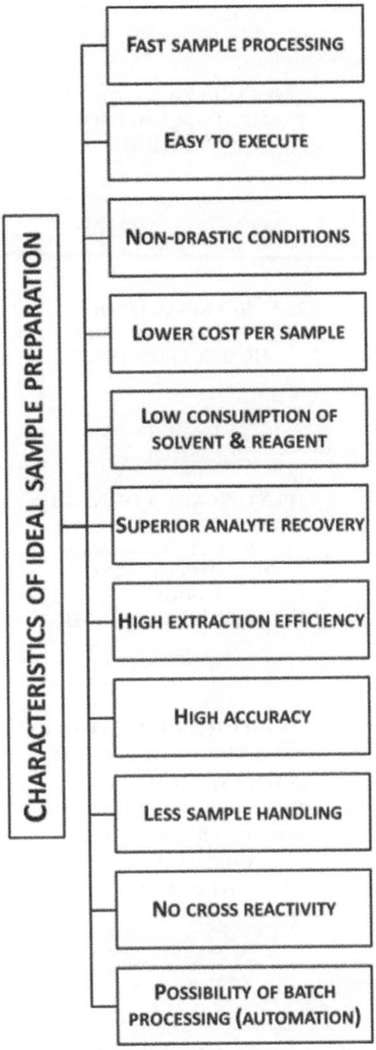

FIGURE 9.2 The desirable characteristics of an ideal sample preparation technique.

Despite its advantages, several challenges are associated with NMR spectroscopy of mixtures (Figure 9.5). The utility of the NMR detector response is maximized when experiments are conducted in quantitative (qNMR) mode. In this mode, the detector response is independent of the chemical nature, structure, and molecular weight of the analytes. Quantitative information can be obtained from NMR spectroscopy if the area under a signal (Ax) is proportional to the number of nuclei (NX) within the NMR probe's view generating the signal (Equation 9.1). The proportionality constant (Cs) is referred to as the spectrometer constant.

$$A_X = C_s \times N_X \quad (9.1)$$

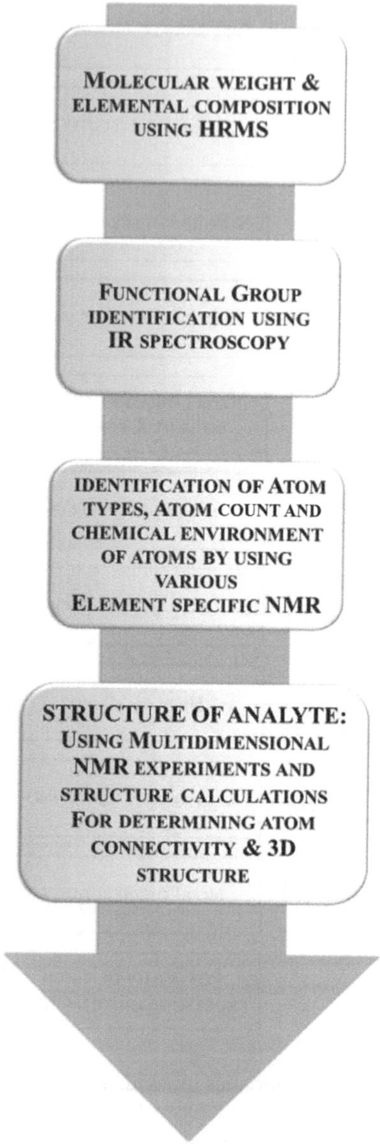

FIGURE 9.3 Structure elucidation of analytes using multiple spectroscopic techniques.

This capability of NMR spectroscopy confers a unique advantage for the analysis of pure analytes as well as the simultaneous detection and quantification of multiple components present. This can be done by using a single experiment in a nondestructive manner. The main advantage of NMR spectroscopy is the deeper insight into molecular and functional aspects of molecules. NMR can handle samples containing

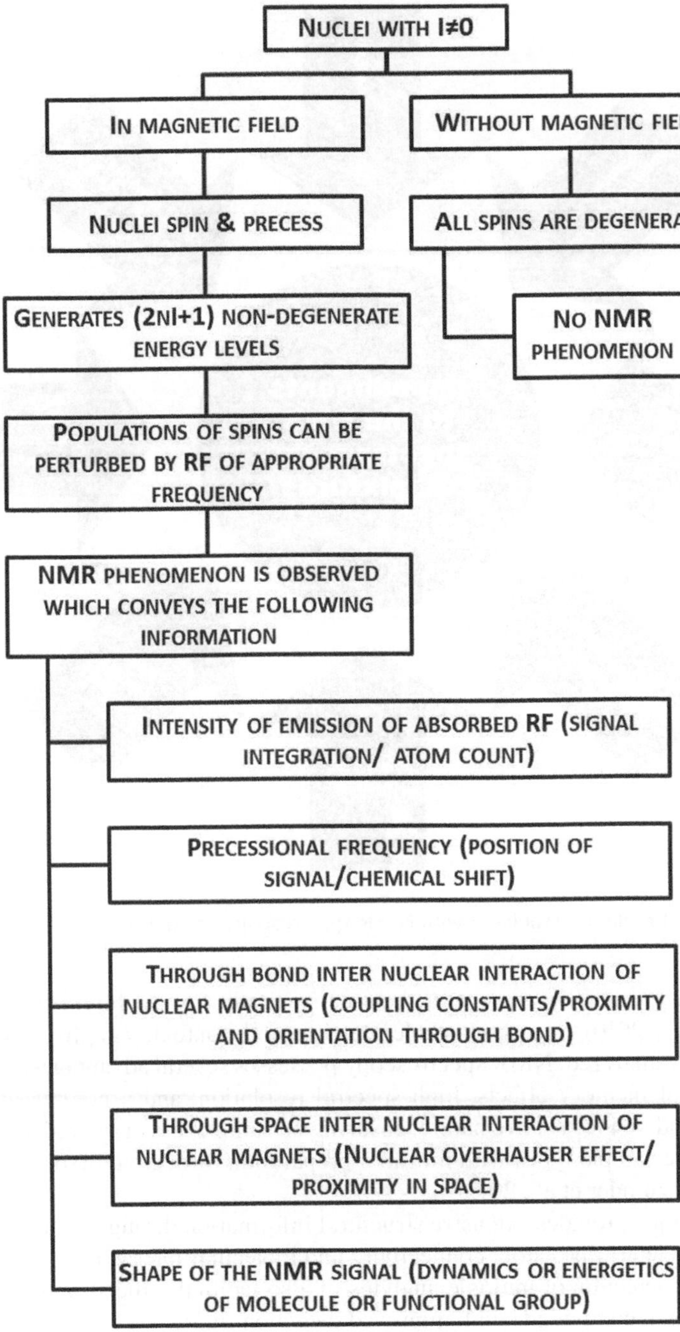

FIGURE 9.4 NMR phenomenon and observables used for the identification of analytes by NMR.

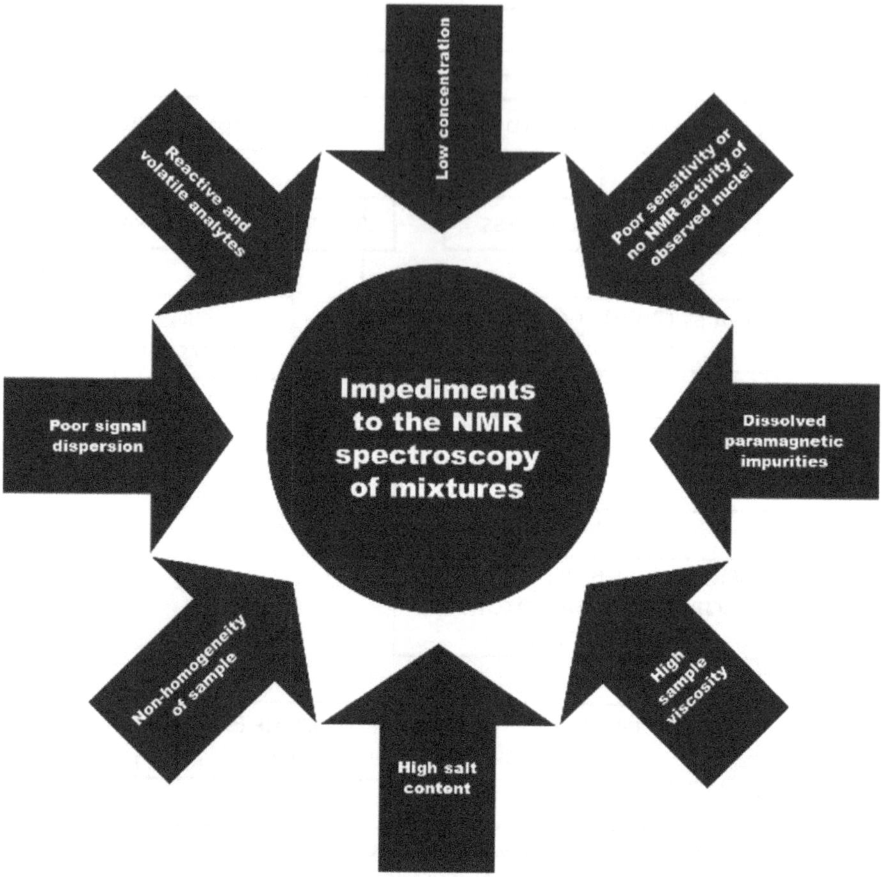

FIGURE 9.5 Problems associated with NMR spectroscopy of mixtures.

analytes that are mostly present in their native state. Solid / liquid (or solution) (Renault et al., 2010) / gaseous, stable/reactive, and nontoxic / highly toxic compounds can be analyzed. NMR spectroscopy possesses several advantages, including the absence of memory effects, high spectral resolution, and a noninvasive, non-aggressive, and nondestructive nature, ensuring no sample loss or degradation. It can handle diverse sample types with minimal preparation through a dissolve-and-scan approach (Mazumder et al., 2013).

The technique provides extensive structural information through isotope-specific observations, determines atom connectivity and molecular interactions, and enables simultaneous detection of multiple analytes. It also facilitates the structure elucidation of unknown compounds with improved detection limits.

Furthermore, NMR offers highly quantitative signal integration, absolute quantification using a single internal reference, excellent reproducibility irrespective of instrument or analyst, and supports high-throughput analysis.

Due to these advantages, scientific literature is replete with examples wherein tissues (and/or their extracts), body fluids, drugs, cosmetics, food beverages, adhesives, polymers, paint, pesticides, explosives, fertilizers, soil, sludge, fuels and lubricants, etc., have been subjected to NMR analysis. Classifying the analytes for an NMR spectroscopist is important before they are subjected to NMR spectroscopy (Chatterjee et al., 2016). The development of an analytical strategy for different types of samples is based on their physical state and solubility of the samples (Figure 9.6).

9.3 DESIRABLE AND UNDESIRABLE PROPERTIES OF THE SAMPLE OF NMR ANALYSIS

With the increasing diversity of samples presented for NMR analysis, it is becoming clear that the generation of reproducible NMR spectra, line shape, and acceptable signal-to-noise ratio for the identification of unknown species (particularly those present at low concentrations) pose major challenges for the analysis of complex, intricate, and diverse mixtures. NMR analysis sensitivity and spectral resolution vary from element to element. Moreover, these difficulties have thrown up new opportunities for researchers working in the field of NMR analysis of mixtures. These problems are compounded when quantitative (or semiquantitative) results have to be obtained from multiple blind samples having diverse matrix compositions, the molecular structure of analytes and matrix components (background chemicals and interferants). Changes in chemical shift, signal-to-noise ratio, and line shape are frequently encountered in such cases. Therefore, automated and high-throughput qNMR spectroscopic methods fail miserably for such mixtures. Therefore, sample preparation methods must be adopted to negate the effect of these interferences/variations presented by the samples. These sample preparation conditions are very important because they determine the quality of spectroscopic data. The factors, viz. pH, matrix components (salt content, background chemicals, and paramagnetic components), the analytes' stability, molecular interactions, viscosity, solubility, freezing point, chemical shifts, resolution, hydrophobicity, acidity/basicity, polarity, etc., are taken into account while performing sample preparation. This necessitates a thorough understanding of the NMR phenomenon, sample nature, and sample preparation method. This cumulative knowledge can be used to extract information about the molecules and their aggregates in the best possible manner. An added advantage could be derived if a single sample preparation method can make the sample/analytes amenable to multiple analytical techniques (Tables 9.1 and 9.2).

9.3.1 Sensitivity, Selectivity, and Resolution in NMR Spectroscopy

NMR sensitivity is influenced by signal-to-noise ratio. An increase in signal strength and a reduction of noise and linewidth of the NMR signals enhance sensitivity (Figure 9.7a and b). This chapter aims to provide a collective sample preparation technique that has been reported to date. Different types of analytes and matrices by NMR spectroscopy have been classified to understand different sample preparation techniques better. The classification-based approach will help to develop new sample preparation methods for new types of samples. Classification has been achieved on the basis of the properties of analytes, background chemicals, and matrices in light of the requirements of different NMR experiments (Figure 9.8).

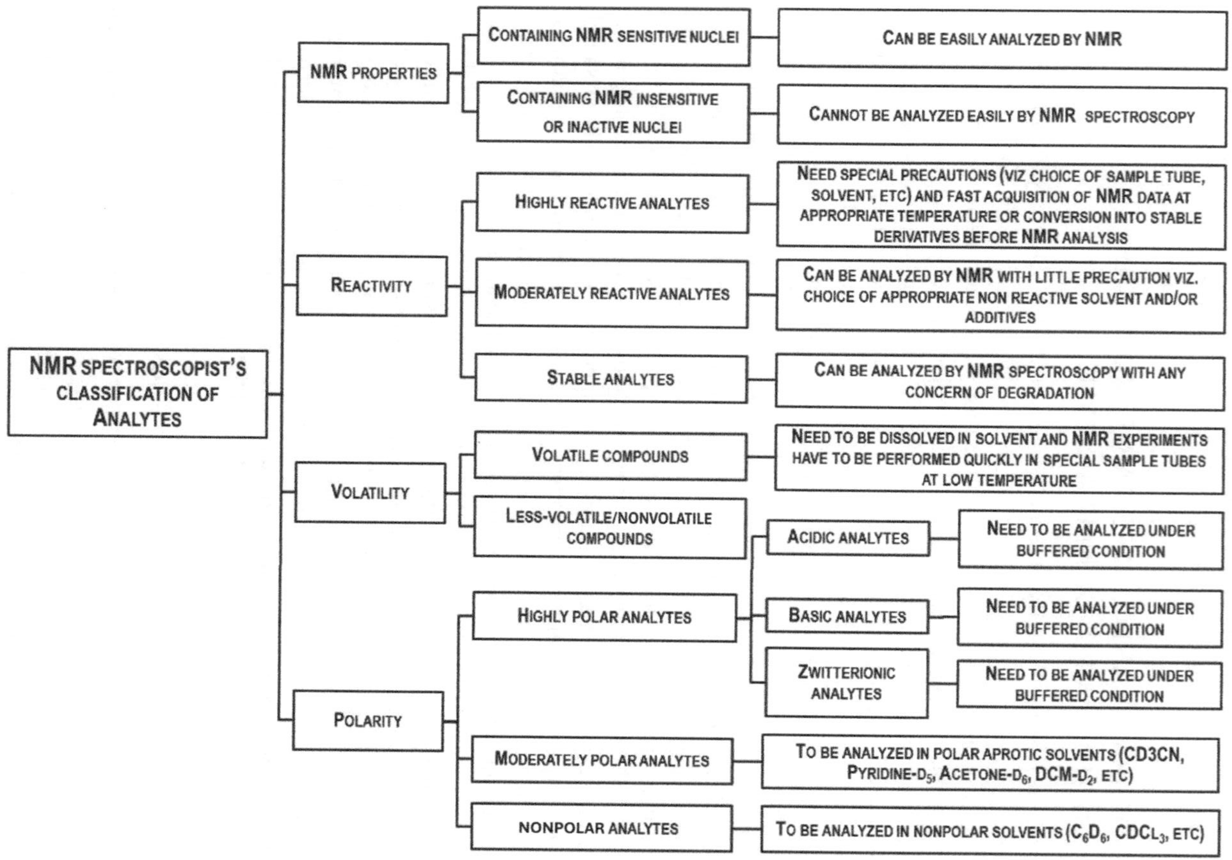

FIGURE 9.6 Classification of analytes and the required conditions of the analysis by NMR spectroscopy.

TABLE 9.1
An Account of NMR Hardware, Pulse Programs, Data Acquisition, and Processing Techniques Used to Boost Signal-to-Noise Ratio

S. N.	Brief Description of NMR Techniques	References
	Signal Enhancement Techniques	
1.	Simultaneous excitation of a wide range of excitation frequencies by delivering short radiofrequency pulses eliminated the need for frequency scanning. This has enabled the accumulation of multiple transients quickly, leading to a ≥ 100-fold increase in the sensitivity of NMR experiments.	Ernst (1987)
2.	The improvement in shimming technology has led to a higher active sample volume. This leads to a more homogeneous magnetic field in less time and a larger number of detectable NMR-active nuclei.	Claridge (2016)
3.	Acquiring more transients has traditionally been used to build upon the signal intensity and average out random noise. Increased experiment time is required for higher transients. Reduction of data collection time to achieve a given S/N is a powerful strategy adopted to enhance NMR sensitivity. Several pulse sequences have been developed for fast pulsing to obtain higher signal-to-noise per unit time. These include Ernst-angle excitation homo-decoupling in combination with multiple strategies to shorten experiment time without significant losses in S/N. These strategies lead to dramatic reductions in recycle delay duration.	Felli (2009), Koos (2017)
4.	As NMR signal-to-noise ratio is proportional to $\sim B_0^{3/2}$, efforts are on to breach the highest magnetic field barrier in order to increase population difference between the higher and lower energy states and thus the S/N.	Markiewicz (2006)
5.	The observable NMR parameters that are typically used are the ^{19}F longitudinal relaxation rate that have been used for improving the sensitivity and dynamic range of these experiments, especially when applied to fragment-based screening of drug candidates.	Dalvit (2007)
6.	Heteronuclear NOE (Nuclear Overhauser Effect) can be employed when sensitive and insensitive nuclei are in close spatial proximity. For instance, 1H saturation, producing steady-state NOE, leads to higher 13C sensitivity. In addition, INEPT (insensitive nuclei enhanced by polarization transfer) pulse scheme can be used to transfer magnetization from sensitive to insensitive nuclei in the presence of scalar coupling between the two. In many heteronuclear pulse sequences, INEPT is followed by reverse INEPT (transfer of magnetization back to the sensitive nuclei for detection). Here, the Preservation of Equivalent Pathways (PEP) method can be used to gain sensitivity improvements by factors up to relative to the typical reverse INEPT.	

(*Continued*)

TABLE 9.1 *(Continued)*
An Account of NMR Hardware, Pulse Programs, Data Acquisition, and Processing Techniques Used to Boost Signal-to-Noise Ratio

S. N.	Brief Description of NMR Techniques	References
	Signal Enhancement Techniques	
7.	Various isotopic labeling strategies have been employed to maximize NMR sensitivity, resolution, and/or selectivity for the NMR spectroscopy of large biomolecules or complex mixtures. These enriched molecules are no longer limited by natural abundance of the isotopes. Additionally, 2H-enrichment is being routinely used in protein NMR spectroscopy for reduction in dipolar cross-relaxation and spin diffusion rates.	Goto et al. (1999)
8.	Low volume probes NMR pulse sequences tailored to these isotope labeling schemes have rendered them maximally useful.	Jones (2012)
9.	In addition, NMR tubes matching the susceptibility of the glass to that of water (e.g., Shigemi tubes)	Voehler et al. (2006)
10.	Reduction in recycle delay using paramagnetic metal ions or their complexes	Cai et al. (2006)
11.	The total data collection time of multidimensional NMR experiments can be significantly shortened by a suite of approaches known as reduced sampling techniques. Hence, in terms of amount of useful data collected per unit time, reduced sampling techniques can be regarded as powerful tools. These approaches include nonuniform sampling (NUS), knowledge-based optimized data undersampling, and non-Fourier methods like Hadamard spectroscopy.	Kazimierczuk et al. (2010)
12.	Interestingly, some studies reported that, depending on the sampling strategy and reconstruction method, NUS may give rise to an actual increase in the S/N ratio and 'detection sensitivity', i.e., the probability of detecting weak peaks, compared to uniformly sampled experiments with an identical total data collection time. However, it needs to be noted that NUS often involves reconstruction procedures that require cautious monitoring of the reliability of the output spectral features.	Felli (2009), Coggins et al. (2010)
	Noise Reduction Techniques	
13.	The improvement in probe technology led to the use of receiver coils that are composed of materials with very small magnetic susceptibility. This has led to minimal distortions in the magnetic field, thereby improving line shapes and signal-to-noise.	Cavanagh (1997), Claridge (2016)
14.	The use of analog filters combined with time-domain oversampling followed by digital filtering led to eliminating aliased noise, contributing to improved sensitivity.	Moskau (2002)
15.	Recently, the introduction of cryogenic probes led to a significant reduction in noise level via cooling of receiver coils and preamplifiers down to approximately 20 K and 80 K, respectively	Molinski (2010)
16.	Use of anti-vibration pads on NMR magnets	

TABLE 9.2
Account of Sample Preparation Techniques Used for Different Types of Samples

S.N.	Sample Type/Sub-Type	Short Description of Sample Preparation Method	References
1.	Extracts of fruit pulp and seeds	SPE cartridges packed with specific sorbents can be used to extract compounds from fruit extracts, pulp, or seeds. Direct dissolution involves directly solubilizing fruit extracts, pulp, or seeds in appropriate NMR-compatible solvents.	Huang et al. (2019)
2.	Fruit juices	The juice was centrifuged for NMR analysis, and the supernatant was mixed with deuterated phosphate buffer containing TSP. After standing and a second centrifugation, the supernatant is transferred into an NMR tube.	Pei et al. (2022)
3.	Strawberry grapes	Approximately 1 kg of grapes were separated into peel, pulp, seed, leaf, and stalk components, lyophilized and placed in Falcon tubes (50 mL), extracted by sonication using pure methanol and centrifuged at 3,500 rpm for 10 min. After removal of the solvent, samples of methanol extract (40.0 mg) was dissolved in 1.0 mL of a solvent system made up of K_2HPO_4 buffer (pH 6.0, 90 mM) in D_2O (containing the internal standard) and methanol-d4 (1:1 v/v) and analyzed by NMR using trimethylsilyl propionic acid sodium salt (TSP, 0.01%, w/v) as internal standard.	Pacifico et al. (2011)
4.	Wine grapes	Randomly harvested intact bunches were rapidly frozen in liquid nitrogen, protected from light and then stored at −80 deg C. Accurately weighed (~500 mg) was ground in a mortar under liquid nitrogen, and seeds were manually removed before weighing. Polar metabolites were extracted using perchloric acid (PCA, 65%). Liquid nitrogen was poured to the mortar containing pulverized frozen berry, until 1 mL of ice-cold 3 M $HClO_4$ was added to the powder. Then, the extract was allowed to thaw on ice for 15 minutes. The next steps were achieved by accurately	Mulas et al. (2011)
5.	Coconut water	A 3.0 mL aliquot of the samples was centrifuged at 605 g for 15 min. A 130 μL of the supernatant was mixed with 14 mM of EDTA in 350 μL of $CD_3OD\text{-}d_4$ containing 1% of sodium-3-trimethylsilylpropionate (TMSP-2,2,3,3-d_4 98% purity), and taken in 5 mm NMR tubes. The EDTA was added to minimize the ionic strength effect on frequency shifts in the NMR spectra.	Sucupira et al. (2017)
6.	Honey, nectar, sap	Each genuine honey sample has to be preheated (50°C for 2 h) until all crystals are dissolved. This has to be followed by stirring for homogenization. It is also essential to reduce the viscosity of the honey for liquid-state NMR analysis. Therefore an appropriate dilution 1.0 to 6.0 has to be performed by using 99.9% D_2O or phosphate buffer solution followed by homogenization using a vortex shaker. The pD has to be adjusted to an appropriate value before transferring it to an NMR tube for analysis.	Schievano (2017), Spiteri (2017)

(*Continued*)

TABLE 9.2 *(Continued)*
Account of Sample Preparation Techniques Used for Different Types of Samples

S.N.	Sample Type/Sub-Type	Short Description of Sample Preparation Method	References
7.	Milk	The metabolomics analysis of milk requires conservation of all the metabolites during sample preparation. Hence, sample preparation is performed by skimming the raw milk samples and passing them through a molecular weight cut-off filter with a 10 kDa threshold in order to remove lipids and proteins.	Sundekilde et al. (2013)
8.	Cheese	After weighing, the cheese sample was added to the NMR nanotube, followed by adding 20 μL of 2 mM TSP-d4 solution in D_2O, and then sealing the nanotube, well mixed and ready for NMR analysis.	Kandasamy et al. (2020)
9.	Ghee and butter	The sample (10 mg) was dissolved in deuterated methanol with 0.5% TMS, sonicated for 1 hour, and left to solubilize for 12 hours at room temperature. After centrifugation for 5 minutes to remove insoluble components, clear supernatants were transferred to NMR tubes for analysis.	Ahmad Fadzillah et al. (2015)
10.	Yoghurt	For NMR analysis, yogurt with 10% D_2O was placed into 5-mm tubes (0.6 mL each), and samples were prepared. The mixture, made from 1 g of Bulgarian yogurt starter and 1 L of milk with D_2O, was fermented at 40°C for 24 hours in the NMR equipment.	Lu et al. (2016)
11.	Alcoholic drinks and beverages	The selection of appropriate solvents is crucial in sample preparation for NMR spectroscopy of alcoholic drinks and beverages. Deuterated solvents, such as deuterated water (D_2O) and deuterated methanol (CD_3OD), are commonly employed to dissolve the samples.	
12.	Nonalcoholic drinks and beverages	Filtration techniques, such as vacuum filtration or centrifugation followed by filtration, are commonly employed to remove solid particles and achieve sample clarity. Dilution with deuterated solvents, such as deuterium oxide (D_2O) or deuterated dimethyl sulfoxide (DMSO-d6), can lower the sample viscosity and improve NMR spectra quality. Adjusting the pH to a consistent level using acid or base solutions can enhance the reproducibility and comparability of NMR data.	Rocha et al. (2020), Silva et al. (2021)
13.	Natural and artificial sweeteners	Samples were prepared in triplicate in DMSO-d6 using a volumetric flask, then transferred to 5 mm NMR tubes and analyzed after 48 hours.	Paniagua-Vega et al. (2019)

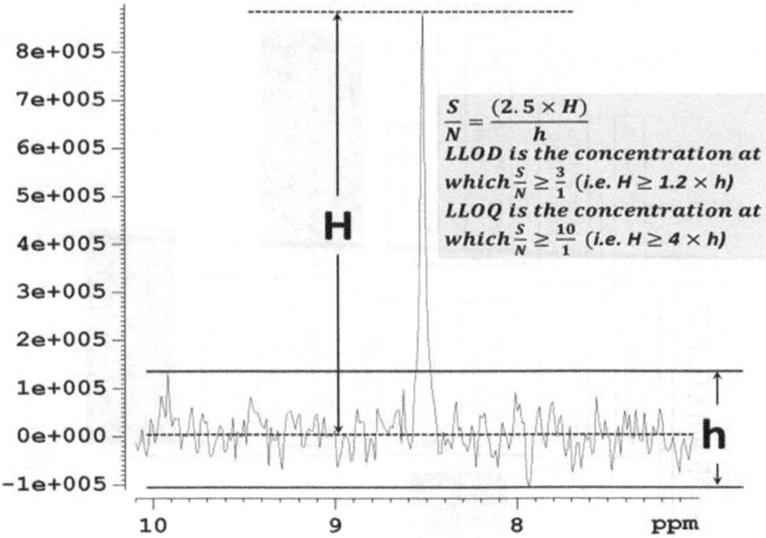

FIGURE 9.7 (A) Determination of signal-to-noise ratio in NMR spectroscopy is performed by using the above formula, and it is used for the determination of LLOD and LLOQ.

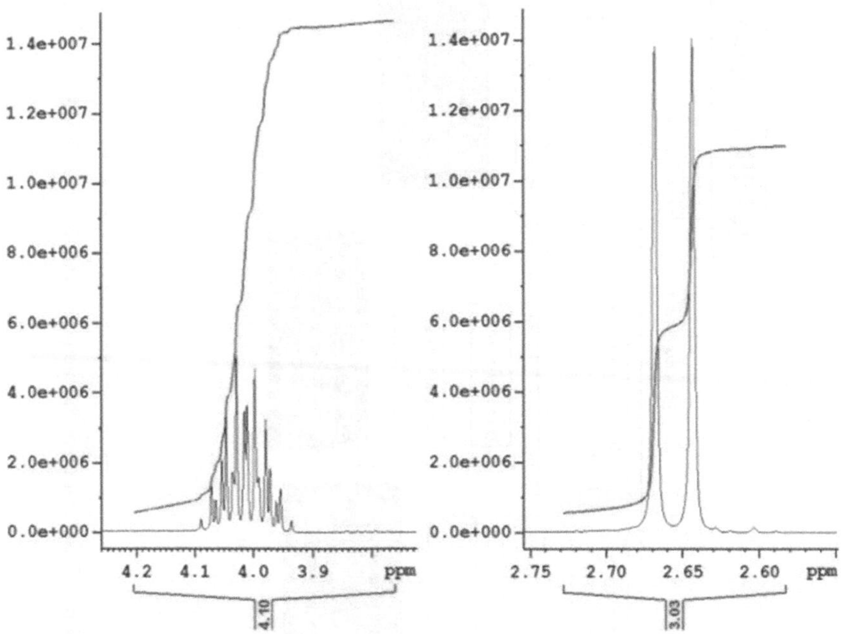

FIGURE 9.7 (B) Even though the proton count is higher in the left signal, greater signal.

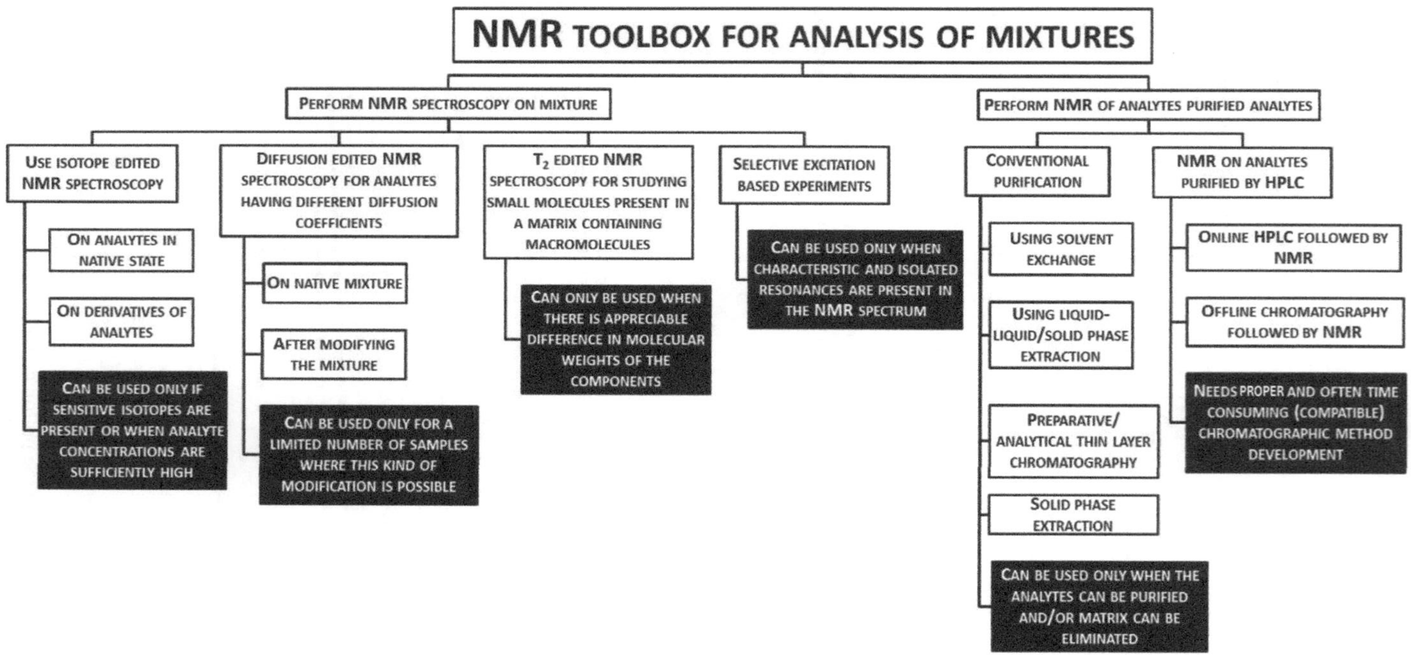

FIGURE 9.8 Classification of the methods available to an NMR spectroscopist for the analysis of mixtures.

9.4 OBJECTIVE OF SAMPLE PREPARATION FOR LIQUID-STATE NMR SPECTROSCOPY

Sample preparation for liquid-state NMR spectroscopy is needed to make the sample homogeneous (by removal of insoluble components), removal or (site-specific) introduction of paramagnetic species, removal of high/ low molecular weight components, selective enrichment of analytes, removal of background/interfering chemicals for easy detection/identification, exchanging matrix with one that facilitates analysis (e.g. by reduction of viscosity), converting unstable analytes into their stable derivatives, introducing one (more) NMR active/sensitive nuclei, converting analytes to a form that is amenable to purification/enrichment, perturbation of intermolecular interactions to study them minutely. An ideal sample preparation method should allow unambiguous identification and/or quantification of analytes with minimal loss of quantity and information about the analyte. The microenvironment of the nuclei affects the NMR spectrum quality. Therefore, unlike other types of spectroscopies, to reveal all information related to the nuclear microenvironment of the nuclei, NMR samples must be prepared in such a way that:

(a) The samples are easy to prepare under reproducible experimental conditions.
(b) The sample should not decompose, evaporate, precipitate, solidify, evolve gases, or cause corrosion of the NMR sample tube.
(c) The samples contain a maximum number of NMR active nuclei.
(d) Concentrations of the analytes do not vary widely in the sample.
(e) Magnetic field homogeneity and stability are maintained during the entire period of analysis.
(f) If physical/chemical changes are to be studied, the timescale of such changes should be sufficiently slow so that they are observable by NMR spectroscopy.
(g) Spins have short T1 (to allow faster repetition of scan) and long T2 (so that the NMR signals are sharp).

9.5 NMR SAMPLING DEVICE AND SAMPLE HANDLING ACCESSORIES

Selection of the right construction material is very important, especially for boron NMR where PTFE insert or a Kel-F tube is used. The NMR tubes are graded according to their concentricity and camber. The tubes with large concentricity values indicate a variety of tube wall dimensions around the circumference (Table 9.1). This leads to non-homogeneity of the magnetic field when the tube is in spinning mode. On the other hand, tubes with large camber values cause not only spinning sidebands and increased shimming times but also wobbling of the spinning sample and possible degradation of the NMR spectrum. These tubes can also mechanically damage the probe's radio frequency coil. However, high-end spectrometers use low-end tubes for routine NMR or scouting experiments. Although the NMR tubes are mostly manufactured in a way that they are completely free of inorganic and organic impurities (viz. lubricants, organic solvents, reagents) that may lead to sample contamination and/or background signals, the tubes should invariably be rinsed with an appropriate organic solvent and air-dried before use. To prevent contamination and/or loss of analytes, the NMR tubes may be sealed by heating them below the open end using a Bunsen burner or a propane torch. If volatile, explosive, or flammable solvents/analytes are present, care should be taken to prevent the contents from catching fire. In such cases, the contents must be cooled by dipping an inappropriate cold bath before sealing.

9.5.1 Use the Correct Volume and Concentration of Analyte Solution

The NMR signal intensity depends upon the number of atoms observed and the molecular weight of the analyte. For half the quantity of material, the data accumulation time is quadrupled. Generally, decent ^{1}H-NMR spectra of organic compounds (except for macromolecules) can be recorded using 5 to 25 mg of a sample.

9.5.2 Screw Cap versus a JY Valve NMR Tube

These tubes are used to protect the sample from atmospheric gases. The JY valved tube has a sealing ability under vacuum and pressure applications. The screw cap tube should only be used to seal samples under ambient pressure and temperature conditions. If very low concentrations are used, the peaks from common contaminants dominate the spectrum, especially when the spectra are recorded on high-field instruments or using highly sensitive (cryo) probes. On the other hand, with increased concentrations, solution viscosity may increase, resulting in broader lines (than one obtained from a more dilute solution). In addition to this, intermolecular associations/dissociations of and freezing point (especially in solvents like DMSO) and boiling point (especially in solvents like dichloromethane) are also affected by analyte concentration. These concentration-dependent changes may necessitate the experiments to be carried out in known and constant concentrations of analytes. As a good rule of thumb, it is better to use higher concentrations, such as to form a saturated solution, if qualitative 13C{1H} experiments are to be performed. The sample should be homogeneous and free from all particulate matter. Homogeneous magnetic and magnetic susceptibility are of utmost importance in generating well-defined energy levels of the nuclear spin states.

9.5.3 Make the Sample to the Correct Depth

NMR signals are generated only from the spins within NMR probe field view (which is smaller than the total sample volume) (Figure 9.9). The shim controls of the spectrometers are used to cancel the distortions of magnetic field homogeneity along the length of the sample magnet, especially at the upper and lower ends of a test sample. The NMR samples must have similar depths and physically resemble the test samples. Samples with longer and shorter sample heights are very difficult to shim, and they require considerable time to record the spectra, leading to the wastage of expensive deuterated solvents.

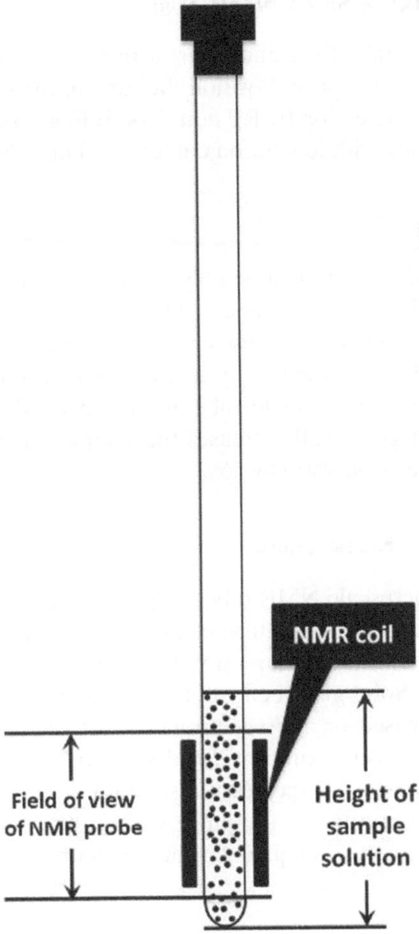

FIGURE 9.9 Schematic diagram of NMR probe setup with sample solution.

9.5.4 NMR Tubes

An NMR tube contains a liquid sample in a cylindrical volume within a magnetic field, rotating at 25 Hz. Its quality is defined by how closely it aligns with the ideal mechanical structure. Although terms like 'Camber' and 'Concentricity' aren't commonly linked to NMR performance metrics like 'Sensitivity' or 'Resolution', understanding Wilmad's structural specifications can help relate them to NMR performance. OD and ID measure a tube's outer and inner diameters, respectively. These diameters are assessed at multiple points along the tube's length and circumference. Wilmad uses fluorometric methods to ensure these dimensions meet stringent standards. Strict adherence to diameter tolerances is crucial for optimal NMR performance.

9.5.5 Si NMR Using a Silica NMR Tube

Place the sample into a tube liner and insert it into the NMR tube; secure it with a rubber O-ring to prevent slippage. Position the tube in the spinner turbine and then place it in a magnet for detection by RF coil. Norell Fluoropolymer NMR tube liner TL-5–7 and quartz tubes with low boron content for boron NMR could be used.

9.5.6 Shigemi Tube

The Shigemi advanced microtube assembly consists of an outer tube and an insert that can be used with NMR tube spinners. They are used when a small solvent volume or less dilution is needed for higher sample concentration. These advanced microtubes have low RF inhomogeneity as they make use of a combination of an improved space-filling 5 mm sample tube and an insert that exactly fits the outer sample tube. A thinner glass wall increases the sample capacity held in this region, enhancing the signal-to-noise ratio by 16%.

9.5.7 Salt-Tolerant NMR Tubes

A Shigemi tube is a microscale NMR tube designed to accommodate smaller sample volumes in ordinary-size NMR spectrometers, where sample availability is limited. The smaller solvent volume helps maintain a higher sample concentration despite the reduced sample depth. Solid glass beneath the sample level compensates to prevent interference, varying based on the spectrometer. After expelling air bubbles, the plunger secures the tube using parafilm. Ideally, the tube and the deuterated solvent are matched to achieve optimal spectrum resolution. The Shigemi Slot Tube offers optimal sensitivity and minimizes the pulse width (90°), making it ideal for challenging samples like proteins in aqueous solutions with high electrolyte concentrations, such as salts.

9.5.8 Screw Cap NMR Tubes

The NMR tube (7 inches) features an open-top, replaceable screw-thread cap with a septum (PTFE/silicone), allowing for sample or reactant addition via syringe and enabling sample storage, making it excellent for air-sensitive samples and in-cell perfusion studies. After use, clean by rinsing with acetone or a suitable solvent, followed by drying with dry air or nitrogen-avoid drying in a hot oven to prevent solvent vapors from impacting spectrum quality. Always cap tubes securely and do not use tubes with chipped or broken tops due to safety risks and potential splintering. Proper cleaning procedures can help to preserve sample integrity and prevent contamination (Fulmer et al., 2010).

Using scratched NMR tubes can lead to broadened lines and reduced resolution due to variations in the magnetic field within the sample. This necessitates frequent spectrometer reshimming for each sample, which is tedious and counteracts the purpose of investing in high-quality tubes designed to avoid such issues. Cleaning NMR tubes can range from straightforward to challenging, depending on the sample type.

Regardless of the cleaning complexity, through rinsing, is crucial to ensure proper tube maintenance.

9.6 SAMPLE PREPARATION METHODS FOR NMR

The ^1H NMR experiments make use of a small sample (typically 1–5 mg) that is dissolved in a deuterated solvent like $CDCl_3$, filling the tube to 4–5 cm depth (determined by field of view of the NMR coil). Protein NMR typically uses H_2O (90%)/ D_2O mixture (10%). The sample can be sonicated or agitated for dissolution, with solids filtered through celite into the NMR tube. Sealing options include polyethene caps, flame sealing, Teflon 'Schlenk' taps, or small rubber septa, often wrapped with parafilm to minimize solvent evaporation. Although there are several methods for preparing samples for NMR spectroscopy, explained below are a select few techniques that have prominently figured in the scientific literature.

9.6.1 Precipitation Followed by Decantation and/or Filtration

This method separates low molecular weight components from macromolecules present in mixtures. The process depends upon playing with the factors that control the solubility of macromolecules. Their solubility in a solvent requires extensive penetration of the solvent molecules in the macromolecules and overcoming the cohesive forces between them. This enables the separation of the solute molecules from one another gradually. Solvent-macromolecular solute stability is due to favorable solute-solvent interactions, resulting in negative free energy of mixing (ΔG), driven by positive entropy (ΔS) and determined by the enthalpy (ΔH), especially for polar, hydrophilic groups. The specific solvent-solute interaction involves more energy change and better intermolecular aggregation. Their solvation, leading to complete solute dispersion, often results in high-viscosity solutions, with macromolecules swelling and dissolving in polar solvents like water, alcohol-water mixtures, dimethyl formamide, and dimethyl sulfoxide. However, cross-linked or network polymers, whether polar or nonpolar, show poor solubility in these solvents. Due to lower symmetry, usually more flexibility and less crystallizability, the branched macromolecules are more soluble. Cross-linking counterparts acquire appreciable changes in shape and size due to their enhanced rigidity and density. Extensive intramolecular/intermolecular hydrogen bonding in linear polymers leads to high crystallinity and insolubility in water and most solvents. Nonpolar linear and branched macromolecules are typically soluble in similar nonpolar solvents. In similar conditions, higher molecular weight polymers have higher solution viscosity. Branching decreases molecular symmetry and increases chain flexibility, enhancing solvent penetration, swelling, and dissolution.

Three different sample preparation techniques are used to remove high molecular weight interferants: precipitation (PT), solid-phase extraction (SPE), and hybrid SPE technique. Among the high molecular weight interferants, proteins, carbohydrates, and polymers are worth mentioning. Their chemical breakdown into low molecular weight fragments makes sample handling easier.

9.6.2 Applications of Precipitation in NMR Spectroscopy

9.6.2.1 Natural Product Analysis

Precipitation is widely employed in isolating and concentrating natural products from complex mixtures. By enhancing the concentration of target compounds, precipitation enables detailed structural characterization using NMR spectroscopy, contributing to natural product discovery and drug development (Hong, 2015; Liu, 2019).

9.6.2.2 Metabolomics

In metabolomics studies, precipitation extracts and concentrated metabolites, combined with NMR spectroscopy, enable the identification and quantification of metabolites, leading to insights in biomarker discovery and metabolic pathway analysis.

9.7 CENTRIFUGATION

Centrifugation is a widely employed sample preparation technique in NMR spectroscopy, offering significant advantages for analyzing molecular structures and dynamics.

A centrifuge is a device that separates particles from solution by applying centrifuge force using a rotor. The two types of centrifuge procedure include preparative (isolate particles) and analytical (measure physical properties)

9.7.1 Applications of Centrifugation in NMR Spectroscopy

9.7.1.1 Cellular and Subcellular Analysis

Centrifugation plays a vital role in isolating and purifying cellular components, including nuclei, mitochondria, and membrane vesicles, for NMR analysis. It enables the investigation of cellular metabolites, biomolecular interactions, and cellular dynamics, contributing to fields like cell biology and biochemistry.

9.7.1.2 Protein Structure and Dynamics

Centrifugation-based techniques, such as analytical ultracentrifugation, offer valuable insights into protein folding, stability, and interactions. Centrifugation aids in NMR-based structural and functional studies of proteins by separating protein species based on their sedimentation rates.

9.7.1.3 Lipidomics

Centrifugation is utilized to isolate and fractionate lipids from complex lipid mixtures. It enables the characterization of lipid classes and their compositions using NMR spectroscopy, contributing to lipidomics research and lipid-related disease studies.

9.8 LIQUID-LIQUID EXTRACTION (LLE)

LLE is a sample preparation technique extensively employed in NMR spectroscopy. Thermodynamics of LLE can be described as the equilibrium distribution of an analyte between two immiscible liquid phases in liquid-liquid extraction. It can be described by distribution ratio (D), given by $[A]_{org} / [A]_{aq}$,

where, $[A]_{org}$ represents analyte concentrations in the organic phase and $[A]_{aq}$ analyte concentrations in aqueous phase.

The partition coefficient (K) is related to D through the equation $K = D / (1 - D)$. Thermodynamic parameters can be determined using the Van't Hoff equation: $\ln(K) = -\Delta G^\circ / (RT) = \Delta S^\circ / R—\Delta H^\circ / (RT)$

where, ΔG° represents standard Gibbs free energy change; ΔH° represents enthalpy change; ΔS° represents entropy change; R represents gas constant; and T is the temperature in Kelvin.

The thermodynamic characterization of LLE provides insights into the energetics and spontaneity of the process.

$$D = [A]_{org}/[A]_{aq}$$
$$K = D/(1 - D)$$

Where K is ratio of the analyte concentrations in the organic and aqueous phases at equilibrium.

The extraction methods are based on extraction by chelate formation, solvation extraction, ion-pair (ion association) formation, and synergistic extraction.

9.8.1 Specific Examples and Applications

9.8.1.1 Extraction of Natural Products

LLE is crucial in isolating natural products from plant extracts or complex biological matrices. By selecting appropriate organic solvents, pH conditions, and extraction techniques, LLE allows for the concentration, purification, and enrichment of natural products, enabling their structural elucidation using NMR spectroscopy.

9.8.1.2 Metabolite Profiling

Liquid-liquid extraction is widely employed in metabolomics studies to separate and concentrate metabolites from biological fluids. By exploiting the differential partitioning behavior of metabolites, liquid-liquid extraction facilitates the identification, quantification, and profiling of metabolites, providing valuable insights into metabolic pathways and disease biomarkers.

9.9 SOLID PHASE EXTRACTION (SPE)

SPE is vital in sample preparation for NMR spectroscopy.

9.9.1 Principles of SPE

SPE involves the selective adsorption of target analytes onto a solid sorbent material while matrix components are washed away. The efficiency of SPE materials is influenced by various factors, including the nature of the analytes, sorbent material properties (such as surface area, pore size, and functional groups), and the morphology of the sorbent (such as particle size and shape). Understanding these aspects is crucial for optimizing the extraction process for NMR spectroscopy. The physicochemical properties of analytes, such as their polarity, molecular weight, and functional groups, dictate their interaction with the solid sorbent material. Hydrophobic analytes tend to interact more favorably with nonpolar sorbents, while polar analytes interact well with polar sorbents. The choice of sorbent material should be tailored to the analytes of interest to achieve optimal extraction efficiency.

9.9.2 Specific Examples and Applications

9.9.2.1 Porous Materials

Porous materials, such as activated carbon, mesoporous silica, or polymer-based materials, offer high surface areas and well-defined pore structures, enabling an efficient extraction of organic pollutants, pesticides, or pharmaceuticals from environmental samples for NMR spectroscopy analysis (Jing et al., 2016; Chen et al., 2020).

9.9.2.2 Functionalized Sorbents

Functionalized sorbents, such as C18-bonded silica or immune affinity materials, exhibit selective interactions with target analytes, allowing for the extraction of specific compounds, including drugs, metabolites, or biomarkers, from complex biological matrices for NMR analysis (Ahmed et al., 2014; Annesini et al., 2020).

9.10 SOLVENT-FREE EXTRACTION

Solvent-free extraction, also known as solvent less extraction, is a sample preparation technique that eliminates the use of liquid solvents, providing an environmentally friendly approach for preparing samples for NMR spectroscopy.

Solvent-free extraction techniques, such as SPME, thermal desorption, and headspace sampling, provide efficient solventless approaches for sample preparation in NMR spectroscopy (Pawliszyn, 2011).

9.10.1 Types of Solvent-Free Extraction Techniques

9.10.1.1 Solid-Phase Microextraction (SPME)

SPME involves extracting analytes from a sample matrix using a solid-phase sorbent coated on a fiber. The sorbent selectively adsorbs or absorbs analytes, allowing their subsequent desorption for NMR analysis. Depending on the application and analyte properties, SPME is performed in different modes such as direct immersion SPME, headspace SPME, or in-tube SPME.

9.10.1.2 Thermal Desorption

Thermal desorption is based on the principle of heating a solid sample to volatilize analytes of interest, which are then collected and introduced into the gas probe of NMR instrument, or it can be absorbed in an appropriate solvent before NMR analysis of the solution in the usual manner. This technique is particularly useful for analyzing volatile or semi-volatile compounds such as environmental pollutants or aroma compounds.

9.10.1.3 Headspace Sampling

Headspace sampling involves the analysis of the vapor phase above a sample matrix. The headspace gas contains volatile compounds in equilibrium with the sample, and it can be directly adsorbed onto an adsorbent. The adsorbent material can then be subjected to extraction into a deuterated NMR solvent and analyzed in the usual manner. Headspace sampling is commonly used to analyze volatile compounds in liquid or solid sample, such as essential oils or food products.

9.10.1.4 Theoretical Principles of Solvent-Free Extraction

Solvent-free extraction relies on the use of physical or mechanical processes to extract analytes from solid samples, eliminating the need for liquid solvents. Techniques such as SPME, headspace sampling, or thermal desorption are commonly employed. The principles underlying solvent-free extraction include analyte partitioning between the sample matrix and the gas phase or sorbent phase, enabling subsequent analysis by NMR spectroscopy.

Partition coefficient (K) is defined as:

$$K = C_{gas} / C_{sample},$$

Where K is the ratio of the concentration of analyte in the gas phase (C_{gas}) to the concentration in the sample matrix (C_{sample}). It describes the partitioning behavior of the analyte between the sample and the gas phase.

9.10.1.5 Applications of Solvent-Free Extraction for NMR Spectroscopy

9.10.1.5.1 Analysis of Volatile Compounds

Solvent-free extraction techniques are commonly employed to analyze volatile compounds, such as aroma compounds, in various matrices. SPME, in particular, has found wide applications in analyzing volatile compounds in beverages, essential oils, and food products. For instance, SPME coupled with NMR spectroscopy is used to analyze the volatile profile of wines, providing valuable information on their composition and quality. Additionally, headspace sampling techniques have been employed to analyze volatiles in food products such as coffee, chocolate, and spices.

9.10.1.5.2 Environmental Monitoring

Solvent-free extraction techniques are crucial in environmental monitoring and analysis of volatile organic compounds (VOCs) in air, water, and soil samples. Thermal desorption coupled with NMR spectroscopy enables the analysis of VOCs emitted from environmental sources, providing insights into air quality, pollution sources, and potential health risks. These techniques have been utilized in studies focusing on monitoring pollutants in urban environments, identifying emission sources, and evaluating remediation strategies.

9.10.1.5.3 Quality Control and Authentication of Natural Products

Solvent-free extraction techniques, such as SPME, are used for quality control and authentication of natural products, including herbs, spices, and medicinal plants. Extracting volatile compounds facilitates the identification of characteristic markers, which can be correlated with the origin, processing, or adulteration of the natural products. NMR spectroscopy combined with SPME has been used to analyze and authenticate essential oils, providing valuable information on their chemical composition and geographical origin.

9.10.1.5.4 Metabolomics and Biomarker Discovery

Solvent-free extraction techniques have gained prominence in metabolomics studies, enabling the analysis of metabolites in biological samples. SPME has been utilized to extract and concentrate metabolites from biological fluids for subsequent NMR spectroscopy analysis. These techniques facilitate detecting and identifying metabolites, leading to insights into metabolic pathways, disease biomarkers, and drug metabolism.

9.11 ISOTOPE TAGGING

Isotope tagging involves incorporating stable isotopes into the analyte molecule, enhancing sensitivity, and providing unique spectral signatures in NMR spectroscopy. Isotope tagging can be achieved through biosynthesis using labeled precursors or through chemical synthesis methods. Commonly used isotopes include ^{13}C, ^{15}N, and ^{2}H. Incorporating isotopes allows for isotopic enrichment, improved spectral resolution, and facilitates tracking metabolic pathways. Isotope labeling increases the sensitivity of NMR spectroscopy by providing amplified signals, enabling the detection of trace compounds, and improving spectral quality. Isotope tagging offers distinct spectral signatures, facilitating peak assignment and determination of molecular structures, particularly in complex mixtures.

9.12 SAMPLE PREPARATION FOR VARIOUS FOOD PRODUCTS FOR NMR SPECTROSCOPY

9.12.1 Fruit Extract, Pulp, and Seeds

9.12.1.1 LLE

LLE is a widely used technique for separating and concentrating target analytes from complex mixtures. By selecting appropriate organic solvents, specific compounds can be extracted selectively from fruit extracts, pulp, or seeds. The partition coefficient (K) can be calculated using the equation $K = D / (1 - D)$, where D is the distribution ratio of the analyte between the organic and aqueous phases. LLE allows for matrix removal and enrichment of target compounds, enhancing the sensitivity and spectral quality (Azmir et al., 2013).

9.12.1.2 SPE

SPE is a sample preparation method utilizing solid sorbents to retain analytes from a liquid sample selectively. It offers advantages such as high selectivity, simplicity, and compatibility with automated systems. SPE cartridges packed with specific sorbents can be used to extract compounds from fruit extracts, pulp, or seeds. The choice of sorbent material and elution solvent allows for customization of the extraction process, enabling efficient removal of interfering components.

9.12.1.3 Direct Dissolution

Direct dissolution involves directly solubilizing fruit extracts, pulp, or seeds in appropriate NMR-compatible solvents. This method is simple and time-efficient, requiring minimal sample preparation steps. However, it may result in overlapping signals and poor spectral resolution due to the presence of matrix components and high molecular weight compounds.

9.12.1.4 Advantages and Limitations

Liquid-liquid extraction provides selective separation, matrix removal, and analyte enrichment. It is advantageous for analyzing low-abundance compounds, metabolites, and natural products in fruit samples. However, effective analyte recovery requires a careful optimization of extraction solvents, pH, and temperature.

SPE offers high selectivity, reproducibility, and automation potential. It enables the extraction of target compounds with reduced interference from matrix components. However, SPE can be time-consuming, and the choice of sorbent and elution solvent must be carefully considered for optimal extraction efficiency.

Direct dissolution is a rapid and simple method requiring minimal sample handling. It is suitable for preliminary screening and qualitative analysis. However, it may result in poor spectral resolution and signal overlap due to the presence of matrix components and high molecular weight compounds.

9.12.2 Edible Oil

9.12.2.1 Direct Analysis

This method directly analyses edible oils without any sample preparation steps. It offers simplicity and rapid analysis, making it suitable for high-throughput screening. However, the presence of impurities, such as water and particulates, can affect the quality of NMR spectra.

9.12.2.2 Solvent Extraction

Solvent extraction involves using organic solvents to extract lipids from the oil matrix. This method effectively removes impurities and enhances the purity of the extracted lipid fraction, leading to improved NMR spectral quality. The distribution ratio (D) and partition coefficient (K) can be calculated to assess the efficiency of the extraction process using appropriate mathematical equations (e.g., $D = [\text{lipid}]_{\text{organic}} / [\text{lipid}]_{\text{oil}}$, $K = D / (1 - D)$). Solvent extraction may require additional time and equipment but provides higher-resolution NMR spectra (Mannina et al., 2003).

9.12.2.3 SPE

SPE utilizes solid-phase sorbents to selectively retain and isolate target compounds from the oil sample. This method allows for the removal of interfering substances and the enrichment of analytes of interest. SPE can be tailored by selecting specific sorbents and elution solvents to improve the detection sensitivity of NMR spectroscopy (Mannina et al., 2003).

9.12.3 Honey, Nectar, and Sap

9.12.3.1 Honey

Sample preparation for honey typically involves the removal of any solid particles or debris, followed by dilution with deuterated solvents suitable for NMR spectroscopy. The dilution ensures that the sugar concentration falls within the linear range of the NMR detector, enabling an accurate quantification of the sugar composition (Schievano et al., 2017).

9.12.3.2 Nectar

Nectar sample preparation involves the collection of nectar samples from flowers or plants. The collected nectar can be centrifuged, transferred supernatant to NMR tubes. Additional steps, such as filtration and dilution, may be required on the basis of the specific characteristics of the nectar samples.

9.12.3.3 Sap

Sample preparation for sap primarily involves the collection of sap from plants such as trees or shrubs. Sap can be extracted using techniques like tapping or incision. Once collected, the sap is typically filtered to remove any debris and transferred to NMR tubes for analysis.

Mathematical Equations: Specific mathematical equations for sample preparation in the context of honey, nectar, and sap for NMR spectroscopy analysis are not commonly applicable. The concentration of specific components, such as sugars or metabolites, can be determined using calibration curves, integration of NMR peaks, or quantification based on known standards.

9.12.3.4 Advantages and Applications

NMR spectroscopy allows characterization and quantification of sugars, organic acids, phenolic compounds, and other constituents in honey. It helps assess the quality, origin, and authenticity of honey, as well as understand its nutritional and medicinal properties. NMR spectroscopy enables the identification and quantification of sugars, amino acids, organic acids, and other metabolites in nectar. It provides insights into the composition, nutritional value, and potential pollinator interactions, contributing to the study of plant-pollinator relationships and ecological dynamics. NMR spectroscopy aids in the analysis of sap composition, including sugars, organic acids, amino acids, and secondary metabolites. It assists in understanding the physiological processes, nutrient transport, and defense mechanisms of plants, especially in studies related to plant physiology, ecology, and plant-microbe interactions (Biswas et al., 2023).

9.12.4 Milk and Milk Products

9.12.4.1 Protein Precipitation

Proteins in milk and milk products can interfere with NMR measurements. Precipitation techniques, such as trichloroacetic acid (TCA) or perchloric acid (PCA) precipitation, are commonly employed to remove proteins. The protein concentration can be determined using the equation:

$$\text{Protein Concentration (mg/mL)} = (A_{280} - A_{320}) / \varepsilon,$$

Where A_{280} and A_{320} are the absorbance values at 280 nm and 320 nm, respectively, and ε is the extinction coefficient.

9.12.4.2 Fat Extraction

Fat is a crucial component in milk and milk products. Its extraction is necessary to obtain fat-free samples for an accurate NMR analysis. Methods such as Soxhlet extraction or liquid-liquid extraction using chloroform/methanol mixtures are commonly employed. The fat content can be calculated using the equation:

$$\text{Fat Content (\%)} = (\text{Weight of extracted fat} / \text{Weight of the sample}) \times 100$$

9.12.4.3 Deuterium Oxide (D_2O) Exchange

The strong water signal in milk samples can overshadow other signals of interest. Deuterium oxide (D_2O) exchange is employed to suppress the water signal. Replacing hydrogen with deuterium significantly reduces the water signal, enhancing the detection of other NMR signals.

9.12.4.4 Sample Filtration

Filtration techniques, such as centrifugation or microfiltration, are used to remove particulate matter and solid residues from milk samples. This step improves the clarity of the sample, reduces spectral artifacts, and enhances the quality of NMR data.

9.12.5 Alcoholic Drinks and Beverages

NMR spectroscopy has emerged as a powerful analytical technique for analyzing alcoholic drinks and beverages, providing valuable insights into their composition and quality.

9.12.5.1 Choice of Solvents

The selection of appropriate solvents is crucial in sample preparation for NMR spectroscopy of alcoholic drinks and beverages. Deuterated solvents, such as deuterated water (D_2O) and deuterated methanol (CD_3OD), are commonly employed to dissolve the samples. These deuterated solvents serve as NMR lock solvents, providing a stable reference signal for chemical shift referencing and ensuring compatibility with the analytes of interest.

9.12.5.2 Quantitative Analysis

Accurate quantification of analytes in alcoholic drinks and beverages can be achieved by employing mathematical equations based on peak integrals in the NMR spectrum. The concentration of a specific compound can be calculated using the equation:

$$C = (I / I_{ref}) * C_{ref},$$

where C represents the concentration of the compound, I is the integral of the compound's peak, I_{ref} is the integral of the reference signal, and C_{ref} is the known concentration of the reference compound. This equation enables the determination of the concentration of various components present in alcoholic drinks and beverages, aiding in quality control and regulatory compliance (Pauli et al., 2012).

9.12.5.3 Specific Examples

9.12.5.3.1 Quantification of Ethanol

Ethanol is the primary component in alcoholic beverages. Its concentration can be accurately quantified using the peak integral of its signal in the NMR spectrum. The ethanol concentration can be determined precisely by applying the aforementioned quantification equation, providing essential information for quality control and regulatory purposes.

9.12.5.3.2 Identification of Flavor Compounds

NMR spectroscopy facilitates the identification and characterization of flavor compounds present in alcoholic beverages. In conjunction with established databases and literature references, specific flavor compounds such as esters, aldehydes, and phenols can be identified by analyzing the chemical shifts and coupling patterns of signals in the NMR spectrum. This enables a better understanding of the beverage's sensory properties and flavor profiles.

9.12.6 Nonalcoholic Drinks and Beverages

Nonalcoholic drinks and beverages, such as fruit juices, soft drinks, and herbal infusions, are popular consumer products that can be analyzed using NMR spectroscopy.

9.12.6.1 Sample Preparation Techniques

9.12.6.1.1 Filtration

Nonalcoholic drinks and beverages often contain particulate matter, such as pulp, seeds, or sediment, which can interfere with NMR measurements. Filtration techniques, such as vacuum filtration or centrifugation followed by filtration, are commonly employed to remove solid particles and achieve sample clarity.

9.12.6.1.2 Dilution

High sugar or high concentration samples can lead to line broadening and reduced spectral resolution in NMR analysis. Dilution D_2O DMSO-d6 can lower the sample viscosity and improve NMR spectra quality.

9.12.6.1.3 pH Adjustment

Nonalcoholic beverages may have varying pH levels, influencing chemical shifts and spectral interpretation. Adjusting the pH to a consistent level using acid or base solutions can enhance the reproducibility and comparability of NMR data (Harris & Lucy 2020).

Dilution Factor (DF):

$$DF = V_1 / V_2,$$

where V_1 is the initial volume of the sample and V_2 is the final volume after dilution.

Concentration Calculation:

$$C_2 = (C_1 \times V_1) / V_2,$$

where C_1 is the initial concentration of the solute, V_1 is the initial volume of the solute, C_2 is the final concentration after dilution, and V_2 is the final volume after dilution.

9.12.6.2 Specific Examples and Applications

9.12.6.2.1 Quantification of Sugars

NMR spectroscopy enables the quantification of sugars in nonalcoholic beverages such as fructose, glucose, and sucrose. The integration of specific peaks in the NMR spectra allows for the determination of sugar concentrations based on established calibration curves or reference compounds.

9.12.6.2.2 Identification of Flavor Components

NMR spectroscopy can be employed to identify and characterize flavor components in nonalcoholic drinks. The distinct chemical shifts and coupling patterns of aromatic compounds, organic acids, and other flavor-active molecules provide valuable flavor profiling and quality control information.

9.12.6.2.2.1 Cinnamon

^1H NMR could be used to investigate the mutagenic and antimutagenic properties of cinnamon bark ethanolic extract (CEE), cinnamon oil (CNO), and cinnamaldehyde (CLD) against a range of foodborne mutagens (MeIQ, IQ, PhIP, Glu-P-1, MeIQx, Glu-P-2, MeAC, Trp-P-1, AFB1) and the direct-acting mutagen sodium azide. Using the in vitro Ames Salmonella/microsome assay, the findings revealed the potential of CEE, CNO, and CLD to influence mutagenic activity, providing insights into their possible protective effects against these mutagens.

In this study, cinnamon (C. cassia) barks were powdered and extracted using 60 g of bark and 300 ml of ethanol in a Soxhlet extractor for 10 hours to prepare the CEE. The extract was centrifuged (7,000 rpm for 2 min), concentrated, and stored at 4°C until use.

NMR spectra were acquired on a Bruker AV III 600 NMR spectrometer with a BBFO probe in non-spinning mode at 24.85 ± 1°C. The data were processed using Topspin 3.1 software. Experiments were conducted on 10 ml samples in 5 mm NMR tubes, using DMSO-d6 for field frequency locking and 0.03% tetramethylsilane for chemical shift calibration.

The R-factor tester strains TA97a, TA98, TA100, TA102, and TA104 of *Salmonella typhimurium* were confirmed for strain-specific markers. For mutagenicity and antimutagenicity assays, a frozen culture was incubated in Oxoid Nutrient Broth No. 2 at 37°C for 12 hours, resulting in approximately 1×10^9 cells/ml.

Male Wistar rats (150–200 g) were pretreated with Aroclor-1254, and the S9 fraction and S9 mix were prepared. A modified liquid-preincubation procedure was used for mutagenicity testing of CEE (0.01 to 1 mg/plate) and CLD/CNO (0.125 to 1 mg/plate). For the antimutagenicity assay, a co-treatment protocol pre-incubated the test chemicals with mutagens and bacterial cells. Percent inhibition of mutagenicity was calculated.

Six bacterial strains, five gram-negative *(P. vulgaris, P. aeruginosa, E. coli, K. pneumoniae, S. typhimurium)* and one gram-positive (*S. aureus*), were sub-cultured on brain heart infusion agar and stored at 4°C. Active cultures for experiments were prepared by transferring cells to brain heart infusion broth and incubating at 37°C for 16–18 hours.

The agar well diffusion method was employed to determine the antibacterial activity of CEE, CLD, and CNO. Overnight cultures (0.1 ml) were mixed with molten top agar, poured onto MHA plates, and allowed to dry. Wells (8 mm diameter) were created, and 60 µl of samples (10, 50, 100, and 200 mg/ml) were added. Ampicillin (10 mg/ml) and tetracycline (10 mg/ml) were positive controls. After 24 hours of incubation at 37°C, plates were examined for inhibition zones, with results averaged from duplicate plates.

NMR spectroscopy confirmed CLD as the major component of cinnamon bark. CEE (0.01–1 mg/plate), CNO (0.125–1 mg/plate), and CLD (0.125–1 mg/plate) were evaluated for mutagenicity and antimutagenicity using the Ames assay against foodborne mutagens and sodium azide. CEE was non-mutagenic in five Salmonella strains and showed antimutagenic activity at higher doses. Additionally, CEE, CNO, and CLD exhibited antimicrobial effectiveness against S*taphylococcus aureus, Proteus vulgaris, S. typhimurium, Klebsiella pneumoniae*, and *Escherichia coli*. These findings suggest potential roles for cinnamon bark extract and CLD in cancer chemoprevention and as natural antimicrobial agents (Vijayan & Mazumdar, 2018).

9.12.7 Fish

^{1}H-NMR is an easy and fast method for identifying fatty acids, especially for similar or processed fish products (Bratu et al., 2013; Standal et al., 2009). To prepare the extraction buffer, K_2HPO_4 (13.9 g) and nicotinamide were dissolved in water (80 ml) and D_2O (20 ml). The D_2O included TMSP (100 mg/L). NaN_3 (2 mg) was added to buffer, and pH was adjusted (7.0). A sample for statistical analysis was prepared by adding phosphate buffer (1,000 μL) to the sample (100 mg) in a sample tube (2.0 mL). The sample without homogenization was taken from muscle tissues, vortexed (30 s), and centrifuged (10 min) at 18.000 g, collected and transferred supernatant to the NMR tube.

In this study, all spectra were acquired by a spectrometer (400 MHz). The pulse sequence (*noesygpprld*) is used to acquire ^{1}H NMR spectra with water suppression, which is critical for samples in an aqueous solution to reduce the dominance of the water signal. The temperature was maintained at 301.8 K to ensure consistency and reproducibility. Multiple scans were collected. HSQC, JRES, and TOCSY spectra were obtained using different pulse sequences.

FIDs were Fourier transformed (0.3 Hz) phased with Topsin for data processing. All spectra aligned to alanine signal (1.49 ppm). Metabolite concentration was determined using citric acid (51.99 mM) solution with known concentration. Data buckets were defined and normalized for multivariate analysis, and tests were performed.

Osmolytes, especially arsenobetaine, were key in differentiating fish species. Its concentration correlated with other trimethylamine osmolytes. High classification accuracies show the potential of ^{1}H NMR spectroscopy for authenticating fish species (Saglam et al., 2024).

9.12.8 Saffron

Saffron is a highly valued, labor-intensive spice, often adulterated with cheaper substances due to its high cost. Effective detection methods using high-resolution ^{1}H NMR and chemometrics can identify adulteration, even at low levels. Recent advancements have improved sensitivity, and benchtop 60 MHz NMR offers a viable alternative for detecting common adulterants.

To prepare a sample for NMR analysis, in this study, saffron was finely grounded using mortar. The mixture was extracted by stirring the sample (10 mg) with 600 μL of DMSO-d6 (3 min), centrifuged, collected, and transferred supernatant NMR tubes (5 mm). DMSO-d6 has the ability to dissolve both hydrophobic and hydrophilic compounds.

Chemometric models (PCA, HCD, and PLS-DA) applied to NMR spectra can identify significant variations (sugar, crocin, and formic acid) in pure saffron and adulterated samples (curcuminoids and linoleic acid) (Sobolev et al., 2022; Petrakis et al., 2014).

9.13 CONCLUSION

Effective sample preparation is crucial for optimizing NMR spectroscopy. It enhances instrument compatibility, improves sensitivity and reproducibility, and mitigates challenges associated with complex sample matrices. Advanced techniques like specialized NMR tube types and precise cleaning protocols are pivotal in maximizing spectral clarity and analytical efficiency. These methods underscore the critical importance of sample handling in advancing scientific research across various disciplines.

ACKNOWLEDGMENT

The authors are thankful to Prof. Raghavendra Prasad Tiwari, Vice Chancellor of Central University of Punjab, Bathinda, India, for providing facilities to write this book chapter.

REFERENCES

Ahmad Fadzillah, N., Che Man, Y. B., Rohman, A., Rosman, A. S., Ismail, A., Mustafa, S., & Khatib, A. (2015). Detection of butter adulteration with lard by employing 1H-NMR spectroscopy and multivariate data analysis. *Journal of Oleo Science*, 64(7), 697–703. https://doi.org/10.5650/jos.ess14255

Ahmed, A. E., et al. (2014). Recent advances in metabolomics and its application in food and nutrition research. *TrAC Trends in Analytical Chemistry*, 57, 83–101.

Annesini, M. C., et al. (2020). Development and validation of a UHPLC–MS/MS method for the quantification of bile acids in human serum. *Journal of Chromatography B*, 1136, 121915.

Azmir, J., Zaidul, I. S. M., Rahman, M. M., Sharif, K. M., Mohamed, A., Sahena, F., Jahurul, M. H. A., Ghafoor, K., Norulaini, N. A. N., & Omar, A. K. M. (2013). Techniques for extraction of bioactive compounds from plant materials: A review. *Journal of Food Engineering*, 117(4), 426–436

Biswas, A., Naresh, K. S., Jaygadkar, S. S., & Chaudhari, S. R. (2023). Enabling honey quality and authenticity with NMR and LC-IRMS based platform. *Food Chemistry*, 416, 135825. https://doi.org/10.1016/j.foodchem.2023.135825

Bratu, A., Mihalache, M., Hanganu, A., Chira, N.-A., Todasca, M.-C., & Rosca, S. (2013). Quantitative determination of fatty acids from fish oils using GC–MS method and ^{1}H-NMR spectroscopy. *UPB Scientific Bulletin, Series B*, 75(2), 23–32.

Cai, S., Seu, C., Kovacs, Z., Sherry, A. D., & Chen, Y. (2006). Sensitivity enhancement of multidimensional NMR experiments by paramagnetic relaxation effects. *Journal of the American Chemical Society*, 128(41), 13474–13478.

Chatterjee, S., Matas, A. J., Isaacson, T., Kehlet, C., Rose, J. K., & Stark, R. E. (2016). Solid-state (13)C NMR delineates the architectural design of biopolymers in native and genetically altered tomato fruit cuticles. *Biomacromolecules*, 17(1), 215–224. https://doi.org/10.1021/acs.biomac.5b01321

Chen, J., et al. (2020). Simultaneous determination of nine atmospheric amines and six inorganic ions by non-suppressed ion chromatography using acetonitrile and 18-crown-6 as eluent additive. *Journal of Chromatography A*, 1627, 461409.

Claridge, T. D. (2016). *High-resolution NMR techniques in organic chemistry* (Vol. 27). Elsevier.

Coggins, B. E., Venters, R. A., & Zhou, P. (2010). Radial sampling for fast NMR: concepts and practices over three decades. *Progress in Nuclear Magnetic Resonance Spectroscopy*, 57(4), 381–419.

Dalvit, C. (2007). Ligand-and substrate-based 19F NMR screening: Principles and applications to drug discovery. Progress in Nuclear Magnetic Resonance Spectroscopy, 51(4), 243–271.

Dona, A. C., Michopoulos, F., Everland, H., & Brown, S. P. (2016). Food metabolite profiling using NMR spectroscopy. *Journal of Agricultural and Food Chemistry*, 64(22), 4149–4158. https://doi.org/10.1021/acs.jafc.6b01601

Ernst, R. R. (1987). Two-dimensional NMR spectroscopy: A powerful tool for the investigation of molecular structure and dynamics. *Chimia*, 41(10), 323–323.

Fulmer, G. R., Miller, A. J. M., Sherden, N. H., Gottlieb, H. E., Nudelman, A., Stoltz, B. M., Bercaw, J. E., & Goldberg, K. I. (2010). NMR chemical shifts of trace impurities: Common laboratory solvents, organics, and gases in deuterated solvents relevant to the organometallic chemist. *Organometallics*, 29(9), 2176–2179. https://doi.org/10.1021/om100106e

Goto, N. K., Gardner, K. H., Mueller, G. A., Willis, R. C., & Kay, L. E. (1999). A robust and cost-effective method for the production of Val, Leu, Ile (δ1) methyl-protonated 15N-, 13C-, 2H-labeled proteins. *Journal of Biomolecular NMR*, 13(4), 369–374.

Harris, D. C., & Lucy, C. A. (2020). *Quantitative Chemical Analysis* (10th ed.). W. H. Freeman and Company. ISBN: 9781319324506 (hardcover); eBook ISBN: 9781319274023.

Hong, (2015). Natural product synthesis at the interface of chemistry and biology. *Chemistry – A European Journal*, 20(33), 10204–10212. https://doi.org/10.1002/chem.201402804.

Huang, Y., Zhan, H., You, X., Yang, Y., Li, C., Cai, S., & Chen, Z. (2019). A pure shift-based NMR method for transverse relaxation measurements on complex samples. *IEEE Transactions on Instrumentation and Measurement*, 69(1), 201–211.

Jones, C. J., & Larive, C. K. (2012). Could smaller really be better? Current and future trends in high-resolution microcoil NMR spectroscopy. *Analytical and Bioanalytical Chemistry*, 402(1), 61–68.

Kandasamy, S., Yoo, J., Yun, J., Kang, H. B., Seol, K.-H., & Ham, J.-S. (2020). HRMAS-NMR-based metabolic fingerprints for discrimination of cheeses based on sensory qualities. *Saudi Journal of Biological Sciences*, 27(6), 1446–1461. https://doi.org/10.1016/j.sjbs.2020.04.043

Kazimierczuk, K., Stanek, J., Zawadzka-Kazimierczuk, A., & Koźmiński, W. (2010). Random sampling in multidimensional NMR spectroscopy. *Progress in Nuclear Magnetic Resonance Spectroscopy*, 57(4), 420–434.

Koos, M. R. (2017). *Novel Methods for Modern NMR Spectroscopy-Optimized Pulses and New Experiments* (Doctoral dissertation, Dissertation, Karlsruhe, Karlsruher Institut für Technologie (KIT), 2016).

Liu, D., Liu, X., & Zhang, X. (2019). Preparative isolation of caffeoylquinic acid isomers from Kuding tea by salt-containing aqueous two-phase extraction and purification by high-speed countercurrent chromatography. *Separation Science Plus*, 2(5–6), 170–177. https://doi.org/10.1002/sscp.201900017

Lu, Y., Hu, F., Miyakawa, T., & Tanokura, M. (2016). Complex mixture analysis of organic compounds in yogurt by NMR spectroscopy. *Metabolites*, 6(2), 19.

Markiewicz, W. D., Miller, J. R., Schwartz, J., Trociewitz, U. P., & Weijers, H. W. (2006). Perspective on a superconducting 30 T/1.3 GHz NMR spectrometer magnet. *IEEE Transactions on Applied Superconductivity*, 16(2), 1523–1526

Mannina, L., Sobolev, A. P., & Segre, A. L. (2003). Olive oil as seen by NMR and chemometrics. *Spectroscopy Europe*, 15(6), 6–14.

Mazumder, A., Kumar, A., & Dubey, D. K. (2013). High resolution 19F {1H} nuclear magnetic resonance spectroscopy and liquid chromatography–solid phase extraction–offline 1H nuclear magnetic resonance spectroscopy for conclusive detection and identification of cyanide in water samples. *Journal of Chromatography A*, 1284, 88–99.

Mazumder, A., Kumar, A., Purohit, A. K., & Kumar Dubey, D. K. (2012). A high-resolution phosphorus-31 nuclear magnetic resonance (NMR) spectroscopic method for the non-phosphorus markers of chemical warfare agents. *Analytical and Bioanalytical Chemistry*, 402, 1643–1652. https://doi.org/10.1007/s00216-011-5561-7

Molinski, T. F. (2010). NMR of natural products at the 'nanomole-scale'. *Natural Product Reports*, 27(3), 321–329.

Monakhova, Y. B., Ilse, M., Hengen, J., el-Atma, O., Kuballa, T., Kohl-Himmelseher, M., & Lachenmeier, D. W. (2014). Rapid assessment of the illegal presence of 1,3-dimethylamylamine (DMAA) in sports nutrition and dietary supplements using 1H NMR spectroscopy. *Drug Testing and Analysis*. https://doi.org/10.1002/dta.1677

Moskau, D. (2002). Application of real time digital filters in NMR spectroscopy. *Concepts in Magnetic Resonance: An Educational Journal*, 15(2), 164–176.

Mulas, G., Galaffu, M. G., Pretti, L., Nieddu, G., Mercenaro, L., Tonelli, R., & Anedda, R. (2011). NMR analysis of seven selections of vermentino grape berry: Metabolites composition and development. *Journal of Agricultural and Food Chemistry*, 59(3), 793–802.

Pacifico, S., D'Abrosca, B., Scognamiglio, M., Gallicchio, M., Potenza, N., Piccolella, S., . . . & Fiorentino, A. (2011). Metabolic profiling of strawberry grape (Vitis× labruscana cv.'Isabella') components by nuclear magnetic resonance (NMR) and evaluation of their antioxidant and antiproliferative properties. *Journal of Agricultural and Food Chemistry*, 59(14), 7679–7687.

Paniagua-Vega, D., Cavazos-Rocha, N., Huerta-Heredia, A. A., Parra-Naranjo, A., Rivas-Galindo, V. M., Waksman, N., & Saucedo, A. L. (2019). A validated NMR method for the quantitative determination of rebaudioside A in commercial sweeteners. *Journal of Food Composition and Analysis*, 79, 134–142. https://doi.org/10.1016/j.jfca.2019.02.009

Pauli, G. F., Chen, S.-N., Simmler, C., Lankin, D. C., Gödecke, T., Jaki, B. U., Friesen, J. B., McAlpine, J. B., & Napolitano, J. G. (2012). Importance of purity evaluation and the potential of quantitative ^1H NMR as a purity assay. *Journal of Medicinal Chemistry*, 57(22), 9220–9231.

Pawliszyn, J. (2011). *Solid Phase Microextraction: Theory and Practice*. John Wiley & Sons.

Pei, Y., He, C., Liu, H., Shen, G., & Feng, J. (2022). Compositional analysis of four kinds of citrus fruits with an NMR-based method for understanding nutritional value and rational utilization: From pericarp to juice. *Molecules*, 27(8), 2579.

Renault, M., Cukkemane, A., & Baldus, M. (2010). Solid-state NMR spectroscopy on complex biomolecules. *Angewandte Chemie International Edition*, 49(45), 8346–8357.

Saglam, M., Paasch, N., Horns, A. L., Weidner, M., & Bachmann, R. (2024). ^{1}H NMR metabolic profiling for the differentiation of fish species. *Food Chemistry Advances*, 4, 100602. https://doi.org/10.1016/j.focha.2023.100602

Schievano, E., Tonoli, M., & Rastrelli, F. (2017). NMR quantification of carbohydrates in complex mixtures. A challenge on honey. *Analytical Chemistry*, 89(24), 13405–13414.

Sobolev, A. P., Ingallina, C., Spano, M., Di Matteo, G., & Mannina, L. (2022). NMR-based approaches in the study of foods. *Molecules*, 27(22), 7906. https://doi.org/10.3390/molecules27227906

Standal, I. B., Axelson, D. E., & Aursand, M. (2009). Differentiation of fish oils according to species by ^{13}C-NMR regiospecific analyses of triacylglycerols. *JAOCS, Journal of the American Oil Chemists' Society*, 86(5), 401–407. https://doi.org/10.1007/s11746-009-1370-y

Sucupira, N. R., Alves Filho, E. G., Silva, L. M. A., De Brito, E. S., Wurlitzer, N. J., & Sousa, P. H. M. (2017). NMR spectroscopy and chemometrics to evaluate different processing of coconut water. *Food Chemistry*, 216, 217–224.

Sundekilde, U. K., Larsen, L. B., & Bertram, H. C. (2013). NMR-based milk metabolomics. *Metabolites*, 3(2), 204–222.

Vijayan, V., & Mazumder, A. (2018). *In vitro* inhibition of food borne mutagens induced mutagenicity by cinnamon (*Cinnamomum cassia*) bark extract. *Drug and Chemical Toxicology*, 41(4), 385–393. https://doi.org/10.1080/01480545.2018.1439056

Voehler, M. W., Collier, G., Young, J. K., Stone, M. P., & Germann, M. W. (2006). Performance of cryogenic probes as a function of ionic strength and sample tube geometry. *Journal of Magnetic Resonance*, 183(1), 102–109.

10 Molecular-Based Techniques

Implication of Genomic and Proteomic Techniques in Food Quality and Safety Analysis

Priya Kalyan, Pratiksha, and Jyoti Rustagi

10.1 INTRODUCTION OF PROTEOMIC AND GENOMIC TECHNIQUES

The term 'foodomics' was introduced in 2009, defined as "a discipline that investigates the fields of food and nutrition by employing and integrating advanced omics technologies to enhance consumer well-being, health, and confidence". This field encompasses a variety of interconnected omics technologies, including transcriptomics (mRNA), nutrigenomics (nutrient interactions), proteomics (proteins), metabolomics (metabolites), and genomics (gene detection) (Cifuentes, 2017). Foodomics is an integrated scientific approach that combines food and nutrition sciences with advanced analytical techniques and bioinformatics to address emerging challenges in the field. The concept of the foodome refers to the complete set of compounds present in a food sample. By applying various omics technologies-such as genomics, proteomics, and metabolomics-foodomics allows us to gain comprehensive molecular-level insights and generate precise quantitative data. This enables a deeper understanding of biological functions and interaction mechanisms within food systems. Ultimately, foodomics opens new avenues for exploring how specific molecular components, like proteins, impact health and disease, leading to more accurate assessments of food-related effects on human biology (Aydoğan, 2024). These technologies are utilized for nutritional assessment, food quality and safety, functional food development, food processing optimization, disease prevention and management, flavor and aroma profiling, and food authentication (Agregán et al., 2021; Ghahary and Abiri, 2021). In this chapter, we will explore the area of proteomics and genomics for food quality and safety analysis.

Genomic technologies extend beyond agriculture, offering advancements in food processing, food safety, and quality assurance. They also support the development

DOI: 10.1201/9781003482611-10

of functional foods and pave the way for new health strategies like 'personalized nutrition', an innovative approach that tailors a person's diet to their specific genetic profile to maximize health and prevent disease (Ortea et al., 2016).

A functional genomics approach to studying the molecular basis of meat quality provides deeper insight into the complex gene expression processes that contribute to meat quality development (Mullen et al., 2006). Genomics in aquaculture is still emerging, with molecular markers developed for many farmed fish species to support genetic improvements. Physical maps and linkages help identify genes linked to key production traits, performance, and development, while genomic reagents like expressed sequence tags have advanced microarray technology (Liu, 2007).

The term 'proteome' was coined by Marc Wilkins in 1995. A proteome is known as a complete set of proteins expressed or produced by a system or organism. Proteomics, a field of study, delves into the interactions, functions, composition, and structures of proteins (Wilkins et al., 1996). Unlike genomics, which primarily focuses on DNA, proteomics provides deeper insights into the structure and function of food. However, proteomics is inherently more complex than genomics due to the dynamic nature of protein composition over time and its susceptibility to various factors that impact food quality.

Proteins serve as indicators of food quality, composition, and origin throughout production stages. Proteome analysis enables the discovery of new marker proteins or peptides, supporting the development of assays to detect adulteration and fraud. With accurate analytical methods, these markers can be reliably identified and validated. Early proteomics in food research focused on exploring various foods-such as meat, genetically modified crops, and beverages like wine-to enhance end-product quality (D'Alessandro and Zolla, 2012). The advancements in omics technologies offer unparalleled benefits for enhancing our understanding of various biological disciplines, particularly in food science. Worldwide, food trends are rapidly evolving, with consumers increasingly focused on their food choices (Afzaal et al., 2022). Today, consumers are not solely concerned with the sensory qualities of food products; but there is a growing demand for safe, nutritious, functional foods that are minimally processed and contain fewer additives (Saeed et al., 2021). In this regard, proteomics holds significant promise for establishing the reliability and safety of food products.

Both genomic and proteomic techniques can serve diagnostic purposes related to food safety and quality. Ensuring food quality and safety is crucial to safeguard consumer health, alleviate financial burdens, and prevent adulteration. To address these challenges, powerful analytical tools are urgently needed. These tools can help tackle food quality and safety issues that are enlisted in Table 10.1.

10.2 DIFFERENTIAL AND FUNCTIONAL PROTEOMICS

10.2.1 Differential Proteomics

Differential proteomics includes qualitative as well as quantitative proteomics. Qualitative proteomics helps in characterization and detection of proteins in different food products, spoiled or contaminated food products, same food products of different origin, and differences in different food processing methods (Carrera,

TABLE 10.1
Applications of Proteomics in Food Safety and Quality Analysis

S. No	Application	Case Studies	References
1	Pathogen detection	Bacteria such as *E. coli*	Fagerquist et al. (2014)
		Staphylococcus aureus	Reinhardt et al. (2013)
		Listeria	Jadhav et al. (2014)
2	Allergens	Allergic potential of various legumes were studied	Perkons et al. (2023)
3	Effect of food processing techniques	Changes in protein profile were identified during soy fermentation	Li et al. (2022)
4	Detection of contaminants	Mycotoxins in cereals	Capriotti et al. (2012)
		Xenobiotics like $HgCl_2$	Zhang et al. (2013)

TABLE 10.2
Major Foodborne Pathogens Detected in Food

Bacteria	Foodborne Species	Disease	Source
Escherichia coli	Shiga toxin producing *E. coli*, including (serovar 0157:H7).	Mild intestinal discomfort, hemolytic uremic syndrome, renal disease and death.	Ruminant animals
Salmonella	*S. enterica* ssp. enterica serovar *Enteriditis* and *Typhimurium*.	Salmonellosis, gastroenteritis (self-limiting)	Food and water
Campylobacter	*C. jejuni, C. coli.*	Guillain–Barré syndrome, Miller Fischer syndrome, mild and self-limiting gastroenteritis	Pig, cattle, cats, water and ready-to-eat foods
Listeria	*L. monocytogenes*	Encephalitis, meningitis, abortion, listeriosis	Milk, meat, egg and dairy products

Source: Deng et al. (2016)

2021). While quantitative proteomics helps in quantifying relative amount of proteins in genetically modified food versus non-genetically modified food, changes due to natural variations, due to technical processes, or due to storage and could also help in biomarker discovery for rapid detection of contaminants or adulterants (Restani et al., 1997; Restani et al., 1999).

10.2.2 Functional Proteomics

Functional proteomics examines how proteins interact with each other and with other molecules. This includes the effects of these interactions (Coombs, 2020). One approach uses protein profiling with activity-based probes to screen for inhibitors and measure enzyme activity (Serim et al., 2012). Another approach, activity-based

TABLE 10.3
Applications of Omics Technologies across Food Systems

Food System	Analytical Approach	Application Summary	Key Outcome	Reference
Dairy	MALDI-TOF MS + lasso regression	Detection of bovine milk adulteration (0.5%–80%) in goat and sheep milk using peptide mass spectra (500–4,000 Da, 4–20 kDa). Low MW spectra + regression modelling enhanced prediction accuracy.	RMSE: 6.34% for goat milk, 6.28% for sheep milk Sensitive even at low adulteration levels	Ryesova et al. (2022)
Meat	DIA-high resolution mass spectrometry (MS) (proteogenomics)	Identification of trace-level meat adulteration using species-specific tryptic peptides from proteins like myosin and hemoglobin. Effective on both raw and cooked meats	Detected 1% pork/horse in mixtures; m/z deviation < 3 ppm; high specificity and sensitivity for regulatory use.	6)
Sea food	DNA Barcoding (FDB and+ MDB)	Authentication of mislabeled seafood (e.g., threadfin/lizardfish, salmon/trout, prawn/fish roe). MDB showed better amplification, especially in degraded/processed samples.	MDB amplified 47% vs. 21% for FDB; highlighted need for improved labeling and traceability systems	Adibah et al. (2020)
Plant-based food	PCR and qPCR (multiplex assay with species-specific primers)	Detection of genetically modified organism (GMO) vs. non-GMO soy fraud. Multiplex PCR protocol detected low-level (0.1%) wheat/soy adulteration in pea flour with high specificity. Real-time qPCR added quantification capability.	Detected 0.1% adulterants across 22 species. Enabled quantitative and traceable authentication in processed plant-based foods	Shin et al. (2021)
Fermented food	Metagenomics + qPCR + Culture-dependent methods	Profiling of microbial communities in *Yucha* (traditional rice-fish fermented food). Identified *Lactobacillus* dominance and enriched metabolic pathways via metagenome and qPCRconfirmation	Found 13 species (147 strains) of *Lactobacillus*; confirmed correlation between microbial diversity and metabolic functionality	Zhang et al. (2016)

proteomics, directly investigates a protein's function within a sample by looking at its activity and how it might be inhibited (Elmore et al., 2021). Mass spectrometry (MS) imaging is a powerful tool for functional proteomics.

10.3 MECHANISM OF PROTEIN IDENTIFICATION USING PROTEOMICS

In the field of proteomics, two primary methods are commonly employed: the bottom-up approach and the top-down approach. Let's explore each method in detail in the following subsections.

10.3.1 Bottom-Up Proteomics

It is a meticulous method that has revolutionized the study of complex protein mixtures. When bottom-up is performed on a mixture of proteins, it is called 'shotgun proteomics', a name coined by the Yates lab because of its analogy to shotgun genomic sequencing (Zhang et al., 2013). At its core, this approach measures proteins indirectly through their constituent peptides. The process begins by the proteolytic enzymatic (such as trypsin) digestion of intact proteins into smaller peptides, which are then separated using high-performance liquid chromatography (HPLC) (Dupree et al., 2020). The separated peptides are subjected to MS. These peptides are ionized and their masses measured. The mass spectra generated are used to deduce the amino acid sequences of the peptides, which can then be mapped back to their parent proteins. This approach is highly sensitive and capable of identifying and quantifying proteins across a wide dynamic range, making it particularly suited for biomarker discovery and the study of post-translational modifications. The process flow of bottom-up proteomics has been explained in Figure 10.1.

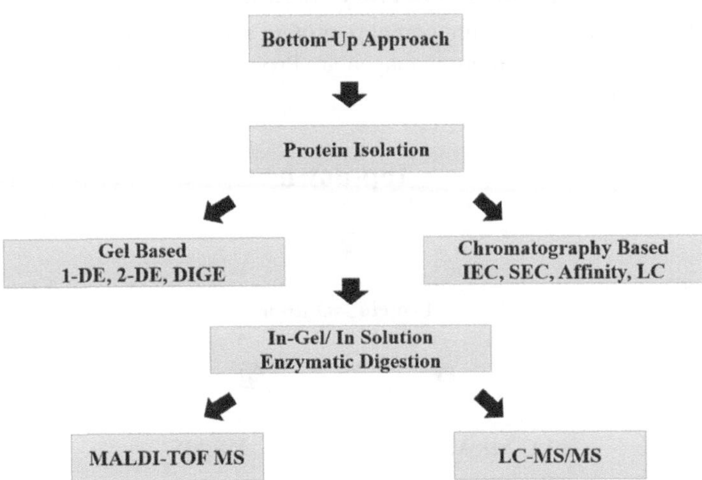

FIGURE 10.1 Bottom-up process flow of proteomics.
Source: Zhang et al. (2013)

10.3.2 Top-Down Proteomics

Unlike the bottom-up approach, top-down proteomics directly measures proteins in their intact form, offering a holistic view of the proteome (Roberts et al., 2024). The intact proteins are directly introduced into a mass spectrometer, where they are ionized and analyzed without prior digestion into peptides for the detailed characterization of proteoforms (Catherman et al., 2014). This method is particularly useful for identifying proteins with a molecular mass greater than 200 kDa. By preserving the native structure of proteins, the top-down approach can provide insights into the functional state of proteins, including their folding, interactions, and dynamics.

The overall workflow for protein identification using proteomics involves the following steps as shown in Figure 10.2:

1. Sample preparation
2. Protein separation and isolation
3. Protein identification
4. Data analysis

10.3.2.1 Sample Preparation

Protein extraction from food is a critical initial step, ensuring the production of purified and concentrated protein samples (Cunsolo et al., 2014). In some cases, optimization is performed during cellular lysis and subcellular fractionation to enhance protein solubilization. Organic solvents and detergents are employed during disruption, which can subsequently be removed through lyophilization. Additionally, selective removal of abundant proteins and targeted enrichment of specific proteins contribute to the overall success of this process.

10.3.2.2 Protein Separation and Fractionation

These methods help in resolving complex protein mixtures into individual components to make the analysis more manageable. Protein separation is usually done by gel-based methods.

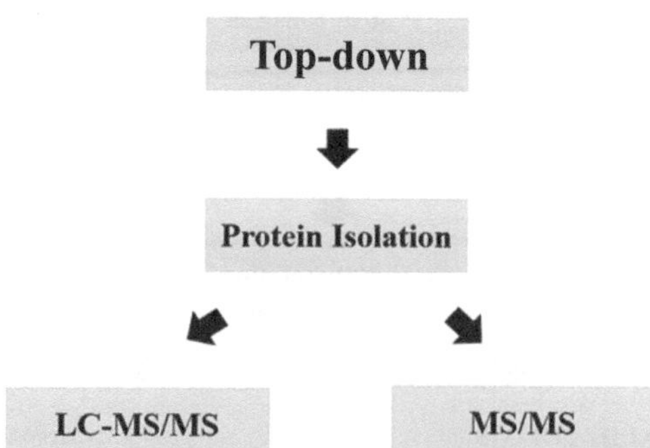

FIGURE 10.2 Top-down process flow of proteomics.
Source: Catherman et al. (2014)

10.3.2.2.1 *One-Dimensional Gel Electrophoresis (1-DE)*

One-Dimensional Gel Electrophoresis (1-DE) represents a foundational technique in the field of proteomics, also known as SDS-PAGE (sodium dodecyl sulfate-polyacrylamide gel electrophoresis), separates proteins (10 kDa to 300 kDa) based on their molecular weight. SDS denatures proteins and imparts a uniform negative charge proportional to their mass. After electrophoresis, the gel can be visualized using visible stains (like silver or Coomassie) or fluorescent stains (like Sypro Ruby, Lava, and Deep Purple) (Brunelle and Green, 2014).

10.3.2.2.2 *Two-Dimensional Gel Electrophoresis (2-DE)*

It is a two-step method for separating proteins. In the first dimension, proteins are initially separated on the basis of their isoelectric point (pI) using isoelectric focusing. In the second dimension, proteins are separated by SDS-PAGE electrophoresis based on their molecular weight. It provides improved resolution compared to 1-DE. One disadvantage to this is that it provides limited reproducibility, and inherent biological variations are also there (Magdeldin et al., 2014).

10.3.2.2.3 *Differential Gel Electrophoresis (DIGE)*

This is a modified version of 2-DE. Different protein samples are labeled with fluorescent dyes (CyTM2, Cy3, and Cy5) prior to isoelectric focusing (Viswanathan et al., 2006). Cy3 and Cy5 are used for two distinct groups of samples, while Cy2 is used as an internal standard for standardization.

10.3.2.3 Protein Identification

Shotgun proteomics has become a widely used method for protein identification. One notable reason is the amenability of common analytical chromatography methods for separation of peptides prior to ESI-or MALDI-MS/MS. MS shines as a powerful tool for protein sequencing and analysis. It can not only identify proteins but also determine their molecular weight. This technique works by ionizing the protein molecules and measuring their mass-to-charge ratio (m/z).

Mass spectrometer has three key parts as shown in Figure 10.3:

1) Ion source: This component ionizes the protein molecules.
2) Analyzer: This part separates the ions based on their m/z ratio.
3) Detector: This component counts and records the ions.

In shotgun proteomics, two primary ionization methods are used to convert peptides into gas-phase ions:

- **MALDI (Matrix-Assisted Laser Desorption/Ionization):** This method uses a laser-absorbing matrix to ionize proteins, typically generating singly charged ions for easier analysis. It has been effectively applied to detect milk adulteration, such as identifying bovine additions in buffalo or goat milk using MALDI-TOF-based peptidomic profiling (Sassi et al., 2015).

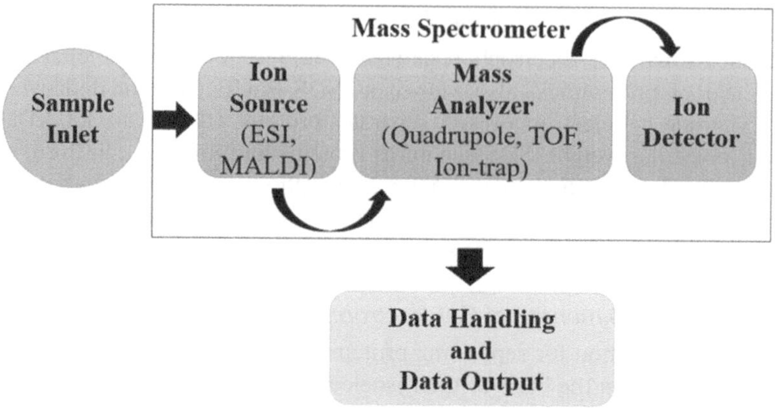

FIGURE 10.3 Components of mass spectrometer.
Source: Pusch and Kostrzewa, (2005)

- **ESI (Electrospray Ionization):** ESI produces multiply charged ions by creating fine, charged droplets that evaporate into gas-phase ions. It offers high reproducibility and compatibility with various MS instruments (e.g., TOF, quadrupole, ion trap), but it is less suitable for molecular imaging, requires more sample, and generates complex spectra (Nadler et al., 2017).

Mass analyzers separate ions by mass-to-charge (m/z) ratio and determine analysis quality. Common types include:

- **TOF:** Fast and accurate across a wide mass range.
- **Quadrupole:** Selective and versatile, often used in tandem MS.
- **Ion Trap:** Sensitive and effective for MS/MS analysis.

Each type of mass analyzer has its specific characteristics and applications, and they are often combined with other components like chromatography systems to enhance the separation and identification of proteins in a sample.

10.3.2.4 Data Analysis

Data analysis in MS-based proteomics is a multistep downstream process that transforms raw data into meaningful biological information. It begins with the preprocessing of spectral data, which includes noise reduction, peak detection, and calibration. The processed spectra are then matched against theoretical spectra from protein databases in a process known as protein identification. This is often done using algorithms like Comet, Tandem, SEQUEST, or Mascot. Software like Max Quant (Tyanova et al., 2015), Trans Proteome Pipeline (TPP) (Deutsch et al., 2023), and Proteome Discoverer are used to visualise the raw MS data like a spectrum called mass spectrum (Orsburn et al., 2021). Perseusa software contains statistical tools for high-dimensional omics data analysis covering normalization, pattern recognition, time-series analysis, cross-omics comparisons, and multiple hypothesis testing of mass spectrometry data (Tyanova et al., 2016). MSstats is a software that includes R package for statistical relative quantification of proteins and peptides in mass-spectrometry-based proteomics. The ultimate goal is to construct a comprehensive map of the proteome, revealing insights into proteins in different food products, spoiled or contaminated food products, same food products of different origin, and differences in different food processing methods. The entire proteome can be screened for potential allergens using in silico bioinformatic tools. Online resources like UniProt and BLAST enable allergenicity prediction. Notably, the UniProt Knowledgebase, linked to Allergome, is a key database, containing 1,303 reviewed (Swiss-Prot) and 3,117 unreviewed (TrEMBL) protein entries related to allergens (López-Pedrouso et al., 2023). Gao et al. (2023) developed a robust method to ensure wheat safety by detecting mildew contamination using UPLC-QTOF-MS combined with chemometric tools. A non-targeted Principal Component Analysis (PCA) model, built from authentic wheat samples, effectively distinguished moldy from normal wheat. Further analysis with OPLS-DA enabled accurate detection even at 5% adulteration, identifying unique biomarkers. This approach highlights the power of metabolomics and data modeling in enhancing food safety.

10.4 MULTI OHMIC APPROACHES IN FOOD SAFETY

Food metabolomics is increasingly applied across the farm-to-table chain for food safety, quality, authenticity, and traceability, enabling the detection of contaminants, assessment of nutritional and sensory attributes, and verification of product origin (Pinu, 2016; Cook and Nightingale, 2018). Given the short shelf life of many food products, metabolomics-based approaches have been employed to identify a wide range of microbial biomarkers associated with varying levels of contamination, demonstrating strong potential for rapid and reliable early-stage detection (Pinu, 2016). Metagenomic approaches-such as metabarcoding, shotgun sequencing, and long-read metagenomics-represent a transformative shift in food safety diagnostics, enabling the comprehensive identification of entire microbial communities within a single food item, ingredient, or environmental sample. These techniques have broad applications across the food industry, including verifying food authenticity and origin and elucidating the microbial composition of probiotics and fermented food products (Billington et al., 2022). Transcriptomics is a powerful molecular biology tool that has significantly advanced our understanding of biological processes relevant to food safety. By analyzing the complete set of RNA transcripts in a cell or organism, transcriptomics enables the investigation of gene expression patterns and regulatory mechanisms. This approach provides valuable insights into how genes are activated or repressed under various conditions, including exposure to foodborne pathogens. Consequently, transcriptomics has contributed to the identification of potential hazards, evaluation of food quality, and development of targeted strategies to improve food safety (Carrera, 2023). Various applications of omics technologies across food systems have been presented in Table 10.3.

10.5 DISCOVERY AND TARGET-DRIVEN PHASE FOR FOOD AUTHENTICITY

Food authenticity is an analytical process that helps to verify if a product complies with the label description. In some cases, compromise with food authenticity does not cause much loss (mislabeled geographic region), but in other cases, it could lead to serious health issues (mislabeled allergen and toxic substances). Even in cases where the health of the consumer is not affected, food fraud can lead to unfair competition in the agri-food market and deception of consumer. Four key elements are identified, being deception, deliberate intent, financial gain, and misrepresentation, which are in line with European Commission's criteria to establish whether a case should be considered as noncompliance or as fraud. High-value products undergo numerous processing steps, and they are many times targeted for food fraudulence. Due to this reason, different approaches are required to address these issues including analytical techniques that could give satisfactory explanation about the quality of food. The most employed method to curb food fraud is targeted method, which focuses on the detection of particular group of molecules, although this process is labor intensive but enables one to detect molecules up to parts per trillion (ppt) levels (Amaral, 2021).

As reported by Grazina et al. (2020), wild salmon attains higher prices in the market compared to the farmed salmon owing to the high polyunsaturated fatty acid content (omega 3 and omega 6), which makes wild salmon prone to the mislabeling practices. Wild salmon and farmed salmon were discriminated on the basis of fatty acid composition sequence, and the lipids were extracted from muscles and analyzed by gas chromatography with flame ionization detector (GC-FID) together with advanced chemometrics.

HPLC is one of the most commonly used techniques for food authentication, and it is great at solving food safety issues. This technique deals with the separation, detection, and quantification of content in a mixture, where the analysis of solute vastly depends on its polarity and solubility in the solvent system. The best feature of this technique is that it can be used to identify and monitor the authentication of certain biomolecules like amino acids, organic acids, sugars, and pigments (Fanelli et al., 2021).

Nuclear magnetic resonance is another high performance and powerful tool with which targeted analysis of metabolites is possible in different food matrices like milk, honey, and coffee. It is relatively an easy technique, and different combinations of compound can be resolved. The extra virgin olive oil (EVOO) is subjected to the addition of low-quality botanical oils, and the chemical composition of these oils is generally different from the original EVOO in terms of triglycerides and fatty acids. The difference among these oils is successfully evaluated with the ^{1}H NMR (nuclear magnetic resonance) analysis (Maestrello et al., 2022).

10.5.1 Next-Generation Sequencing and Whole Genome Sequencing

As food-borne illnesses continues to be a threat, the identification of disease-causing bacteria is critical to find the exact vehicle of transmission and to remove contaminated food from surroundings. Earlier molecular subtyping was an instrumented tool for surveillance and food-borne illness investigation, and its utility was demonstrated by wide use of pulse field gel electrophoresis (PFGE). In recent past, next-generation sequencing (NGS) for whole genome sequencing (WGS) has been gaining tremendous amount of attention by food regulatory and health agencies to detect, investigate, and control outbreaks caused by microorganisms (Deng et al., 2016). It is a laboratory technique that has potential to change how we detect and identify microbial hazards in food chain (WHO, 2023). NGS, coupled with advanced bioinformatics, is transforming food microbiology. Once limited to research, NGS is now widely used in diagnostics, outbreak tracing, antimicrobial resistance monitoring, forensics, and verifying food authenticity (Jagadeesan et al., 2019). Whole Genome Sequencing (WGS) is increasingly replacing traditional microbiological methods by enabling rapid, cost-effective identification and characterization of pathogens, including serotyping, antimicrobial resistance, and virulence profiling, all within a single workflow (Brown et al., 2019). WGS provides immediate information on identification and detection of pathogen, which is critical in management of conventional outbreak management.

10.5.1.1 Pulse Field Gel Electrophoresis

It is employed for subtyping of food-borne pathogens. It is a DNA fingerprinting method that is based on restriction digestion of purified genomic DNA (Xia et al., 2009). As reported by Sandt, 2006, PFGE played a critical role in identifying outbreak causing isolates of pathogen *Salmonella* and successfully distinguished it from unrelated isolates. Pulse field gel electrophoresis was discovered by Schwartz in 1984. It allows resolving DNA mix from 20 kb up to 5 Mb. The principle of DNA separation is based on alternating the direction of two applied electric fields. The size of the molecule decides the reorientation time. Smaller molecules reorient faster than larger sized molecules. The movement of larger molecule from the gel is delayed with respect to smaller molecules, and consequently the separation of molecules occurs (Muñoz *et al.*, 2016). The major disadvantage of PFGE is that it is a laborious method, and result achieved can be inconsistent because of different standardized protocols. The same has been explained with the help of Figure 10.4.

This method is regarded as a standard method for subtyping *L. monocytogenes*, known for its high discriminatory power, accuracy, and reproducibility in detailed structural comparison and molecular typing of the bacterium (Park et al., 2016). An epidemiological study used PFGE on *Salmonella enteritidis* isolates from a food poisoning outbreak, revealing identical PFGE patterns among isolates from 19 patients, identifying a single causative strain. Variations in PFGE patterns in a few isolates suggested potential chromosomal mutations or multiple strains from various sources, with further analysis supporting mutation as the cause. These findings indicate the importance of PFGE analysis on multiple colonies per patient in epidemiological studies of *S. enteritidis* (Murase et al., 1996). A total of 495 *Listeria monocytogenes* isolates, temporally and geographically matched from human clinical cases, food sources, ruminant farms, and urban and natural environments, were analyzed to assess the diversity of PFGE types. The two-enzyme PFGE method using AscI and

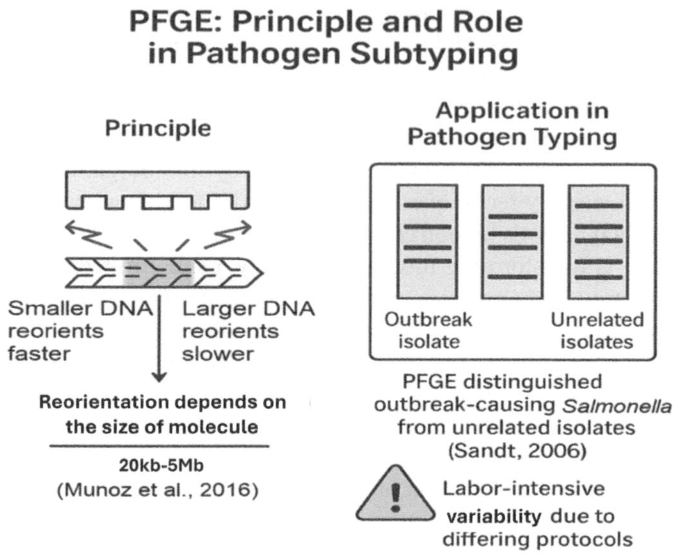

FIGURE 10.4 PFEG: Principle and role in pathogen subtyping.

ApaI differentiated 310 unique PFGE types and demonstrated greater overall discriminatory power (Fugett et al., 2007).

10.5.1.2 Multiple Locus Variable Number Tandem Repeat Sequence Analysis (MLVA)

The primary aim of any subtyping method is to characterize bacteria beyond the species level and effectively group individual isolates. Quick and reliable subtyping is essential for laboratory-based surveillance; as indistinguishable subtypes are more likely to share a common source. This principle underpins the use of molecular subtyping for monitoring bacterial pathogens, detecting outbreaks, and responding to them. MLVA has been instrumental in detecting and investigating food-borne disease outbreaks globally (Nadon et al., 2013). Being a PCR-based typing method for classification, identification, and typing of microorganism, and a rapid alternative to PFGE as it is simple and reproducible and can discriminate closely related bacterial strains. MLVA has been used to type bacteria like *S. Typhimurium,* S. Enteritidis, *E. coli* O157, and *Vibrio parahaemolyticus*. Recently, it has been recommended as a supplement to PFGE for subtyping *S. Enteritidis*, as it can link rapidly evolving bacterial strains to outbreaks even when PFGE results appear identical (Liu et al., 2016). VNTR (variable number of tandem repeats) are sequence of 20–100 nucleotides that vary in copy number in genome of different isolates; they are known to arise from a phenomenon known as DNA strand slippage during replication and are known to perform no function (Pitt and Barer, 2012). The primers of PCR are choosing outside the VNTR region, and the analysis of PCR product is performed with the aid of capillary electrophoresis. MLVA is subjected to subtyping different food-borne pathogens like *E.coli* O157:H7 (Lindstedt et al., 2003) and *Salmonella* (Lindstedt et al., 2004). MLVA method is directly applied to the *Listeria monocytogenes* positive sample without pure culture isolation to subtype the different strains of culture of different serotypes (Chen et al., 2011). Sobral et al. (2012), in their study, subjected a total of 145 isolates of *S. aureus* to MLVA subtyping, The isolates were obtained from humans', animals', and food poisoning products' diseased condition. Forty-six percent of clonal complexes (CCs) in animal isolates were common in humans. Isolates responsible for food poisoning showed CC distribution resembling human isolates, which suggests human origin (Sobral et al., 2012). The same has been explained with the help of Figure 10.5.

10.5.1.3 Multi Locus Sequence Typing (MLST)

The MLST approach offers accurate species and strain identification while providing insights into population genetics and gene flow. It is more precise than conventional immunological assays, which may not reflect genealogical relationships, leading to

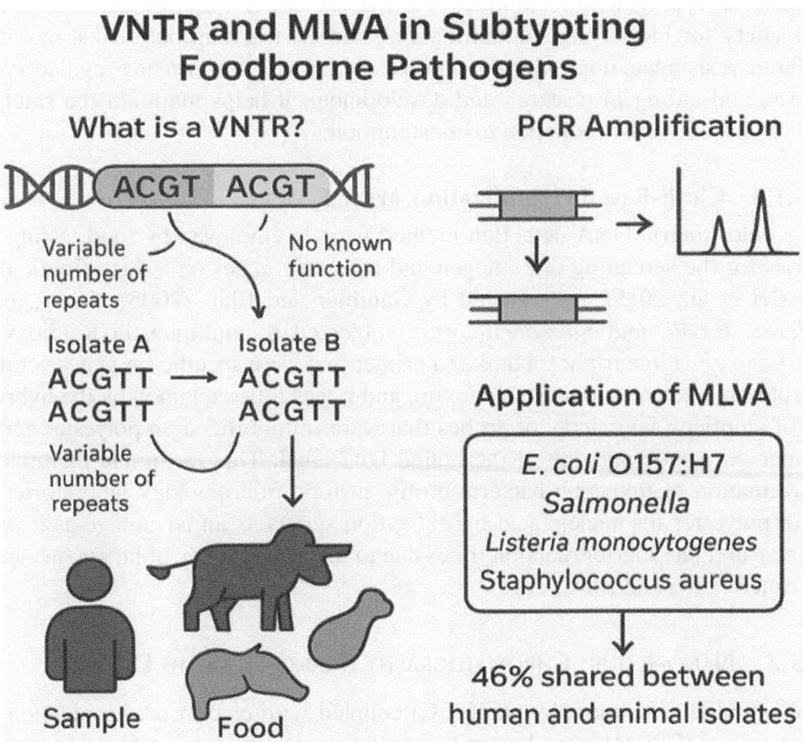

FIGURE 10.5 VNTR and MLVA in subtyping food-borne pathogens.

potential misdiagnoses. MLST has become a cornerstone technique for molecular typing of pathogenic microorganisms, enabling high-resolution genealogical data that enhances an understanding of pathogen evolution and epidemiology. However, its main limitation is the need for a reference genome to select housekeeping loci. Additionally, some pathogens may lack diversity in their genomes or housekeeping genes, reducing MLST's discriminatory power. The method typically analyzes a small fraction of the genome (six to seven housekeeping genes), which is expected to represent overall genetic diversity. Housekeeping genes are preferred markers due to their slow evolution, providing reliable data for accurate typing and phylogenetic analysis (Pérez losada et al., 2013). MLST typing is based on housekeeping gene sequence, and these genes in bacteria are conserved. Each gene-recorded sequence is given a number that defines the isolate, and the isolates sharing the related sequence are grouped together into CCs. The characterization of hypervariable region is possible as reported in the case of Campylobacter. A total 28 isolates of *C. jejuni* and 19 isolates of *C. coli* were collected from samples of meat (chicken, chicken liver, chicken gizzard, pork, and turkey). MLST of isolates revealed 21 different sequence types (STs) that belonged to 8 different CCs. In conclusion, MLST showed campylobacter STs from chicken liver was related to STs of isolates associated with humans (Noormohamed and Fakhr, 2014).

Bacterial typing with techniques like MLVA, MLST, and PFGE is essential in food safety for identifying contamination sources, tracking outbreaks, monitoring antibiotic resistance, improving food processing practices, ensuring regulatory compliance, and aiding in research and development. It helps maintain the safety and quality of food from production to consumption.

10.5.1.4 Cloth-Based Hybridization Array System

It is a colorimetric DNA detection method usually employed by food testing laboratories for the screening of pathogen and toxigenic genes on a microfluidic device (Geissler et al., 2015). As reported by Gauthier and Blais (2005), the bacteria *V. cholerae*, *E.coli*, and *Salmonella* were subjected to multiplex PCR (dioxygenin (DIG)-deoxyuridine triphosphates and primer that were specific for genes verotoxin, enterotoxin, and heat-labile stable toxin), and it was followed upon by the hybridization of amplicon with array of probes that were immobilized on polyester and then immunoenzymatic reaction of the bound DIG label. This technique facilitates the determination of toxigenic bacteria profile in food microbiology laboratories. The use of polyester for nucleic acid hybridization serves as an advantage as it is cost-effective and has shorter reaction times due to the accessibility of larger surface area (Martinez-Perez and Blais, 2010).

10.5.2 Nano-Liquid Chromatography (Nano-LC) with ESI

Nano-liquid chromatography (nano-LC) coupled with electrospray ionization mass spectrometry (ESI-MS) has become one of the most widely employed techniques in advanced proteomics research. A major limitation in current proteomic studies is the reduced sensitivity caused by the necessary dilution of protein samples-often more than tenfold-prior to enzymatic digestion and analysis, which negatively impacts detection efficiency. The application of highly sensitive and modern nano-LC systems is crucial for enhancing the performance of food and plant proteomics, offering advantages such as minimal sample requirement and decreased chemical waste generation. As more robust and commercially accessible nano-LC platforms become available, they enable deeper characterization and improved understanding of complex biological processes. Measuring protein abundance and identifying specific proteoforms in food and plant matrices further contribute to the field of foodomics, potentially offering therapeutic insights. Continuous advancements in nano-LC instrumentation-such as the development of innovative nano-column technologies—have significantly improved analytical sensitivity at nanoscale flow rates over the past two decades. These developments, alongside progress in MS, have elevated the capabilities of nano-LC/MS in proteomic analyses. Consequently, the adoption of nano-LC/MS is steadily growing within food science, particularly for detailed protein profiling and functional analysis. Nano-LC is more sensitive than normal flow technique; therefore, it is the preferred choice in bottom-up proteomics (Ayodğan, 2024).

10.6 AI IN OMICS-BASED FOOD SAFETY

Machine learning, with its ability to process large datasets, can uncover hidden patterns in the food production and distribution chain, enhancing risk prediction and control (Soroushianfar et al., 2025). AI can enhance the detection of foodborne pathogens and indicator organisms (e.g., coliforms, *E. coli*); classification, identification, and differentiation of bacterial isolates; characterization of bacteria for AMR, host specificity, and virulence; modeling of microbial growth; and analysis of population dynamics (Qian et al., 2023). AI has also aided pathogen characterization and identification through analytical and diagnostic tools. Weis et al. (2020) reviewed studies where ML algorithms analyzed spectral data from MALDI-TOF MS (Singhal et al., 2015) to enhance species identification. The agri-food sector has actively integrated artificial intelligence (AI) technologies to enhance various operations, including crop yield forecasting, quality control of products, traceability improvements, and innovative product development (Qian et al., 2023). Machine learning and deep learning have enhanced protein–protein interaction (PPI) analysis by improving accuracy, speed, and scalability. Their growing application in plant PPI networks is advancing our understanding of biological systems, with significant implications for agriculture and global food security (Sarkar, 2025).

10.7 ROLE IN FOOD SPOILAGE AND MICROBIAL AND PROCESSING ANALYSIS

Public health organizations have concerns about ensuring food safety due to the growing demand for minimally processed ready-to-eat goods and the popularity of street foods. Various foods, like fruits, vegetables, and ready-to-eat items, can include foodborne infections that are consumed without additional care. Foodborne infections may arise from improper handling of food. Furthermore, consuming raw or undercooked foods including meat, poultry, and seafood has been linked with foodborne diseases. Therefore, it is crucial to analyze food for the presence of foodborne pathogens to ensure a safe food supply and reduce the incidence of foodborne diseases. With a significant increase in occurrence over the preceding 20 years, foodborne illnesses have become a major global public health concern. Many regions have reported an increase in instances, but the fact of underreporting, especially in poorer nations, makes it difficult to estimate the global incidence of these diseases. Estimates of foodborne illnesses provide a valuable foundation for developing and implementing future public health interventions. In 2019, the Centers for Disease Control and Prevention (CDC) estimated the average annual number of foodborne illnesses, hospitalizations, and deaths in the United States caused by seven major foodborne pathogens. Six of these pathogens were responsible for approximately 9.9 million domestically acquired foodborne illness cases. Collectively, illnesses caused by all seven pathogens resulted in an estimated 53,300 hospitalizations and 931 deaths, highlighting the significant public health burden associated with foodborne diseases (Centers for Disease Control and Prevention, 2025). Furthermore, about a quarter of the population is now at a higher risk for foodborne diseases (Law et al., 2015). Some of the major foodborne pathogens detected in food have been presented in Table 10.2. Animal goods, especially eggs and egg products, and pig meat were among the primary food items linked to strong-evidence outbreaks, accounting for 10% of all outbreaks. Additional sources were fish and fish products (7%), cheese (8%), broiler meat (9%), bovine meat (4%), milk and dairy products (5%), and crustaceans (3%). *Salmonella* spp. in eggs was one of the top five food-pathogen combinations for sickness cases and hospitalizations in 2015 and was connected to the most reported foodborne outbreaks (Bintsis, 2017). The WGS illumina sequencing employs fluorescently labeled dinucleotides (dNTPs) to decipher the sequence of DNA fragments. Three illumina model series are known: MiniSeq (most affordable sequencer), MiSeq (rapid sequencer but gives low outputs), and NextSeq (high range output). Sequencing of *L. monocytogenes* is performed using NextSeq and MiSeq (Kwong et al., 2016). Nanopore technology another sequencing technology that reads single nucleotide, have added advantage of not requiring sequence replication, instead a protein nanopore is inserted at the base of polymer membrane which is electrically resistant. Electric current flows through the aperture of the nanopore. As single nucleotide passes through the nanopore, it brings changes in the electric current, variation in the current flowing through the pore is measured, and specific sequence is identified on these particular disturbances. This technology's sequencers for commercial use are pocket-size MinION Mk 1 system (real-time sequencer for DNA, RNA, and protein analysis, with quick preparation time and the cheapest method for WGS till date) and PromethION system (high output) (Quainoo et al., 2017). Quick et al. (2015) assessed the performance of MinION in sequencing of an outbreak pathogen *S. enterica*. Poultry and products derived from poultry are major source of foodborne pathogens like *Listeria*, *Campylobacter*, and *Salmonella etc.* Chicken is the source of foodborne illness in United States (CDC). The low cost, ease of handling, and operation of these technologies mean that they can further contribute to improve the quality of product, poultry health, and genetic disease resistance.

Food frequently spoils due to microbial or enzymatic activity. This spoilage not only impacts individual health but also has social, economic, and public health implications. Employing proteomics to analyze the nutritional quality of food involves identifying molecular markers for specific spoilage or pathogenic microorganisms (Champomier et al., 2002). *Pseudomonas fluorescens* is commonly associated with food spoilage, particularly in cold storage conditions. Its ability to form biofilms allows it to resist standard disinfection methods, enhancing its persistence in the fresh food supply chain. A comparative proteomic analysis was conducted by Quintieri et al. (2019) to explore the metabolic pathways involved in biofilm formation by the blue-cheese-pigmenting *P. fluorescens* strain ITEM 17298 after incubation at 15°C and 30°C. The same approach was also used to examine the effects of bovine lactoferrin hydrolysate (HLF) as an antibiofilm agent. Proteomic studies for meat quality evaluation have been conducted on various types of meat, including beef, pork, lamb, and chicken (Hollung et al., 2007). The synergistic application of proteomics tools, combined with bioinformatics analysis, is highly significant. Bioinformatics-driven data analysis marks a real advancement in this field, as it is essential for identifying key features of the meat proteome that are correlated with quality indices (Gobert et al., 2014).

Various processing techniques can be applied individually or in combination, depending on the type of food. These treatments can either enhance or diminish the nutritional value of the food. Over the past decade, proteomics has been successfully employed to study proteins and their modifications in food before and after processing. This has provided valuable insights into the molecular changes at the protein level associated with each type of treatment (Piras et al., 2016). Proteomic methods have been effectively employed to study microbe-host interactions, pathogenic mechanisms, toxin biosynthesis, and microbial responses to environmental factors. Consequently, proteomics offers valuable insights that can help develop strategies to prevent mold spoilage and mitigate the risks associated with mycotoxins in food (Álvarez et al., 2023). In their study on proteomics for fish species authentication, Mazzeo and Siciliano (2016) outlined various methods for identifying fishery products using proteomic approaches. They concluded that MALDI-TOF molecular profiling can identify fish species within minutes, while MS-based proteomics techniques not only aid in species identification but also detect major fish allergens like β-PRVBs. Thus, they emphasized that MS-based methods offer the potential for rapid and reliable authentication.

Recent advances in proteomics have gained significant relevance in assessing milk quality for cheese production. Hinz et al. (2012) emphasized the seasonal variations in milk quality for cheese making, demonstrating through 2D electrophoresis how proteolysis at different lactation stages can influence the quality of cheddar cheese. Beta-cypermethrin (β-CY) is a pesticide that can remain in food and pose health risks. Certain food-related bacteria break down β-CY during fermentation. A strain called *Gordonia alkanivorans* GH-1, found in fermented Pixian Doubanjiang, was able to break down 77.11% of β-CY (at 45 mg/L) within 24 hours of using its enzymes. Proteomics analysis showed that an enzyme called esterase breaks the ester bond in β-CY, forming 3-phenoxybenzoic acid, which is then further broken down by other enzymes like oxidoreductases. Gene expression studies confirmed that these enzymes were more active during the process. When one of these enzymes, Est280, was produced in *E. coli*, it improved the breakdown of β-CY by 48.43%, proving its key role (Peng et al., 2024).

10.8 RECENT INNOVATION IN PROTEOMICS AND GENOMICS FOR FOOD SAFETY AND QUALITY

The primary goal of food analysis has consistently been to guarantee food safety and quality while safeguarding consumers from adulteration (Gallo & Ferranti 2016).

10.8.1 Detection of Pathogens

Proteomic approaches aid in identifying microorganisms by analyzing variations in their proteome, which facilitates the detection of various types of pathogens in food (Pavlovic et al., 2013; Shiny Matilda et al., 2020). MALDI-TOF MS, MALDI-TOF, and HPLC ESI MS/MS are effective proteomic-based techniques that have been employed in various studies for the detection, quantification, and identification of different microbes, their toxins, and various allergens in food (Afzaal et al., 2022). Toxins cannot be eliminated by standard food processing methods. However, proteomic techniques such as LS-MS/MS have been employed for detecting toxins like mycotoxins and aflatoxins in food (Callahan et al., 2006). MALDI-TOF MS was used for detecting 24 different food spoilage bacteria and foodborne pathogens, including genera such as *Staphylococcus, Yersinia, Proteus, Escherichia, Lactococcus, Listeria, Pseudomonas, Morganella,* and *Salmonella* (Mazzeo et al., 2006). Several studies have demonstrated the use of MALDI-TOF MS for identifying foodborne pathogens, particularly bacteria. Pathogens like *Listeria* and *Escherichia coli* (Jadhav et al., 2014) can be efficiently detected using this technique, which is both time-saving and cost-effective (Singhal et al., 2015).

10.8.2 Analyzing Allergens

Proteomics and bioinformatics are widely regarded as essential tools for analyzing allergens and studying protein structural changes caused by industrial processes, which are critical for food quality and safety. Ample research demonstrates that this approach is pivotal for high-throughput screening of potential allergens. Additionally, a comprehensive understanding of food allergens could enable various industrial protein applications, including the development of allergen biosensors (Zhou et al., 2019). The use of MS-based proteomic techniques in food science is on the rise, utilizing both targeted and untargeted approaches, which is anticipated to become the predominant technology for food allergen analysis in the coming years. Typically, qualitative and quantitative proteomic analyses involve a two-step process: liquid chromatographic separation followed by MS identification, making them especially valuable for analyzing complex and processed foods (López-Pedrouso et al., 2023).

Shotgun proteomics enables the study of complex protein mixtures by detecting specific peptides produced after trypsin-mediated proteolysis (Monaci et al., 2018; Abril et al., 2023). This approach allows for comprehensive proteome analysis, which can help identify new potential allergens using bioinformatic tools. Online resources such as UniProt and the Basic Local Alignment Search Tool (BLAST) facilitate allergenicity predictions. A key database for protein allergen identification is the UniProt Knowledgebase, which is linked to Allergome and comprises 1,303 reviewed entries from UniProtKB/Swiss-Prot and 3,117 unreviewed entries from UniProtKB/TrEMBL (Mari et al., 2006).

10.8.3 Traceability of Food Origin and Food Authentication

In the global food supply chain, monitoring the traceability and authenticity of origin and food processing is essential to meet consumer expectations. Tracking food information throughout the supply chain is critical for enhancing consumer trust and purchase intent. Meat products, in particular, represent a high-risk category for adulteration or substitution, with numerous cases reported (Li et al., 2020). To address these concerns, technological advancements, including isotope analysis, chemometrics, near-infrared (NIR) spectroscopy, and DNA barcoding, are being developed to improve traceability (Badia-Melis et al., 2015). DNA barcoding serves as a regulatory tool for verifying identity and authenticity, helping to minimize microbiological and toxicological risks linked to food consumption (Dawan and Ahn, 2022). From a molecular perspective, DNA-based methods and rapid evaporative ionization mass spectrometry (REIMS) can effectively identify various animal species in minced meat products (Cavin et al., 2018). Protein also serves as a key target for traceability using techniques such as MS, HPLC, and immunoassays. In this context, proteomic fingerprinting, which detects structural differences in proteins and DNA, is a valuable tool for both quantitative and qualitative analyses (Esteki et al., 2019). Traceability of animal-derived foods is vital for ensuring authenticity. Along with stable isotope and multielement analysis, MS-based proteomics is increasingly used to determine the geographical origin, especially for verifying the authenticity of 'blue foods'. Food and feed traceability remains a key concern, especially in complex supply chains. Metabolomics, closely linked to environmental factors, is effective in verifying product origin. For example, it reveals abiotic effects in plants and helps authenticate food based on species biogeography. Ultra-performance LC–MS has identified 20 biomarkers in corn used for feed and starch. Similarly, metabolomics can protect the unique geographical identity of Ethiopia's Coffea arabica, known for its distinct flavor and market value (Zhang et al., 2024).

10.8.4 Dairy, One Health, and the Global Fight Against Antibiotic Resistance

The World Health Organization (WHO) recommends that 25–33% of dietary protein for malnourished children comes from dairy sources due to its nutritional and health benefits. Advances in genomics have highlighted the beneficial interaction between milk bioactive compounds and gut bacteria like *Bifidobacterium,* which supports gut health, cognitive development, and immunity in children. While dairy products play a crucial role in nutrition and health, the widespread use of antibiotics has contributed to the rise of antibiotic-resistant bacteria, now a global health threat. WHO's 'Global Priority List of Antibiotic-Resistant Bacteria' underscores the urgent need for new antimicrobials. Adopting a One Health approach in dairy production—linking human, animal, and environmental health—can enhance food safety, prevent disease, and support global food security. This includes using clean inputs, with responsible antibiotic use and proper waste disposal, and ensuring that dairy products are free from contaminants. Such integrated practices are vital for reducing zoonotic diseases and promoting sustainable health worldwide (Garcia et al., 2019).

10.9 LIMITATION AND CHALLENGES OF PROTEOMICS AND GENOMICS IN FOOD SAFETY AND QUALITY ANALYSIS

While these techniques offer benefits for food analysis, they have certain limitations, including their dependency on the complexity of the matrix, the requirement for high protein concentration ranges, and the need for multiple processing steps (de Oliveira Vieira et al., 2021). As suggested by most studies, a primary challenge in using proteomics for food authentication lies in extracting proteins from complex matrices and dealing with the low concentration of target proteins. Therefore, improvements in extraction and enrichment processes are needed to enhance assay sensitivity and reproducibility. Another limitation that food analysis laboratories might encounter is the limited availability of reference materials, given the vast range of possible analytes, concentrations, and matrix combinations found in real samples (Ortea et al., 2016). The complexity of the matrix, the high protein concentration ranges required, and the multiple steps involved present some limitations to applying proteomics in studying foodborne molds. To address these challenges, model systems have been developed, and advancements from other scientific fields, such as library-free data-independent acquisition analyses, ion mobility implementation, and the evaluation of post-translational modifications, are expected to be gradually adopted in this area to help prevent unwanted mold contamination in food products (Álvarez et al., 2023). Food proteomics is still in its early stages compared to other life science fields, and proteomic databases for many species remain incomplete. However, as methodologies and applications advance, food proteomics is poised to become an increasingly vital analytical tool for various aspects of food quality assessment (Mora et al., 2018).

10.10 FUTURE SCOPE AND CONCLUSION

Although the use of protein and peptide biomarkers through proteomic technologies is still less common for assessing food authenticity than established methods like immunoassays or DNA analyses, they hold significant potential due to their robustness, sensitivity, multiplexing capabilities, and high-throughput discriminatory power (Ortea et al., 2016). Given the major impact of pre-analytical factors on proteomic study outcomes, it's important to provide thorough details on sample treatment, covering aspects like processing and storage. To improve result reproducibility, standardizing sample treatment protocols and carefully reporting technical details are equally essential. Progress in high-resolution MS instruments is expected to further increase assay sensitivity and specificity. Targeted MS/MS acquisition methods can help close the gap between preliminary discovery studies and final validation experiments, where exhaustive sampling is needed to address variability comprehensively. The development of validated SRM LC–MS methods, combined with the use of certified reference materials, could lead to more reliable analytical results that can be consistently compared across laboratories. Additionally, creating new bioinformatic tools, and applying current ones to food proteomes and matrices, will be key to advancing integrative and systems biology in food technology and nutrition.

REFERENCES

Abril, A. G., Calo-Mata, P., Böhme, K., Villa, T. G., Barros-Velázquez, J., Pazos, M., & Carrera, M. (2023). Shotgun proteomics analysis, functional networks, and peptide biomarkers for seafood-originating biogenic-amine-producing bacteria. *International Journal of Molecular Sciences*, *24*(9), 7704.

Afzaal, M., Saeed, F., Hussain, M., Shahid, F., Siddeeg, A., & Al-Farga, A. (2022). Proteomics as a promising biomarker in food authentication, quality and safety: A review. *Food Science & Nutrition*, *10*(7), 2333–2346.

Agregán, R., Echegaray, N., Nawaz, A., Hano, C., Gohari, G., Pateiro, M., & Lorenzo, J. M. (2021). Foodomic-based approach for the control and quality improvement of dairy products. *Metabolites*, *11*(12), 818. https://doi.org/10.3390/metabo11120818

Álvarez, M., Andrade, M. J., Núñez, F., Rodríguez, M., & Delgado, J. (2023). Proteomics as a new-generation tool for studying moulds related to food safety and quality. *International Journal of Molecular Sciences*, *24*(5), 4709.

Amaral, J. S. (2021). Target and non-target approaches for food authenticity and traceability. *Foods*, *10*(1), 172.

Aydoğan, C. (2024). New advances in nano-liquid chromatography for proteomics analysis. In N. Grinberg & P. W. Carr (Eds.), *Advances in chromatography* (pp. 79–115). CRC Press, Boca Raton.

Badia-Melis, R., Mishra, P., & Ruiz-García, L. (2015). Food traceability: New trends and recent advances. A review. *Food Control*, *57*, 393–401.

Billington, C., Kingsbury, J. M., & Rivas, L. (2022). Metagenomics approaches for improving food safety: A review. *Journal of Food Protection*, *85*(3), 448–464.

Bintsis, T. (2017). Foodborne pathogens. *AIMS Microbiology*, *3*(3), 529.

Brown, E., Dessai, U., McGarry, S., & Gerner-Smidt, P. (2019). Use of whole-genome sequencing for food safety and public health in the United States. *Foodborne Pathogens and Disease*, *16*(7), 441–450.

Brunelle, J. L., & Green, R. (2014). One-dimensional SDS-polyacrylamide gel electrophoresis (1D SDS-PAGE). *Methods in Enzymology*, *541*, 151–159. https://doi.org/10.1016/B978-0-12-420119-4.00012-4

Callahan, J. H., Shefcheck, K. J., Williams, T. L., & Musser, S. M. (2006). Detection, confirmation, and quantification of staphylococcal enterotoxin B in food matrixes using liquid chromatography–mass spectrometry. *Analytical Chemistry*, *78*(6), 1789–1800

Capriotti, A. L., Caruso, G., Cavaliere, C., Foglia, P., Samperi, R., & Laganà, A. (2012). Multiclass mycotoxin analysis in food, environmental and biological matrices with chromatography/mass spectrometry. *Mass Spectrometry Reviews*, *31*(4), 466–503. https://doi.org/10.1002/mas.20351

Carrera, M. (2021). Proteomics and food analysis: Principles, techniques, and applications. *Foods*, *10*(11), 2538.

Carrera, M. (2023). Food safety—Transcriptomics and proteomics. *International Journal of Molecular Sciences*, *24*(24), 17127.

Catherman, A. D., Skinner, O. S., & Kelleher, N. L. (2014). Top-Down proteomics: facts and perspectives. *Biochemical and biophysical research communications*, *445*(4), 683–693. https://doi.org/10.1016/j.bbrc.2014.02.041

Centers for Disease Control and Prevention Burden of Foodborne Illness: Findings. [(accessed on 4th June 2025)]; Available online: https://www.cdc.gov/food-safety/php/data-research/foodborne-illness-burden/?CDC_AAref_Val=https://www.cdc.gov/foodborne burden/2011-foodborne-estimates.html#cdc_report_pub_study_section_3-what-did-we-estimate

Cavin, C., Cottenet, G., Cooper, K. M., & Zbinden, P. (2018). Meat vulnerabilities to economic food adulteration require new analytical solutions. *Chimia*, *72*(10), 697–697.

Champomier-Verge's, M. C., Maguin, E., Mistou, M. Y., Anglade, P., & Chich, J. F. (2002). Lactic acid bacteria and proteomics: Current knowledge and perspectives. *Journal of Chromatography B*, *771*, 329–342.

Chen, S., Li, J., Saleh-Lakha, S., Allen, V., & Odumeru, J. (2011). Multiple-locus variable number of tandem repeat analysis (MLVA) of *Listeria monocytogenes* directly in food samples. *International Journal of Food Microbiology*, *148*(1), 8–14.

Cifuentes, A. (2017). Foodomics, foodome and modern food analysis. *Trends in Analytical Chemistry*, *96*, 1. https://doi.org/10.1016/j.trac.2017.09.001

Cook, P. W., and Nightingale, K. K. (2018). Use of omics methods for the advancement of food quality and food safety. *Animal Frontiers: The Review Magazine of Animal Agriculture*, *8*(4), 33–41.

Coombs, K. M. (2020). Update on Proteomic approaches to uncovering virus-induced protein alterations and virus-host protein interactions during the progression of viral infection. *Expert Review of Proteomics*, *17*(7–8), 513–532. https://doi.org/10.1080/14789450.2020.1821656

D'Alessandro, A., & Zolla, L. (2012). We are what we eat: Food safety and proteomics. *Journal of Proteome Research*, *11*(1), 26–36.

Dawan, J., & Ahn, J. (2022). Application of DNA barcoding for ensuring food safety and quality. *Food Science and Biotechnology*, *31*(11), 1355–1364.

Deng, X., den Bakker, H. C., & Hendriksen, R. S. (2016). Genomic epidemiology: Whole-genome-sequencing–powered surveillance and outbreak investigation of foodborne bacterial pathogens. *Annual Review of Food Science and Technology*, *7*(1), 353–374.

de Oliveira Vieira, K. C., da Silva, H. R. A., Rocha, I. P. M., Barboza, E., & Eller, L. K. W. (2021). Foodborne pathogens in the omics era. *Critical Reviews in Food Science and Nutrition*, *62*, 6726–6741.

Deutsch, E. W., Mendoza, L., Shteynberg, D. D., Hoopmann, M. R., Sun, Z., Eng, J. K., & Moritz, R. L. (2023). Trans-proteomic pipeline: Robust mass spectrometry-based proteomics data analysis suite. *Journal of Proteome Research*, *22*(2), 615–624.

Dupree, E. J., Jayathirtha, M., Yorkey, H., Mihasan, M., Petre, B. A., & Darie, C. C. (2020). A critical review of bottom-up proteomics: the good, the bad, and the future of this field. *Proteomes*, *8*(3), 14.

Elmore, J. M., Griffin, B. D., & Walley, J. W. (2021). Advances in functional proteomics to study plant-pathogen interactions. *Current Opinion in Plant Biology*, *63*, 102061. https://doi.org/10.1016/j.pbi.2021.102061

Esteki, M., Regueiro, J., & Simal-Gándara, J. (2019). Tackling fraudsters with global strategies to expose fraud in the food chain. *Comprehensive Reviews in Food Science and Food Safety*, *18*(2), 425–440.

Fagerquist, C. K., Zaragoza, W. J., Sultan, O., Woo, N., Quiñones, B., Cooley, M. B., & Mandrell, R. E. (2014). Top-down proteomic identification of Shiga toxin 2 subtypes from Shiga toxin-producing Escherichia coli by matrix-assisted laser desorption ionization-tandem time of flight mass spectrometry. *Applied and Environmental Microbiology*, *80*(9), 2928–2940. https://doi.org/10.1128/AEM.04058-13

Fanelli, V., Mascio, I., Miazzi, M. M., Savoia, M. A., De Giovanni, C., & Montemurro, C. (2021). Molecular approaches to agri-food traceability and authentication: An updated review. *Foods*, *10*(7), 1644.

Fugett, E. B., Schoonmaker-Bopp, D., Dumas, N. B., Corby, J., & Wiedmann, M. (2007). Pulsed-field gel electrophoresis (PFGE) analysis of temporally matched Listeria monocytogenes isolates from human clinical cases, foods, ruminant farms, and urban and natural environments reveals source-associated as well as widely distributed PFGE types. *Journal of Clinical Microbiology*, *45*(3), 865–873.

Gallo, M., & Ferranti, P. (2016). The evolution of analytical chemistry methods in foodomics. *Journal of Chromatography A*, *1428*, 3–15.

Gao, B., Lu, W., Jin, M., & Chen, Y. (2023). Non-targeted metabolomics of moldy wheat by ultra-performance liquid chromatography–quadrupole time-of-flight mass spectrometry. *Frontiers in Microbiology*, *14*, 1136516.

Garcia, S. N., Osburn, B. I., & Cullor, J. S. (2019). A one health perspective on dairy production and dairy food safety. *One Health*, *7*, 100086.

Gauthier, M., & Blais, B. W. (2005). Cloth-based hybridization array system for detection of toxin genes associated with major foodborne pathogenic bacteria. *Journal of Rapid Methods & Automation in Microbiology*, *13*(4), 243–256.

Geissler, M., Clime, L., Hoa, X. D., Morton, K. J., Hébert, H., Poncelet, L., . . . & Veres, T. (2015). Microfluidic integration of a cloth-based hybridization array system (CHAS) for rapid, colorimetric detection of enterohemorrhagic escherichia coli (EHEC) using an articulated, centrifugal platform. *Analytical Chemistry*, *87*(20), 10565–10572.

Ghahary, A., & Abiri, R. (2021). Foodomics; principles, challenges, and applications: A promising tool for food analysis. *Acta Scientific Nutritional Health (ISSN: 2582–1423)*, *5*(3).

Gobert, M., Sayd, T., Gatellier, P., Sant´e-Lhoutellier, V. (2014). Application to proteomics to understand and modify meat quality. *Meat Science*, *98*, 539–543.

Grazina, L., Rodrigues, P. J., Igrejas, G., Nunes, M. A., Mafra, I., Arlorio, M., . . . & Amaral, J. S. (2020). Machine learning approaches applied to gc-fid fatty acid profiles to discriminate wild from farmed salmon. *Foods*, *9*(11), 1622.

Hinz, K., O'Connor, P. M., O'Brien, B., Huppertz, T., et al. (2012). Proteomic study of proteolysis during ripening of Cheddar cheese made from milk over a lactation cycle. *Journal of Dairy Research*, *79*, 176–184.

Hollung, K., Veiseth, E., Jia, X., Færgestad, E. M., & Hildrum, K. I. (2007). Application of proteomics to understand the molecular mechanisms behind meat quality. *Meat Science*, *77*, 97–104.

Jadhav, S., Sevior, D., Bhave, M., & Palombo, E. A. (2014). Detection of Listeria monocytogenes from selective enrichment broth using MALDI-TOF Mass Spectrometry. *Journal of Proteomics*, *97*, 100–106. https://doi.org/10.1016/j.jprot.2013.09.014

Jagadeesan, B., Gerner-Smidt, P., Allard, M. W., Leuillet, S., Winkler, A., Xiao, Y., . . . & Grant, K. (2019). The use of next generation sequencing for improving food safety: Translation into practice. *Food Microbiology*, *79*, 96–115.

Kwong, J. C., Mercoulia, K., Tomita, T., Easton, M., Li, H. Y., Bulach, D. M., . . . & Howden, B. P. (2016). Prospective whole-genome sequencing enhances national surveillance of Listeria monocytogenes. *Journal of Clinical Microbiology*, *54*(2), 333–342.

Quintieri, L., Zühlke, D., Fanelli, F., Caputo, L., Liuzzi, V. C., Logrieco, A. F., . . . & Riedel, K. (2019). Proteomic analysis of the food spoiler Pseudomonas fluorescens ITEM 17298 reveals the antibiofilm activity of the pepsin-digested bovine lactoferrin. *Food microbiology*, *82*, 177–193.

Law, J. W. F., Ab Mutalib, N. S., Chan, K. G., & Lee, L. H. (2015). Rapid methods for the detection of foodborne bacterial pathogens: Principles, applications, advantages and limitations. *Frontiers in Microbiology*, *5*, 770.

Li, Y., Han, Y., Zhao, Q., Tang, C., Zhang, J., & Qin, Y. (2022). Fermented soy and fish protein dietary sources shape ileal and colonic microbiota, improving nutrient digestibility and host health in a piglet model. *Frontiers in Microbiology*, *13*, 911500. https://doi.org/10.3389/fmicb.2022.911500

Li, Y. C., Liu, S. Y., Meng, F. B., Liu, D. Y., Zhang, Y., Wang, W., & Zhang, J. M. (2020). Comparative review and the recent progress in detection technologies of meat product adulteration. *Comprehensive Reviews in Food Science and Food Safety*, *19*(4), 2256–2296.

Lindstedt, B. A., Vardund, T., Aas, L., & Kapperud, G. (2004). Multiple-locus variable-number tandem-repeats analysis of Salmonella enterica subsp. enterica serovar Typhimurium using PCR multiplexing and multicolor capillary electrophoresis. *Journal of Microbiological Methods*, *59*(2), 163–172.

Lindstedt, B. A., Heir, E., Gjernes, E., Vardund, T., & Kapperud, G. (2003). DNA fingerprinting of Shiga-toxin producing Escherichia coli O157 based on multiple-locus variable-number tandem-repeats analysis (MLVA). *Annals of Clinical Microbiology and Antimicrobials*, *2*(1), 12.

Liu, Y., Shi, X., Li, Y., Chen, Q., Jiang, M., Li, W., & Hu, Q. (2016). The evaluation and application of multilocus variable number tandem repeat analysis (MLVA) for the molecular epidemiological study of Salmonella enterica subsp. enterica serovar Enteritidis infection. *Annals of Clinical Microbiology and Antimicrobials*, *15*, 1–9.

Liu, Z. J. (2007). Fish genomics and analytical genetic technologies, with examples of their potential applications in management of fish genetic resources. *America*, *5*, 145–179.

López-Pedrouso, M., Lorenzo, J. M., Alché, J. D. D., Moreira, R., & Franco, D. (2023). Advanced proteomic and bioinformatic tools for predictive analysis of allergens in novel foods. *Biology*, *12*(5), 714.

Maestrello, V., Solovyev, P., Bontempo, L., Mannina, L., & Camin, F. (2022). Nuclear magnetic resonance spectroscopy in extra virgin olive oil authentication. *Comprehensive Reviews in Food Science and Food Safety*, *21*(5), 4056–4075.

Magdeldin, S., Enany, S., Yoshida, Y., Xu, B., Zhang, Y., Zureena, Z., Lokamani, I., Yaoita, E., & Yamamoto, T. (2014). Basics and recent advances of two dimensional-polyacrylamide gel electrophoresis. *Clinical Proteomics*, *11*(1), 16. https://doi.org/10.1186/1559-0275-11-16

Mari, A., Scala, E., Palazzo, P., Ridolfi, S., Zennaro, D., & Carabella, G. (2006). Bioinformatics applied to allergy: Allergen databases, from collecting sequence information to data integration. The Allergome platform as a model. *Cellular Immunology*, *244*, 97–100.

Martinez-Perez, A., & Blais, B. W. (2010). Cloth-based hybridization array system for the identification of Escherichia coli O157: H7. *Food Control*, *21*(10), 1354–1359.

Mazzeo, M. F., Sorrentino, A., Gaita, M., Cacace, G., Di Stasio, M., Facchiano, A., . . . & Siciliano, R. A. (2006). Matrix-assisted laser desorption ionization-time of flight mass spectrometry for the discrimination of food-borne microorganisms. *Applied and environmental microbiology*, *72*(2), 1180–1189.

Mazzeo, M. F., & Siciliano, R. A. (2016). Proteomics for the authentication of fish species. *Journal of Proteomics*, *147*, 119–124. https://doi.org/10.1016/j.jprot.2016.03.007

Monaci, L., De Angelis, E., Montemurro, N., & Pilolli, R. (2018). Comprehensive overview and recent advances in proteomics MS based methods for food allergens analysis. *Trends in Analytical Chemistry*, *106*, 21–36.

Mora, L., Gallego, M., & Toldrá, F. (2018). New approaches based on comparative proteomics for the assessment of food quality. *Current Opinion in Food Science*, *22*, 22–27.

Mullen, A. M., Stapleton, P. C., Corcoran, D., Hamill, R. M., & White, A. (2006). Understanding meat quality through the application of genomic and proteomic approaches. *Meat Science*, *74*(1), 3–16.

Muñoz-Atienza, E., Araújo, C., del Campo, R., Hernández, P. E., Herranz, C., & Cintas, L. M. (2016). Safety assessment and molecular genetic profiling by pulsed-field gel electrophoresis (PFGE) and PCR-based techniques of Enterococcus faecium strains of food origin. *LWT-Food Science and Technology*, *65*, 357–362.

Murase, T., Nakamura, A., Matsushima, A., & Yamai, S. (1996). An epidemiological study of salmonella enteritidis by pulsed-field gel electrophoresis (PFGE): Several PFGE patterns observed in isolates from a food poisoning outbreak. *Microbiology and Immunology*, *40*(11), 873–875.

Nadler, W. M., Waidelich, D., Kerner, A., Hanke, S., Berg, R., Trumpp, A., & Rösli, C. (2017). MALDI versus ESI: The impact of the ion source on peptide identification. *Journal of Proteome Research*, *16*(3), 1207–1215. https://doi.org/10.1021/acs.jproteome.6b00805

Nadon, C. A., Trees, E., Ng, L. K., Møller Nielsen, E., Reimer, A., Maxwell, N., . . . & the MLVA Harmonization Working Group, C. (2013). Development and application of MLVA methods as a tool for inter-laboratory surveillance. *Eurosurveillance*, *18*(35).

Noormohamed, A., & Fakhr, M. K. (2014). Molecular typing of Campylobacter jejuni and Campylobacter coli isolated from various retail meats by MLST and PFGE. *Foods*, *3*(1), 82–93.

Orduna, A. R., Husby, E., Yang, C. T., Ghosh, D., & Beaudry, F. (2017). Detection of meat species adulteration using high-resolution mass spectrometry and a proteogenomics strategy. *Food Additives & Contaminants: Part A*, *34*(7), 1110–1120.

Orsburn, B. C. (2021). Proteome discoverer: A community enhanced data processing suite for protein informatics. *Proteomes*, *9*(1), 15. https://doi.org/10.3390/proteomes9010015.

Ortea, I., O'Connor, G., & Maquet, A. (2016). Review on proteomics for food authentication. *Journal of Proteomics*, *147*, 212–225.

Park, S., Jung, H., Lee, M., Choi, H., Kim, J., Jung, J., . . . & Jung, K. (2016). Detection of Listeria monocytogenes in foods and characterization by PFGE. *Advances in Microbiology*, *6*(4), 343.

Pavlovic, M., Huber, I., Konrad, R., & Busch, U. (2013). Application of MALDI-TOF MS for the identification of food borne bacteria. *The Open Microbiology Journal*, *7*, 135. https://doi.org/10.2174/18742 85801307010135

Peng, C., Tang, J., Yu, X., Zhou, X., Wang, M., Zhang, Y., . . . & Cai, T. (2025). Biodegradation of various phthalic acid esters at high concentrations by Gordonia alkanivorans GH-1 and its degradation mechanism. *Environmental Technology & Innovation*, *38*, 104066.

Pérez-Losada, M., Cabezas, P., Castro-Nallar, E., & Crandall, K. A. (2013). Pathogen typing in the genomics era: MLST and the future of molecular epidemiology. *Infection, Genetics and Evolution*, *16*, 38–53.

Perkons, I., Varunjikar, M. S., & Rasinger, J. D. (2023). Development and testing of proteomics tools and databases for the species and tissue-specific identification of processed animal protein (PAP) in aquafeed. *EFSA Journal*, *21*(S1), 1–15. https://doi.org/10.2903/j.efsa.2023.e211013

Pinu, F. R. (2016). Early detection of food pathogens and food spoilage microorganisms: Application of metabolomics. *Trends in Food Science & Technology*, *54*, 213–215.

Piras, C., Roncada, P., Rodrigues, P. M., Bonizzi, L., & Soggiu, A. (2016). Proteomics in food: Quality, safety, microbes, and allergens. *Proteomics*, *16*(5), 799–815. https://doi.org/10.1002/pmic.201500369

Pitt, T. L., & Barer, M. R. (2012). *Classification, Identification and Typing of Micro-Organisms* In *Medical Microbiology*. Greenwood, D., Barer, M., Slack, R., and Irving, W (Eds.). Churchill Livingstone. (18th ed., pp. 24–38).

Pusch, W., & Kostrzewa, M. (2005). Application of MALDI-TOF mass spectrometry in screening and diagnostic research. *Current pharmaceutical design*, *11*(20), 2577–2591. https://doi.org/10.2174/1381612054546932

Qian, C., Murphy, S. I., Orsi, R. H., & Wiedmann, M. (2023). How can AI help improve food safety? *Annual Review of Food Science and Technology*, *14*(1), 517–538.

Quainoo, S., Coolen, J. P., van Hijum, S. A., Huynen, M. A., Melchers, W. J., van Schaik, W., & Wertheim, H. F. (2017). Whole-genome sequencing of bacterial pathogens: The future of nosocomial outbreak analysis. *Clinical Microbiology Reviews*, *30*(4), 1015–1063.

Quick, J., Ashton, P., Calus, S., Chatt, C., Gossain, S., Hawker, J., & Loman, N. J. (2015). Rapid draft sequencing and real-time nanopore sequencing in a hospital outbreak of Salmonella. *Genome Biology*, *16*, 1–14.

Quintieri, L., Zühlke, D., Fanelli, F., Caputo, L., Liuzzi, V. C., Logrieco, A. F., Hirschfeld, C., Becher, D., Riedel, K., Laura, Q. (2019). Proteomic analysis of the food spoiler Pseudomonas fluorescens ITEM 17298 reveals the antibiofilm activity of the pepsin-digested bovine lactoferrin. *Food Microbiol. 82*, 177–193.

Reinhardt, T. A., Sacco, R. E., Nonnecke, B. J., & Lippolis, J. D. (2013). Bovine milk proteome: Quantitative changes in normal milk exosomes, milk fat globule membranes and whey proteomes resulting from Staphylococcus aureus mastitis. *Journal of Proteomics*, *82*, 141–154. https://doi.org/10.1016/j.jprot.2013.02.013

Restani, P., Fiocchi, A., Beretta, B., Velonà, T., Giovannini, M., & Galli, C. L. (1997). Meat allergy: III—Proteins involved and cross-reactivity between different animal species. *Journal of the American College of Nutrition*, *16*(4), 383–389. https://doi.org/10.1080/07315724.1997.10718701

Restani, P., Gaiaschi, A., Plebani, A., Beretta, B., Cavagni, G., Fiocchi, A., Poiesi, C., Velonà, T., Ugazio, A. G., & Galli, C. L. (1999). Cross-reactivity between milk proteins from different animal species. *Clinical and Experimental Allergy: Journal of the British Society for Allergy and Clinical Immunology*, *29*(7), 997–1004. https://doi.org/10.1046/j.1365-2222.1999.00563.x

Rysova, L., Cejnar, P., Hanus, O., Legarova, V., Havlik, J., Nejeschlebova, H., . . . & Bozik, M. (2022). Use of MALDI-TOF MS technology to evaluate adulteration of small ruminant milk with raw bovine milk. *Journal of Dairy Science*, *105*(6), 4882–4894.

Saeed, F., Afzaal, M., Hussain, M., & Tufail, T. (2021). Advances in assessing product quality. In *Food losses, sustainable postharvest and food technologies* (pp. 191–218). Academic Press.

Sarkar, D. (2025). Role of artificial intelligence in increasing agricultural productivity and ensuring global food security through study of plant protein-protein interaction networks and interactomes. *Journal of Plant Science & Research*, *12*(2), 279.

Sassi, M., Arena, S., & Scaloni, A. (2015). MALDI-TOF-MS platform for integrated proteomic and peptidomic profiling of milk samples allows rapid detection of food adulterations. *Journal of Agricultural and Food Chemistry*, *63*(27), 6157–6171.

Serim, S., Haedke, U., & Verhelst, S. H. (2012). Activity-based probes for the study of proteases: Recent advances and developments. *ChemMedChem*, *7*(7), 1146–1159. https://doi.org/10.1002/cmdc.201200057

Shin, J., Kim, M. J., & Kim, H. Y. (2021). Development of triplex PCR for simultaneous detection of soybean and wheat. *Food Science and Biotechnology*, *30*, 801–806.

Shiny Matilda, C., Madhusudan, I., Gaurav Isola, R., & Shanthi, C. (2020). Potential of proteomics to probe microbes. *Journal of Basic Microbiology*, *60*(6), 471–483. https://doi.org/10.1002/jobm.20190 0628

Singhal, N., Kumar, M., Kanaujia, P. K., & Virdi, J. S. (2015). MALDI-TOF mass spectrometry: An emerging technology for microbial identification and diagnosis. *Frontiers in Microbiology*, *6*, 791.

Sobral, D., Schwarz, S., Bergonier, D., Brisabois, A., Feßler, A. T., Gilbert, F. B., & Vergnaud, G. (2012). High throughput multiple locus variable number of tandem repeat analysis (MLVA) of Staphylococcus aureus from human, animal and food sources. *PLoS ONE*, *7*(5), e33967.

Soroushianfar, M., Asgari, G., Afzali, F., Falahat, A., Soroush, M., Baghahi, M., . . . & Nazari, E. (2025). Application of bioinformatics and machine learning tools in food safety. *Current Nutrition Reports*, *14*(1), 67.

Tyanova, S., Temu, T., Carlson, A., Sinitcyn, P., Mann, M., & Cox, J. (2015). Visualization of LC-MS/MS proteomics data in MaxQuant. *Proteomics*, *15*(8), 1453–1456.

Tyanova, S., Temu, T., Sinitcyn, P., Carlson, A., Hein, M. Y., Geiger, T., Mann, M., & Cox, J. (2016). The Perseus computational platform for comprehensive analysis of (prote)omics data. *Nature Methods*, *13*(9), 731–740. https://doi.org/10.1038/nmeth.3901

Viswanathan, S., Unlü, M., & Minden, J. S. (2006). Two-dimensional difference gel electrophoresis. *Nature Protocols*, *1*(3), 1351–1358. https://doi.org/10.1038/nprot.2006.234

Weis, C. V., Jutzeler, C. R., & Borgwardt, K. (2020). Machine learning for microbial identification and antimicrobial susceptibility testing on MALDI-TOF mass spectra: A systematic review. *Clinical Microbiology and Infection*, *26*, 101310–101317.

WHO. (2023). WHO issues rapid communication on use of targeted next-generation sequencing for diagnosis of drug-resistant tuberculosis. Accessed on May 2025. https://www.who.int/news/item/25-07-2023-who-issues-rapid-communication-on-use-of-targeted-next-generation-sequencing-for-diagnosis-of-drug-resistant-tuberculosis

Wilkins, M. R., Sanchez, J. C., Gooley, A. A., Appel, R. D., Humphery-Smith, I., Hochstrasser, D. F., & Williams, K. L. (1996). Progress with proteome projects: Why all proteins expressed by a genome should be identified and how to do it. *Biotechnology & Genetic Engineering Reviews*, *13*, 19–50. https://doi.org/10.1080/02648725.1996.10647923

Xia, X., Zhao, S., Smith, A., McEvoy, J., Meng, J., & Bhagwat, A. A. (2009). Characterization of Salmonella isolates from retail foods based on serotyping, pulse field gel electrophoresis, antibiotic resistance and other phenotypic properties. *International Journal of Food Microbiology*, *129*(1), 93–98.

Zhang, J., Wang, X., Huo, D., Li, W., Hu, Q., Xu, C., . . . & Li, C. (2016). Metagenomic approach reveals microbial diversity and predictive microbial metabolic pathways in Yucha, a traditional Li fermented food. *Scientific Reports*, *6*(1), 32524.

Zhang, Q. H., Huang, L., Zhang, Y., Ke, C. H., & Huang, H. Q. (2013). Proteomic approach for identifying gonad differential proteins in the oyster (*Crassostreaangulata*) following food-chain contamination with $HgCl_2$. *Journal of Proteomics*, *94*, 37–53. https://doi.org/10.1016/j.jprot.2013.08.018

Zhang, S., Chen, J., Gao, F., Su, W., Li, T., & Wang, Y. (2024). Foodomics as a tool for evaluating food authenticity and safety from field to table: A review. *Foods*, *14*(1), 15.

Zhou, J., Qi, Q., Wang, C., Qian, Y., Liu, G., Wang, Y., & Fu, L. (2019). Surface plasmon resonance (SPR) biosensors for food allergen detection in food matrices. *Biosensors and Bioelectronics*, *142*, 111449.

Zhang, Y., Fonslow, B. R., Shan, B., Baek, M. C., & Yates III, J. R. (2013). Protein analysis by shotgun/bottom-up proteomics. *Chemical reviews*, *113*(4), 2343–2394.

11 Principle and Application of Surface-Based Analysis in Food Quality and Safety Analysis

Milad Tavassoli, Behnam Bahramian, Reza Abedi-Firoozjah, Ali Ehsani, and Ehsan Sadeghi

11.1 INTRODUCTION

Over the past ten years, the food industry's growth has presented a challenge with regard to food contamination and food fraud (Moradi et al., 2021; Sadeghi et al., 2023). In the trade of food items, it is imperative for all nations to adhere to global standards for food safety. Due to this, organizations worldwide, both public and private, are interested in finding affordable and effective methods for testing food safety (Abedi-Firoozjah, et al., 2024a; Gordon, 2017). Risks related to food products, like toxins, microbial contamination, hazards during food processing, presence of heavy metals, antibiotic residues, and pesticides, pose a significant danger to food safety (King, et al., 2017). Concerns about the inadequate safety of food items can have adverse effects on both economic and health sectors. Therefore, ensuring food safety is crucial and has emerged as a significant global issue. Employing technological methods to diagnose and monitor food safety is vital. Hence, in order to guarantee the safety of food items, it is essential to create precise, rapid, and reliable techniques for advancing food safety detection technology. The utilization of surface-centered analysis, which encompasses techniques like surface-enhanced Raman spectroscopy (SERS), surface plasmon resonance (SPR), surface-based biosensors, has risen in recent times for identifying contamination and adulteration in food items. Surface-centered analysis techniques have gained significance in identifying contamination in food items because of their nondestructive detection, high sensitivity, rapid and continuous detection in small sample sizes, cost-effectiveness, and low usage of reagents and energy (Yu Jiang & Wu, 2019; Shankaran, Gobi, & Miura, 2007).

Surface-based sensing analysis has created a lot of excitement in the chemical science field, encompassing various topics such as plasmonics, nanoscience, surface chemistry, analytical methods, and interfacial science (Guo et al., 2021; Jackman, Ferhan, & Cho, 2019). Particular interest lies in utilizing the simple equipment requirements and exceptional surface detection capabilities of these sensors to study

DOI: 10.1201/9781003482611-11

biomacromolecules and biological nanoparticles. Our recent work in exploring surface-based sensing systems for food analysis science is comprehensively detailed in this book chapter, along with our thoughts on potential future research directions. We start by introducing the basic design principles and measurement methods that form the foundation of surface-based sensing systems, and then we carefully examine recent examples of their application in our book chapter projects. Additionally, we investigate the potential for integrating these sensors with other surface-sensitive measurement techniques. Looking ahead, the prospects for the continued use of surface-based nanoplasmonic sensors in biointerfacial science applications are promising, with numerous opportunities for innovation from both theoretical and practical perspectives.

11.2 PRINCIPLE OF SURFACE-BASED ANALYSIS IN FOOD QUALITY AND SAFETY ANALYSIS

Investigating the interactions between food and its environment at the interface is important for evaluating food quality and safety. Analyzing the surfaces and interfaces of food components provides valuable information about their quality, safety, and characteristics (Dale, Fischl, & Sereno, 1999). Surface-based methods used in food analysis offer important insights into the makeup, arrangement, and actions of food elements. This enables the detection of impurities and additives as well as the supervision of quality factors. By concentrating on the exterior features of food substances, researchers can gain a deeper understanding of the connections between food constituents and their surroundings, resulting in more precise evaluations of food integrity and security.

11.2.1 Surface-Enhanced Raman Spectroscopy (SERS)

In general, the mechanism of this method is based on the two bases of chemical or charge-transfer enhancement and electromagnetic field enhancement, and the way these methods work is different. Most researchers have identified the SPR mechanism in the electromagnetic field enhancement method (Wang et al., 2022a). This method enhances inelastic and incident light scattered by Raman molecules absorbed by plasmonic metal nanostructures. In this method, both the inelastic and incident lights that are scattered are absorbed by the Raman molecule and amplified by the plasmonic metal nanostructure. These lights scattered by the plasmonic surface of the metal nanostructure caused an increase in the Raman signal. This shows that the nanoparticles are illuminated by light at a specific wavelength, causing an oscillating dipole. The amplification obtained in this method depends on the polarization of the light, the localization of the molecules in the complex, the excitation wavelength, and the morphology of the assembled nanoparticles. In the method of chemical enhancement or charge transfer with the effect of charge transfer, nanomaterials intensify with incident light (Jiang et al., 2021). By absorbing molecules on the surface of the noble metal nanostructure, new electronic forms are created. The generated electronic forms act as resonant intermediate forms in Raman scattering (Figure 11.1).

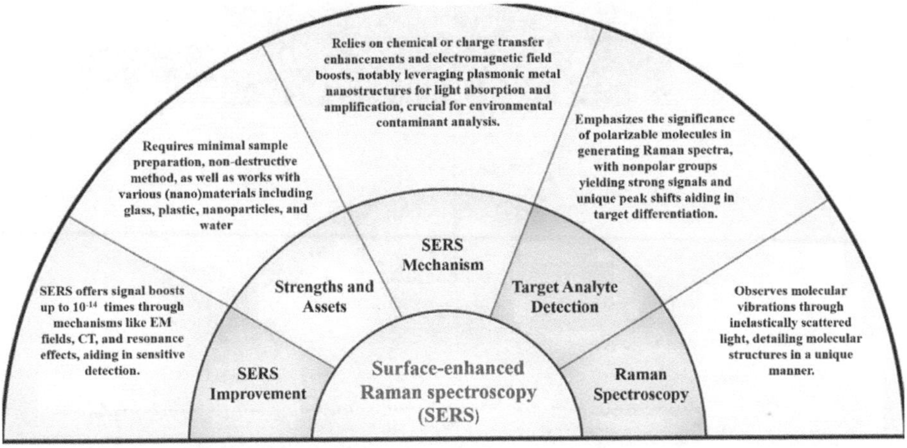

FIGURE 11.1 Schematic diagrams illustrating the benefits of Raman spectroscopy, the amplified capabilities of SERS, the fundamentals of analyte detection, the mechanisms of signal improvement, and their pivotal role in expanding SERS utility for food safety/quality monitoring and analysis.

Charge transfer becomes thrilling when the metal possesses a fermi level located in between the highest occupied molecular orbital and the lowest unoccupied molecular orbital in terms of energy. In Albrecht's classification, if laser excitation closely aligns with a permissible electronic transition and a single excited state is engaged in a charge transfer process, solely the symmetric modes undergo resonant enhancement. When two excited states of the chromophore interact with these modes, resonance Raman effects occur on nonsymmetric vibrational modes. This phenomenon is elucidated by the process known as the B mechanism in Albrecht's system or the Herzberg-Teller mechanism (Xia et al., 2014).

11.2.2 Surface Plasmon Resonance

SPR is an optical occurrence taking place at the surface of a thin film (Yola, Atar, & Eren, 2014). When the incident wavelength remains constant, variations in the resonance angle are primarily influenced by the concentration of the substance bonded to the surface of the film. The unique interaction between the substance immobilized on the film surface and the analyte results in modifying the response signal. Utilizing SPR proves to be a potent and delicate technique for scrutinizing molecular interactions (Paliwal et al., 2015). SPR has the capability to observe real-time interactions of biological molecules without the need for labeling. The fundamental concept of SPR is illustrated in Figure 11.2. When light reflects off the interface between a metal and a medium, a phenomenon called SPR occurs. This happens when the evanescent wave generated during total internal reflection of light stimulates the release of free electrons on the metal surface, leading to the formation of surface plasmons. When the frequency and wave number of the incident light align with the resonant frequency of the metal, a resonance phenomenon occurs. This results in

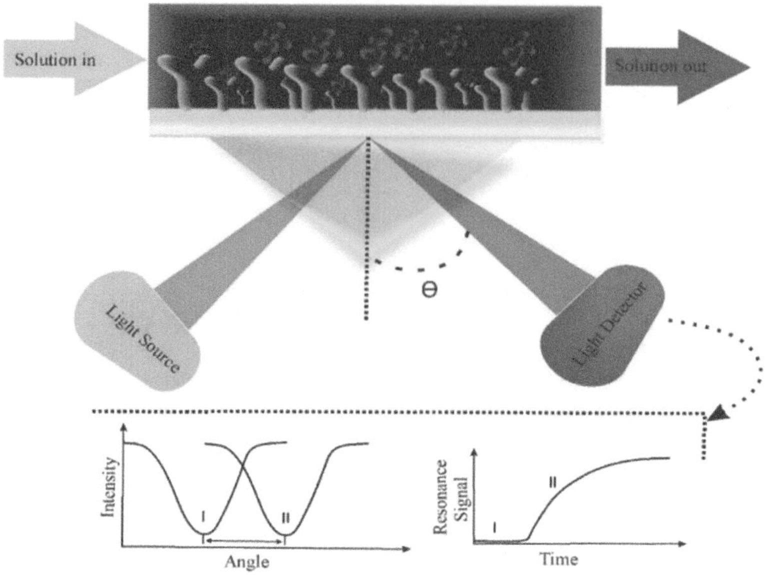

FIGURE 11.2 A visual representation of the traditional Kretschmann optical setup used in SPR biosensing systems.

the absorption of the incoming light, a noticeable decrease in reflected light energy, and the emergence of a distinct peak in the reflection spectrum, indicating the point of minimum reflection intensity. By observing changes in this peak, we can detect alterations in the quantity or arrangement of molecules attached or adsorbed on the metal surface (Schasfoort, 2017; Wei et al., 2022).

11.2.3 Localized SPR

The vibrant hues of noble metal particles have long been a source of fascination, historically employed as colorful additives in stained glass windows. In contemporary times, the adjustable optical characteristics of metal nanoparticles and their manipulability through spectroscopic methods have once again placed them at the forefront of both fundamental and applied scientific pursuits (Chen & Ming, 2012). Notably, gold and silver nanoparticles have garnered significant attention for their role as highly sensitive optical sensors that operate without the need for labels and in minimal volumes (Mayer & Hafner, 2011; Willets & Van Duyne, 2007). These plasmonic nanoparticles function as converters that translate alterations in the nearby refractive index into shifts in the spectral properties of the localized surface plasmon resonance (LSPR) band. The utilization of the LSPR-shift mechanism offers a versatile approach for gauging bonding affinities and reaction rates triggered by molecules impacting the refractive index surrounding these metallic nanostructures (Chen &

Ming, 2012). Through the attachment of specific recognition components (be they chemical or biological ligands) onto these nanostructures, the resulting nanosensor gains selectivity and precision in detecting the target analyte (Willets & Van Duyne, 2007). Given its capabilities, LSPR spectroscopy with metallic nanoparticles emerges as a potent tool for conducting chemical and biological sensing assays. Furthermore, the LSPR phenomenon plays a crucial role in facilitating electromagnetic field amplification, pivotal for SERS-based sensors and comparable surface-enhanced spectroscopic methodologies (Mayer & Hafner, 2011).

Scientists have found that materials with specific dielectric properties can support SPR sensors via coherent electron oscillations triggered by light. Plasmonics explores these interactions, leading to diverse applications, aided by nanoscale metal structure advancements (Willets & Van Duyne, 2007). Table 11.1 outlines SPR and LSPR sensor system differences.

In the intriguing realm of surface plasmon polaritons, minuscule waves of electron density known as plasmons journey along the boundary between a metal and dielectric material, traversing short to relatively long distances horizontally before dissipating vertically at a distance of approximately 200 nanometers. When these plasmons interact with a molecular layer on the surface, their resonance properties undergo alterations. These modifications can be observed through changes in the angle of incidence, shifts in the plasmon's wavelength, and through the use of advanced imaging techniques (Mayer & Hafner, 2011). In the case of localized surface plasmons, there is interaction with particles that are smaller than the incident wavelength, leading to the emergence of a plasmonic oscillation referred to as the LSPR. Similar to the SPR, the LSPR responds to changes in the local dielectric environment. These changes are often monitored through assessments of wavelength shifts, and there is the potential for angle-resolved sensing (Willets et al., 2007). Research interest in understanding propagating and localized surface plasmons

TABLE 11.1
A Concise Comparison of SPR and LSPR According to Their Different Properties

Properties	SPR	LSPR
Propagation	Propagates along metal-dielectric	Local oscillation around nanoparticle
Distance	Tens to hundreds of microns	Localized to nanoparticle vicinity
Sensitivity	Higher to bulk refractive index	Higher to short-range refractive index
Sensing	Angle resolved, wavelength shift	Wavelength-shift primarily for environmental
Modes	Imaging	Changes, angle-resolved sensing possible
Benefits	Established in commercial use	High sensitivity, tunable wavelengths

Abbreviations: SPR, surface plasmon resonance, LSPR; localized surface plasmon resonance heightened with the discovery of SERS, where diverse enhancements on rough metal surfaces signify intricate interplays of chemical and electromagnetic field mechanisms (Mayer & Hafner, 2011; Nishi et al., 2015). While SPR is common, LSPR offers similar advantages with added benefits for sensing. Both methods offer data on binding processes, with LSPR excelling in short-range refractive index changes due to its smaller sensing area.

11.2.4 Microfluidic Detectors

Bio or nanosensors represent technologies that enable the development of dependable, quick, and cost-effective microfluidic analysis systems (Gao et al., 2022; Lace & Cleary, 2021). Arrays of sensors exploit the merging of numerous recognition components on a solitary micro detector, facilitating the identification of a wide range of molecular substances within intricate samples. Progress in lab-on-a-chip technology has propelled microfluidic methods, allowing swift and portable detection of various beneficial or detrimental elements with substantial sensitivity, efficiency, and minimal sample usage (Atalay et al., 2011).

The rise of sophisticated microfabrication methods has revamped the realm of analytical chemistry research. This transformation involves downsizing traditional analytical methods like bioassays, separation techniques, and chemical synthesis into compact systems capable of precise manipulation of minute liquid quantities and conducting analyses swiftly and with unparalleled efficiency (Atalay et al., 2011; Gao et al., 2022). One key feature of downsized systems is their significant surface-to-volume ratios, which improve molecular diffusion and heat transfer. Microfluidic analytical tools encompass a range of devices that carry out chemical and biological processes on diminutive systems spanning from millimeters to a few square centimeters in size. Therefore, microfluidic analysis tools encompass microfluidic chips and compact systems like sensors and arrays (often termed biochips), all within the realm of miniaturization for high-throughput applications. Microarray biosensors have emerged within this context for screening multiple analytes in food samples efficiently (Atalay et al., 2011; Chen, Yao, & Li, 2023). Table 11.2 describes the specific properties and advantages of microfluidic-based analytical devices for food safety/quality monitoring and analysis.

Conversely, microfluidics technology focuses on managing and analyzing fluids within compact systems, providing a framework for merging diverse operations to create microfluidic analytical devices. Over time, these systems have the potential to surpass traditional robotic fluid handling systems in performance. Furthermore, microfluidic technologies enable extensive and rapid analysis by combining numerous steps, multiplexing, and conducting parallel analyses on a single device (Atalay et al., 2011; Gervais & Delamarche, 2009). The typical channel dimensions of microfluidic analytical devices range from 10 mm to 200 mm (Atalay et al., 2011). While many microfluidic analytical instruments have traditionally utilized silica-based materials, advancements in fabrication techniques have facilitated the production of affordable disposable microfluidic devices from materials like polymers and, surprisingly, paper.

TABLE 11.2
Specific Properties and Benefits of Microfluidics-Based Analytical Devices for Food Safety/Quality Monitoring and Analysis

Characteristic	Description
Miniaturization	Reduced size of devices, enabling the analysis of small sample volumes
Integration	Ability to integrate multiple functions (e.g., sample preparation, analysis, detection) on a single platform
Low reagent consumption	Utilizes small amounts of reagents, reducing costs and minimizing waste
Rapid analysis	Enables quick analysis due to reduced sample volumes and enhanced mass transfer properties within microchannels
High sensitivity	Enhanced sensitivity to detect low concentrations of analytes due to reduced background noise and enhanced signal-to-noise ratios
Automation	Capable of automation and parallelization, improving throughput and reducing human labor
Portability	Compact design allows for portable and point-of-care applications, suitable for on-site testing
Multiplexing	Capability to analyze multiple parameters simultaneously in a single device, enhancing efficiency and data collection
Continuous flow operation	Enables continuous and steady sample flow, leading to stable and reproducible results
Versatility	Versatile platform adaptable for various applications including healthcare, environmental monitoring, and food safety
References	Atalay et al. (2011), Crevillén et al. (2007), Gao et al. (2022), Gervais and Delamarche (2009), Mogensen and Kutter (2009)

Microfluidic devices meet the essential criteria sought by the food industry and quality regulatory bodies to uphold food quality and safety across the entire supply chain (Atalay et al., 2011; Chen et al., 2023; Lace & Cleary, 2021). Consequently, there exists significant enthusiasm for employing microfluidic technologies in biotechnological realms like monitoring and analyzing fermentation processes in sectors such as pharmaceuticals and food for real-time assessments (Atalay et al., 2011).

Capillary manipulation in microfluidic systems utilizes surface tension modifications for fluid transport. Capillary-driven devices, like lateral-flow tests, are popular due to their simplicity and effectiveness, with Gervais and Delamarche (2009) creating a one-step immunodiagnostic tool. This device, depicted in Figure 11.3a, operates using capillary forces, needing only a microliter of sample to initiate an immunoassay monitored by fluorescence. Analyte-biorecognition interactions (Figure 11.3b) occur as the sample moves through structured mesh to prevent obstructions, encountering fluorescently labeled detection antibodies (dAbs) that bind to the target. These complexes then bind to immobilized capture antibodies (cAbs) on the device's cover, with their quantity measured using external fluorescence detection (Gervais & Delamarche, 2009).

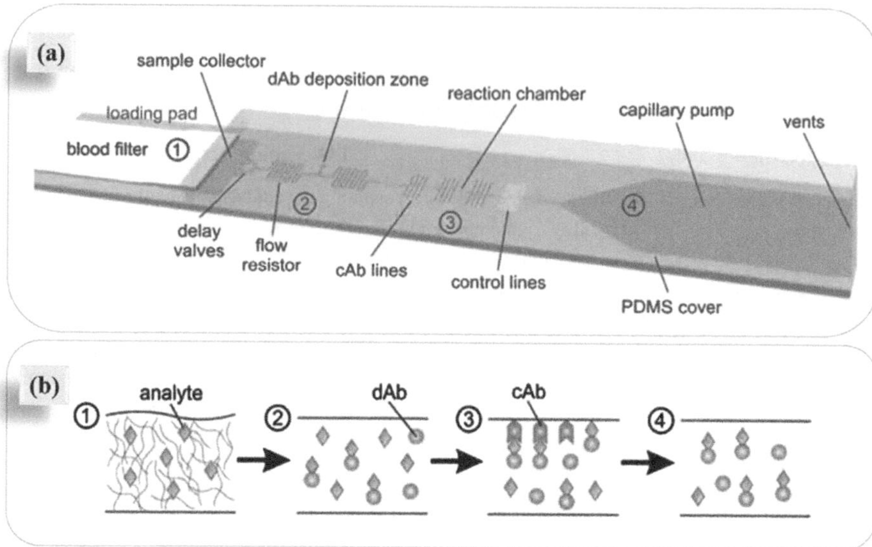

FIGURE 11.3 Schematic diagrams illustrating a capillary-powered microfluidic device designed for carrying out a one-step immunosensing system (a). The interaction of the target analyte, detection antibodies (dAbs), and capture antibodies (cAbs) throughout the microfluidic device (b).

Source: Reprinted from Gervais & Delamarche (2009) with permission

11.3 DIFFERENT TECHNIQUES INVOLVED IN SURFACE-BASED ANALYSIS IN FOOD

11.3.1 SERS-Molecular Labeling System

Aptamers, typically comprising dozens of nucleotide sequences, are generated through systematic evolution of exponential enrichment (SELEX) techniques, where these single-stranded DNA or RNA molecules are carefully selected from random oligonucleotide pools. Aptamers exhibit strong specificity and high affinity toward particular ligands or small molecules. Moreover, their production is straightforward and cost-efficient, making them widely favored in real-world applications (Hassan et al., 2021; Szlag et al., 2018). Given their cost-effectiveness, ease of chemical modification, and high specificity for particular molecules, aptamers are extensively employed in the creation of target-specific biological sensors. The use of SERS-based aptamer systems mitigates the challenge of non-target molecule binding that often hinders label-free nanoparticle-based SERS approaches (Pu et al., 2019).

Moreover, aptamers can substitute antibodies in swiftly identifying small molecules and have already been applied to various targets (Nie et al., 2018; Wang et al., 2010). With these attributes in mind, the SERS-molecular labeling system offers a versatile foundation for identifying various components within the food matrix through the utilization of specific aptamers (Yang et al., 2017). In a research study conducted in 2018, Song and colleagues used cDNA-modified Fe_3O_4@Au magnetic nanoparticles (cDNA-MGNPs) as capture probes and enhanced substrates. Additionally, they employed 5,5-dithiobis-(2-nitrobenzoic acid) (DNTB) in conjunction with OTA aptamer-modified Au@Ag NPs as probes for molecular recognition and Raman signals. To validate the effectiveness of this approach, the researchers tested spiked wine and coffee samples. The recovery rates ranged from 80% to 110%. A nanostructure containing aptamer-functionalized gold nanoparticles within a polydimethylsiloxane matrix (Apt-Au-PDMS film) served as a SERS substrate. This film, along with the self-assembled 4-mercaptobenzoic acid (4-MBA)-functionalized gold aptamer, was designed to selectively identify and attach to *Vibrio parahaemolyticus*, creating a SERS complex consisting of the target, the targeting agent, and Raman signal molecules (Shen et al., 2019).

11.3.2 SERS-Immunochromatography System

The immunochromatography assay (ICA) is a method for detection based on membranes, which relies on the interaction between antigens and antibodies. It is utilized for identifying a range of substances like drugs and microorganisms, among others. This method enables the isolation of the reaction product from the unreacted product without the need for extra precipitation or washing procedures (Dzantiev et al., 2014). In ICA-SERS technology, SERS is integrated into the final detection stage, setting it apart from conventional ICA. This approach offers the benefits of straightforward, quantitative, and uncomplicated procedures when contrasted with standard immunochromatography methods (Deng et al., 2019; Li et al., 2019). Deng and colleagues (2019) introduced a sandwich-style substance featuring mercaptobenzoic acid located between the core and shell of Ag@SiO2 NPs, with the monoclonal antibody situated on the surface of Ag@SiO_2 NPs. This novel material was employed in an ICA-SERS-based immunoassay to detect diclofenac residue in water. Figure 11.4 outlines the methodology's structure. The Raman substance utilized for SERS was fixed on the test line of the ICA strip, enhancing sensitivity and simplifying operational procedures. Likewise, Au@Ag nanorods have been identified as a SERS base and integrated into the construction of the ICA for the identification of Sudan I within food samples (Deng et al., 2019c). For instance, Li et al. devised a SERS-driven ICA technique to simultaneously detect two pyrethroid pesticides by incorporating two distinct recognition lines (Li et al., 2019).

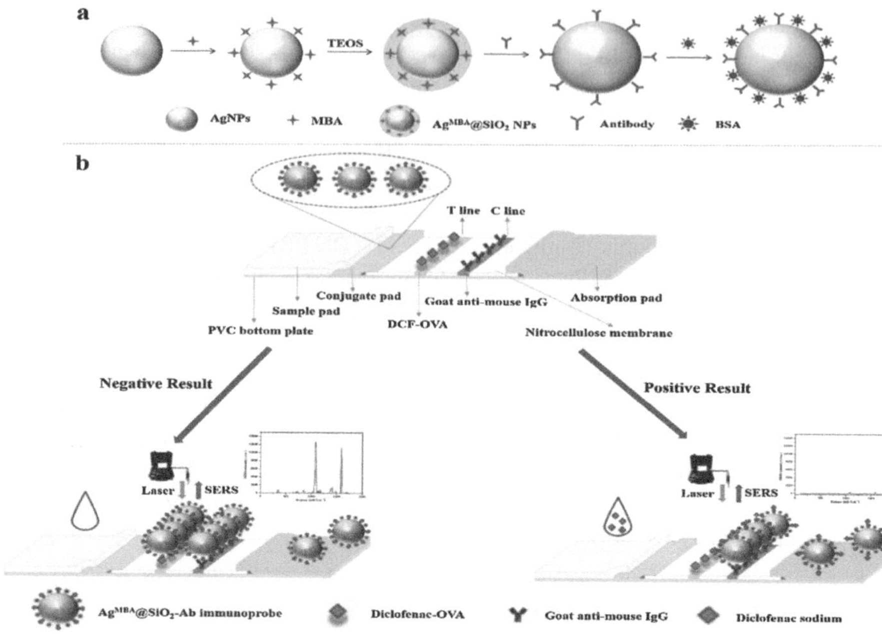

FIGURE 11.4 (a) Schematic preparation of immunoprobe ($Ag^{MBA}@SiO_2$-Ab NPs); (b) Assemble of the ICA strip and the scheme of the SERS-based ICA for diclofenac.

Source: Deng et al. (2019)

11.3.3 SERS-Microfluidics System

Microfluidic technology pertains to the manipulation, management, and analysis of intricate fluids at a miniature scale (Sackmann et al., 2014). The incorporation of SERS into a microfluidic system can address challenges linked to the uniformity and reproducibility of measurements within intricate matrices. This combined SERS-microfluidic setup is now utilized for precise sample preparation, demonstrating remarkable promise for in-field assessments of food contaminants (Pu et al., 2017). Furthermore, the integration of SERS into microfluidics essentially combines sample preparation, reaction, separation, and detection processes. In an interesting project, Zhu and colleagues merged paper-based microfluidics with a SERS setup, offering cost-effective production and enhanced sensitivity. This integration enables the rapid detection of thiram pesticides in tea samples (Zhu et al., 2017). In 2017, Durucan and coauthors conceived an automated centrifugal microfluidic SERS assessment system for processing samples and identifying substances in liquid mediums. The system successfully extracted and identified melamine within a brief timeframe of under 10 minutes, achieving a limit of detection (LOD) of 10 ppm (Durucan et al., 2017). Likewise, Krafft and colleagues created a three-dimensional microfluidic device that integrates sample handling alongside SERS analysis. This microfluidic system featured channels linked at varying heights through nanoporous polycarbonate track etching (PCTE) membranes. The setup utilized a specialized electrokinetic mechanism to purify the sample by selecting and transporting desired components while eliminating impurities. This microfluidic apparatus was specifically employed to identify a concentration of 1 ppm of melamine within milk (Krafft et al., 2020).

11.3.4 SERS-Molecularly Imprinted Polymers System

Molecular imprinting represents an emerging approach in molecular recognition technology focused on creating a polymer containing a distinct template molecule known as a molecularly imprinted polymer (MIP). MIPs are typically derived from templates, functional monomers, and cross-linking agents. Before solvent cross-linking occurs, functional monomers interact with templates through non-covalent (like hydrogen bonding) and covalent bonds to form complexes (Gui et al., 2018). Critical elements essential for crafting MIPs include the copolymerization of functional monomers, cross-linking involving template molecules, template removal, and target molecule recognition based on MIPs' distinct dimensional, chemical, and stereoscopic binding site characteristics (Huang et al., 2018). MIPs offer numerous benefits including exceptional selectivity and sensitivity, cost-effectiveness, chemical robustness, straightforward preparation, and resilience in organic solvents (Wackerlig & Lieberzeit, 2015). Presently, MIPs find applications across various scientific domains. Cai and his team synthesized MIPs specifically to detect trace benzimidazoles present in mutton and beef (Cai et al., 2019). MIP-SERS sensors were developed using gold nanoparticles (AuNPs) as the SERS substrate. This sensor showed outstanding performance in detecting patulin, with a LOD of 5.37×10^{-12} M and a recovery rate ranging from 96% to 101% in samples of blueberry sauce, grapefruit sauce, and orange juice (Wu et al., 2019).

11.3.5 SERS-Colorimetric System

The distinct optical properties of colloids made of gold and silver in the visible light spectrum have been thoroughly studied. These properties, along with the color changes associated with colloid clustering, have been extensively explored for developing color-dependent sensors for detecting various substances (Wei et al., 2015). One example is the work by Zhang and colleagues, who developed a method using specialized reporting nanoprobes, Au@Ag nanoparticles modified with 4-mercaptophenylboronic acid (4-MPBA), which enabled a dual optical and spectroscopic readout technique for assessing alkaline phosphatase activity. This system demonstrated notable capabilities for both visual analysis and accurate detection (Zhang et al., 2017). Apart from exploiting nanoparticle aggregation, the color reaction can also play a role in a combined colorimetric and SERS setup. Song et al. developed a method that combines colorimetry and SERS for the identification of mercury ions in water. They synthesized peroxidase-like Au@AgPt nanoparticles by selectively depositing platinum on the edges of hexahedral gold nanoparticles. The nanoparticles were modified with mercaptohexanol (MCH) to create the dual-mode probe (Au@AgPt@MCH). This nanoprobe facilitated the oxidation reaction of 3,3′,5,5′-tetramethylbenzidine (TMB) and H_2O_2, leading to a change in color and revealing the SERS fingerprint of the oxidized TMB (Song et al., 2020a) (Figure 11.5).

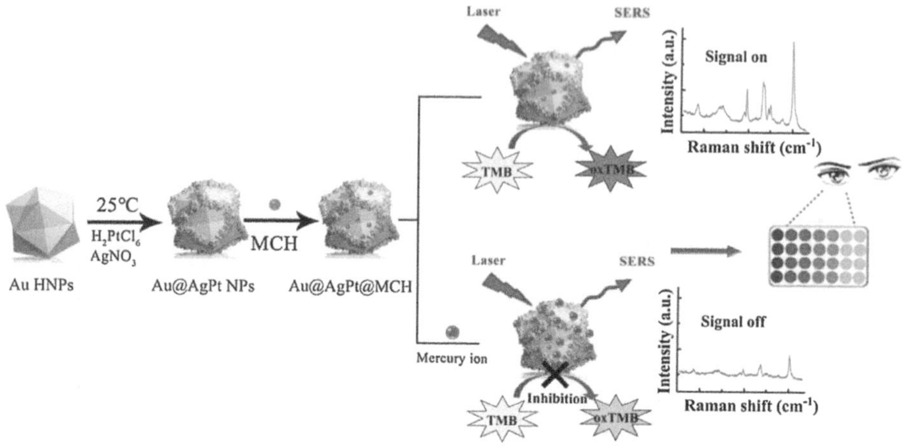

FIGURE 11.5 The detection of mercury ion is achieved using a dual-mode approach that combines colorimetric and SERS techniques with SERS-active peroxidase-like Au@AgPt NPs.

Source: Adapted from Song et al. (2020) with permission

The SERS-colorimetric probe was engineered by merging Pt's strong catalytic capabilities with Au's effective SERS properties. This system functioned by detecting alterations in both SERS signals and catalytic behavior following the introduction of Hg^{2+}. The Au@AgPt NPs catalyzed the redox reaction between H_2O_2 and TMB, converting TMB into its oxidized form, causing a switch in color from transparent to blue. When Hg^{2+} was introduced, the SERS-colorimetric probe (Au@AgPt@MCH) specifically trapped the Hg^{2+} ions before facilitating a chemical interaction where Pt atoms reacted with Hg^{2+}, producing Hg atoms and Pt^{2+}. The utilization of Pt atoms within the nanoparticles diminished the catalytic efficacy of the colorimetric probe, hindering the oxidation process. However, through this mechanism, the presence of Hg^{2+} in the sample was effectively identified by the dual-mode SERS-colorimetric probe (Harshita et al., 2023).

11.3.6 SERS-Imaging System

Raman technology, when integrated with a digital imaging technique, offers a means to acquire spectral data, chemical makeup, and spatial details simultaneously (Dybas et al., 2018). SERS imaging technology has proven effective in detection and visual monitoring tasks. In contrast to conventional SERS methods, SERS imaging merges spectral and spatial data, offering the distinct advantage of visualization (Li et al., 2017). The method offers notable benefits in intricate component systems. When data is collected through imaging, the interference in the field of view is reduced because each pixel in the image contains chemical information that is specific to a restricted sampling area (Stewart et al., 2012). Chen and coworkers conducted a study in 2018 using SERS microscopic imaging to explore how pesticide residues on the exteriors of fruits and leaves can be detected. The SERS imaging analysis of omethoate and chlorpyrifos provided a clearer insight into the distribution of pesticide residues (Chen et al., 2018). The approved technique of combining SERS imaging with multivariate data analysis made it possible to gather spectral and spatial details about the presence of thiophanate-methyl (TM) and its metabolite carbendazim residues on the outer layer of red bell peppers (Li et al., 2017).

11.4 CHALLENGES AND LIMITATION OF SURFACE-BASED ANALYTICAL TECHNIQUES IN FOOD ANALYSIS

The utilization of SPR in commercial settings is limited by factors such as cost, upkeep requirements, and its affordability. Given these challenges, there is a pressing demand for the creation of a cost-effective SPR system that delivers precise and accurate results, addressing the current requirement for more accessible SPR technology in various applications (Prabowo et al., 2018). A key limitation of conventional SPR technology is its inability to simultaneously detect and analyze multiple allergens in food samples (Zhou et al., 2019). Yet, this challenge can be resolved by incorporating SPR into diverse sensing systems, enabling strong biomolecular identification characterized by exceptional specificity and extended durability. Numerous research endeavors are underway to boost efficacy, while simultaneously shrinking the size of SPR sensors to reduce production expenses and enhance portability. Nevertheless, the effectiveness of SPR biosensors in intricate samples remains subpar, largely because of the disruption caused by nonspecific bindings on output signals, necessitating their improvement (Shankaran et al., 2007). Recent studies on the application of SPR in analyzing complex samples have revealed its limited sensitivity and analytical utility, posing a significant obstacle to its advancement (Chen et al., 2017). Constraints in both sensitivity and the design of SPR sensors currently restrict the utilization of SPR within the food industry. Enhancements in these aspects could position SPR as a valuable tool for detecting food allergens and foodborne pathogens (Lan et al., 2008).

After assessing the sources consulted, we have outlined the latest progress of SERS regarding food safety and quality. While SERS proves to be a potent detection technique, its implementation within the food sector remains largely experimental, presenting limited opportunities for practical use. Certain obstacles continue to hinder the widespread application of SERS in this domain (Panneerselvam et al., 2018). Given its evolving nature in detection, the cost of employing an SERS-based detection system is notably high, necessitating a need for cost reduction. The substantial initial investment required for SERS methodologies, encompassing control systems, signal processing, detection systems, data acquisition systems, and laser light sources, can be financially prohibitive for medium- and small-sized initiatives. This difficulty impedes the straightforward application of this method in routine food industry analyses. Presently, portable and cost-efficient handheld SERS spectrometers with enhanced capabilities are accessible for real-time identification of food impurities directly at production sites. Despite these advancements, the quantitative analysis of hazardous compounds in food samples using SERS techniques remains challenging due to their intricate nature. Through the integration of spectral data analysis, chemometrics techniques, smart-nanosubstrates, and micro-Raman spectrometers, SERS has the potential for swift quantitative and qualitative evaluations of various substances in food samples. Yet, further investigations are necessary to standardize variables to ensure consistent outcomes in intricate food matrices. Specificity and sensitivity represent typical constraints of methodologies reliant on SERS. The improvement of detection specificity can be accomplished by using functionalized

SERS substrates or utilizing specific target capture agents such as aptamers or MIPs. Additionally, increasing the resolution of the Raman spectrum in the device can enhance both the sensitivity of detection and the range of its practical uses (Qiu et al., 2018). Moreover, the current availability of commercialized SERS substrates is limited. These substrates fall short of meeting market demands due to their limited shelf life. Consequently, there is a necessity for target-specific, durable, economically viable, and dependable substrates to bolster the technique's reliability and utility. Moreover, there are persisting challenges in utilizing the method for analyzing diverse food samples like liquids (e.g., milk, fruit juice), semi-solids (e.g., yogurt, fruit gel), or solids (e.g., cereals, vegetables, fruits, and meats). The behavior of SERS signals varies considerably based on the composition of the sample. SERS shows promise in distinguishing compounds located in close proximity to the substrate. As non-target analytes generate considerable background noise in the spectrum, this can diminish the sensitivity toward the desired target analyte. Therefore, enhancing the efficiency of this technique requires the creation of advanced SERS substrates with enhanced hotspots. At times, alterations in SERS configurations can occur notably because of complex formations between the target molecules and the capturing probe. Thus, precise recognition of signals related to the target molecule is paramount. Moreover, the advancement of portable Raman instruments based on SERS hinges on the progress made in developing SERS substrates (Kalachyova et al., 2018).

11.5 APPLICATIONS OF SURFACE-BASED ANALYTICAL IN FOOD SAFETY

11.5.1 SERS

Numerous studies have discussed the utilization of SERS technology in assessing food quality and safety. Enhancing sensitivity and stability is crucial for optimizing any detection system. A recent trend involves integrating SERS with other techniques like separation methods to enhance the precision of food quality assessment. Typical targets for detection in food samples comprise pesticides, heavy metals, metal ions, antibiotics, and microorganisms, among others. Table 11.3 presents a summary of SERS detection methods related to food, as well as the SERS substrates commonly employed.

11.5.2 Microorganisms

Microbial pathogens play a role in the production of numerous food items. Nevertheless, the presence of harmful microbes leads to food spoilage and decay, posing health risks. Consequently, the significance of microbial detection has increased in monitoring food safety (Li et al., 2020). In contrast to typical food impurities, precise identification of foodborne pathogens necessitates the utilization of biomolecule-linked substances. Utilizing labels in detection entails employing antibodies or aptamers to selectively attach to targeted microbes, enhancing detection sensitivity and specificity, although this might extend the detection duration (Abedi-Firoozjah et al., 2024a). Xu and colleagues introduced a technique for identifying *Salmonella typhimurium* through a DNA-assembled aptasensor utilizing Au nanodimers. This method was effectively validated in spiked milk

TABLE 11.3
A Summary of the Selected Some Surface-Based Sensing Systems for Food Safety/Quality Analysis

Analytes	SERS Substrates	Method	LOD	Linearity	Real Sample	Ref.
Salmonella Typhimurium	GNP modified with 5,5'-dithiobis(succinimidyl-2-nitrobenzoate) (GNP-DSNB)	SERS-immunosensor	10 cells/mL	10^1-10^7 cells/mL	Cottage cheese, egg white and canned mixed fruit juice	(Chattopadhyay, Sabharwal, Jain, Kaur, & Singh, 2019)
E. coli 25922 *S. aureus* *S. typhi* *E. coli K12*	AgNPs	SERS-laser induced breakdown spectroscopy (LIBS)	5×10^3–5×10^7 CFU/mL	5×10^3–5×10^7 CFU/mL	Lake water, tap water, and drinking water	Liao et al. (2018)
Chlorpyrifos Imidacloprid	AgNPs coated glass bead	SERS	10 ng/mL	NS	Apple	(Tang, Chen, & Ju, 2019)
Lysozyme	Apt/AgNPs	SERS	0.35 nM	0.35×10^{-2} nM to 419.3 nM	Steel food-handling surface	(Boushell, Pang, & He, 2017)
Hg^{2+}	AuNPs	SERS-molecular labeling	0.1 nM	10–100 ng/mL	Drinking water	Zou et al. (2019)
Sulfapyridine	MWCNTs-AuNPs	SERS	8.8 ng/mL	10–100 ng/mL	Milk	Moreno et al. (2019)
S. aureus	Apt/MSNs-Ag NFs	SERS	17 CFU/mL	4.7×10^1 to 4.7×10^8 CFU/mL	Fish	Zhu et al. (2021)
E. coil BL21 and *S. aureus*	PEI-modified Fe_3O_4@Au/Au@Ag NPs	SERS	10^3 CFU/mL	10^3 to 10^7 CFU/mL	Tap water and milk	Wang et al. (2016)
Yeast, Saccharomycesc-erevisiae, Dekkera bruxellensis and Wickerhamomyc-eanomalus	AgNPs	SERS	NS	NS	Beer	(Lemma, Wang, Arstila, Hytönen, & Toppari, 2019)
Neomycin	AuNPs	SERS- lateral flow immunoassay (LFI)	0.216 pg/mL	0.04 ng/mL and 0.216 pg/mL	Milk	(Shi et al., 2018)

(*Continued*)

TABLE 11.3 (*Continued*)
A Summary of the Selected Some Surface-Based Sensing Systems for Food Safety/Quality Analysis

Analytes	SERS Substrates	Method	LOD	Linearity	Real Sample	Ref.
Enrofloxacin and amoxicillin	AgNPs/MMF/R6G	SERS	10^{-9} M	10^{-3} to 10^{-9} M	Milk	Fu et al. (2021)
Ampicillin and nitrofurantoin	Au nanoparticles/graphene oxide	SERS	0.01 ng/mL	500 ng/mL and 5 ng/mL	Milk	(Y. Wang, Chen, & Jiang, 2021)
Thiabendazole, thiram, endosulfan, and malathion	Polydimethylsiloxane	SERS	5 μg/kg	44–88 μg/kg	Strawberry	(Asgari, Wu, Aghvami, Zhang, & Lin, 2021)
Kanamycin	Au/Ag NPs	SERS	142 pg/mL	6.67×10^{-10}–2×10^{-7} g/mL	Milk	(Jiang, Sun, Pu, & Wei, 2020)
Ciprofloxacin and chloramphenicol	TiO_2/Au/Ag nanoparticles	SERS	10^{-9} M and 10^{-8} M	NS	Water	Jiao et al. (2022)
Tetracycline	Lanthanide metal ions/ carbon quantum dots	Microfluidic sensor	5 μM	1 ~ 60 μM	Milk	Zhu et al. (2023)
Tetracycline	Ovalbumin-stabilized gold nanocluster	Microfluidic sensor	0.09 μg/mL	NS	Spiked milk and water samples	Zhang et al. (2021)
Salmonella and *Brucella*	Magainin 1/RPA/aptamer	Microfluidic sensor	50 CFU (Salmonella) and 10^2 CFU (Brucella)	10^4 to 50 CFU	Urine	(Dao et al., 2018)
E. coli O157:H7 and *Staphylococcus aureus*	Polyethylene glycol/ nanoporous alumina membrane	Microfluidic sensor	10^2 CFU/mL	10^2 CFU/mL to 10^5 CFU/mL	NS	(Tian, Lyu, Shi, Tan, & Yang, 2016)
Carbamate pesticides	AuNPs	Microfluidic sensor	3.102 ppb	0.01 ppm to 10 ppm	Cabbage	Zhao et al. (2021)
Methyl parathion	Graphene-gold nanoparticles	Microfluidic sensor	0.000646 μM	0.001 μM to 200 μM	Ureum and lettuce	Zhang et al. (2024)
Lactoferrin	Antibody@gold electrode	Surface-based biosensor	4.9 pg/mL	0.01–1000 ng/ml	Milk	Huang et al. (2018)

samples, demonstrating a detection limit of 35 CFU/mL (Xu et al., 2018). Bozkurt and his team used a SERS sandwich immunoassay method, making use of alkaline phosphatase's enzymatic activity on 5-bromo-4-chloro-3-indolyl phosphate to detect *E. coli*. Label-free SERS detection, a rapid technique in which microbial molecules interact directly with SERS-active substrate to generate signals, was compared to label-assisted detection (Bozkurt et al., 2018). Wang and collaborators researched label-free SERS detection to identify various bacteria within a 10-minute timeframe. The effectiveness of this method was confirmed in tap water and milk samples (Wang et al., 2016).

11.5.3 Pesticides

Controlling crop diseases with pesticides enhances overall agricultural productivity, boosting stable grain production and sustainable rural incomes. Yet, even minimal pesticide residues can pose significant risks to human health. As a result, the demand for detecting these residues is growing. To identify trace amounts of pesticides, high-sensitivity detection technologies are essential. SERS technology has been employed for detecting pesticides, meeting the need for precise trace-level detection (Xu et al., 2020). Zhu et al. created a SERS platform using a hybrid film of hydrophilic Ag nanocubes and graphene oxide. This film was placed on a small Cu foil on a hydrophobic Teflon surface. The analyte solution droplet was added to the hybrid film, leading to the concentration of analyte molecules through evaporation. The system's efficiency was confirmed through the detection of 4-chlorobiphenyl and methyl parathion in Chinese cabbage (Zhu et al., 2019). Alsammarraie and colleagues introduced a swift SERS technique for analyzing and measuring pesticides extracted from lemon, carrot, and mango juices utilizing an Au nanosubstrate. This setup successfully identified and quantified thiabendazole at concentrations as low as 50 μg/L in all three varieties of fruit juices (Alsammarraie et al., 2018).

11.5.4 Metal Ions

Metal ions are commonly found in nature and can pose a risk to human health due to their ability to accumulate in living organisms. Li et al. developed a method for detecting potassium ions (K^+) and tested it on rice samples (Li et al., 2019). In another study, Dugandžić et al. proposed a technique using SERS to detect copper ions (Cu^{2+}) in complex samples, such as white wine, by employing gold nanoparticles (AuNPs) with dipyridine ligands. The concentration of Cu^{2+} was determined indirectly by monitoring changes in the spectral properties of the pyridine ring as it interacts with Cu^{2+}(Dugandžić et al., 2019). Additionally, Li and coworkers outlined a dual detection method (metal-enhanced fluorescence (MEF) and SERS) for detecting mercury ions (Hg^{2+}) based on rhodamine derivatives grafted onto core-shell nanocubes consisting of gold nanostars coated with silver (AuNS@Ag). This system shows promise for detecting adulterated milk (Li et al., 2018).

11.5.5 Antibiotics

The issue of antibiotic residues in food, notably in animal-derived products, is a significant concern. Ingesting these residues through items like eggs, meat, and milk can lead to antibiotic buildup in humans, fostering resistance to medications, organ damage, and potentially cancerous developments. Various methods exist to identify antibiotic residues in foodstuffs (Chen et al., 2017). SERS, known for its quick and noninvasive nature, provides an improved method for identifying antibiotic residues in food samples. Oxytetracycline belongs to the class of broad-spectrum tetracycline-type antibiotics. Dhakal and coworkers have developed an innovative approach for performing on-site screening of full-fat milk to identify small quantities of tetracycline using SERS with a portable Raman spectrometer. They utilized cyclodextrin-capped silver colloids as the SERS substrate for the detection system (Dhakal et al., 2018). Additionally, Li and colleagues presented a speedy and highly sensitive SERS approach for the detection of tetracycline (TTC) utilizing a magnetic nanosphere-targeting aptasensor. The efficiency and feasibility of this detection system were confirmed through testing with artificially adulterated TTC milk, achieving a recovery rate of 98–120% (Li et al., 2017). Furthermore, in another investigation, Chen and coworkers utilized a two-step pretreatment process for SERS-based penicillin G detection in milk. They utilized silver nanoparticles as the SERS substrate and successfully achieved an LOD of 2.54×10^{-9} mol/L for penicillin G under optimal conditions (Chen et al., 2017).

11.5.6 SPR

Table 11.4 depicts the outcomes of mycotoxin identification through different methods. Extensive research has focused on employing SPR technology for mycotoxin detection due to its rapid and straightforward detection capabilities, which offer a distinct advantage in this field.

Aflatoxin (AFT) is a cancer-causing and mutation-inducing substance produced by *Aspergillus parasiticus* and *Aspergillus flavus*, with AFB1 currently acknowledged as the most powerful natural carcinogen. In a research conducted by Wu et al. (2018), they presented an aptamer-based SPR biosensor to simultaneously detect AFB1 and AFB2. This biosensor, based on aptamers and SPR technology, demonstrated remarkable specificity toward these toxins and remained unaffected by OTA, ochratoxin B (OTB), ZEN, and T_2 toxin. Its applicability extends to detecting AFB traces in agricultural produce, vinegar, and various food items. Wei et al. (2019) developed a SPR sensor chip using self-assembled monolayers to continuously monitor AFB1, ZEN, OTA, and DON. Figure 11.6 illustrates the production process of the AFB1, ZEN, OTA, DON antigen and antibody, as well as the SPR sensor chip. The outcomes obtained through this approach are in close agreement with those acquired using the HPLC-MS/MS technique. This method not only exhibits high sensitivity but also opens up the potential for high-throughput mycotoxin detection (Wei et al., 2019). Joshi and his colleagues created a portable iSPR detection system for DON, ZEA, T2 toxin, OTA, FB1, and AFB1, which allows for quick detection and field screening of different mycotoxins. However, there is a need to improve the sensitivity for OTA and AFB1. OTA is mainly produced by *Penicillium* and *Aspergillus fungi* and is commonly found in various food and feed sources, including fruits, grains, beverages, and coffee. The unique advantages of SPR technology in detection and analysis have generated significant interest, leading to increased adoption by researchers for OTA detection. They immobilized the toxin directly onto the nanostructured iSPR chip for competitive inhibition immunoassays. The chip is also

TABLE 11.4
A Summary of Detection of Mycotoxins by SPR and LSPR Method

Target	Sensing Platform	LOD	Linearity	Real Sample	Reference
[A] SPR					
AFB_1	AuNPs	0.003 nM	0.01 to 50 nM	Wheat	Bhardwaj et al. (2020)
Lysozyme	PMAPA/Au chip	0.66 nM	1 nM to 50 nM	Egg white	Saylan et al. (2017)
AFB_1	AuNPs/ antibody	0.59 ng/mL	10.37–103.31 ng/mL	Corn and wheat	Wei et al. (2019)
OTA	AuNPs/ antibody	1.27 ng/mL	10.37–103.31 ng/mL	Corn and wheat	Wei et al. (2019)
Lactoferrin	Abd/AuNPs	2.8×10^{-7} M	0.5×10^{-6}–3.50×10^{-6} M	Cow and goat milk	Tomassetti et al. (2013)
Lysozyme	AgNPs/glutamic acid	1.5 nM	3 nM to 150 nM	Egg white	Shrivas et al. (2019)
OTA	N-methacryloyl-(L)-phenylalanine and 2-Hydroxyethyl methacrylate/MIP	0.028 ng/mL	0.1 ng/mL and 20 ng/mL	Dried fig	Akgönüllü et al. (2022)
ZEN	AuNPs/ antibody	7.07 ng/mL	10.37–103.31 ng/mL	Corn and wheat	Wei et al. (2019)
Lactoferrin	Graphene/AuNPs/ antibody	0.12 mg/mL	1 to 10 mg/mL	Milk	Billakanti, Fee, Lane, Kash, & Fredericks, (2010)
Penicilin-G	AuNPs	0.00017 ppb	0.01 to 0.25 ppb	Milk	Safran et al. (2021)
Lysozyme	PMOXA/dopamine @MIP	153 nM	0.19 µg/mL and 0.58 µg/mL	Egg white	Liangyu Zhu, Hu, Shen, & Wang, (2023)
Glycopeptide antibiotics	AuNPs	17.7 ng/mL	10 ng/mL to 125 ng/mL	Milk	Altintas, (2018)
Benzylpenicillin	MIP-GO	0.021 ppb	1–100 ppb	Spiked and unspiked milk	Çelik, Saylan, Göktürk, Yılmaz, & Denizli, (2023)

(Continued)

TABLE 11.4 (*Continued*)
A Summary of Detection of Mycotoxins by SPR and LSPR Method

Target	Sensing Platform	LOD	Linearity	Real Sample	Reference
Ciprofloxacin, moxifloxacin and ofloxacin	Nano-MIPs	0.54 nM	1.95–31.25 nM	Milk and river water	Sullivan, Henderson, Hand, & Turner, (2022)
SNZ, SMZ, and ATZ	MIP/AuNPs	0.095, 0.031 and 0.091 nM	0.10 to 6.64 nM	NS	Saylan, Akgönüllü, et al., (2017)
[B] LSPR					
Acetamiprid	Cy3-labeled aptamer/AuNPs	166.7 pg/mL	NS	Vegetables	Wang et al. (2022a)
Enrofloxacin	PDA-MIP	61.1 ng/mL	25–1000 ng/mL	Chicken meat	Wang, et al. (2021)
Enrofloxacin	AuNPs-Aptamer	0.001 ng/mL	0.01–100 ng/mL	Water	Wang, et al. (2023)
Ampicillin, amoxicillin, penicillin G, oxacillin, and carbenicillin	AuNPs	1 nM	1 nM to 1 mM	Chinese lettuce leaf extract	Aghamirzaei et al. (2021)
Hg^{2+}	Gold nanorods	11 pM	1 µM to 10 µM	River water, tube well water, drinking water, and industrial effluent water	(Jayabal et al., 2015)
Hg^{2+}	Gold nanorods	0.38 nM	1 nM to 1 mM	Tap water, lake and river	(Park, Byun, Yim, & Kim, 2016)

Abbreviations: AFB1, aflatoxin B1; OTA, ochratoxin A; ZEN, zearalenone; DON, deoxynivalenol; PMOXA, poly(2-methyl-2-oxazoline); MIP, molecularly imprinted polymeric; GO, graphene oxide; SNZ, cyanazine; SMZ, simazine; ATZ, atrazine

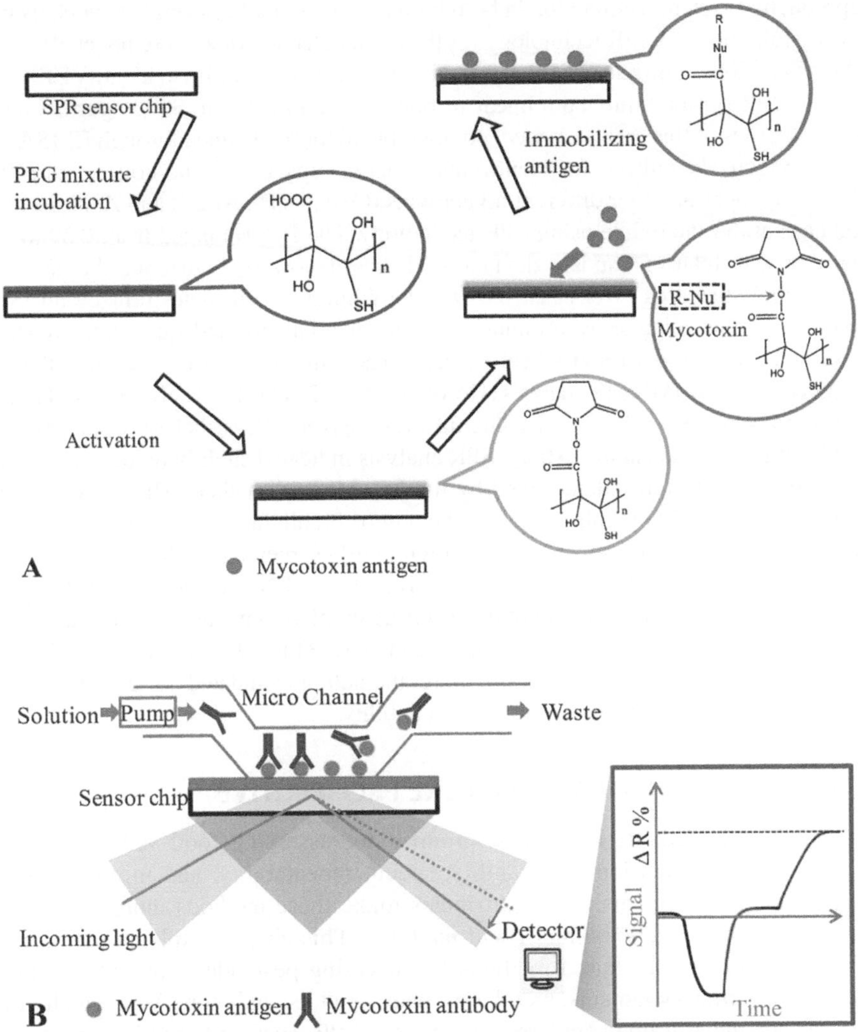

FIGURE 11.6 A: Illustration showing the modification process of an SPR sensor chip. Diagram B: Outline depicting the optical configuration and sensor chip within the SPR operational setup. Signal indicated through alterations in reflectivity percentage (ΔR%).

Source: Wei et al. (2022)

reusable. Unfortunately, this method is not suitable for on-site detection because of the high OTA concentration in beer (Joshi et al., 2026a). Akgonullu et al. (2022) showed how OTA contamination in figs can be identified using an OTA-imprinted SPR sensor, with detection time of less than 8 min. They utilized unlabeled MIP films to develop a straightforward and sensitive method for detecting OTA (Akgönüllü et al., 2022). ZEN, which is predominantly synthesized by *Fusarium graminearum and Fusarium culmorum*, is a secondary metabolite. It is frequently found in various cereal crops, including maize, wheat, and rice. ZEN is known to act as an endocrine disruptor, contributing to reproductive system dysfunction in both humans and animals. Several researchers have successfully utilized SPR technology for the rapid detection of ZEN (Choi et al., 2009). Choi et al. (2009) utilized an SPR sensor along with an MIPPy film to identify ZEN. The MIPPy-SPR sensor exhibited a linear response within the 0.3 to 3000 $ng{\cdot}mL^{-1}$ range. The outcomes of this approach were comparable to those obtained through ELISA, and it was effectively utilized for identifying maize samples. Wei and colleagues (2019) effectively identified four different mycotoxins (OTA, DON, AFB1, and ZEN) in wheat and corn at the same time using SPR technology. The LODs ranged from 0.59 to 3.26 $ng{\cdot}mL^{-1}$ with relative standard deviations (RSDs) below 10%, meeting the necessary food safety standards (Wei et al., 2019). Dorokhin and colleagues utilized an iSPR-based multiplex microarray immunosensor to qualitatively and quantitatively detect ZEN and DON in corn and wheat extracts. Their findings closely matched the data obtained from LC–MS/MS analysis (Dorokhin et al., 2011). The LODs for ZEN in corn and wheat samples were 16 and 10 ng/mL, respectively. Nevertheless, when assessing ZEN and DON levels in the extract, SPR analysis indicated slightly reduced concentrations compared to those determined by the LC–MS/MS method. The detection technique aligns with EU regulations, and the author highlighted the significant potential of the iSPR sensor chip detection system for swiftly screening 40 distinct mycotoxins. Specifically, out of the four mycotoxins (AFB1, OTA, ZEN, and DON) identified using the SPR sensor chip with the mentioned self-assembled monolayer (Wei et al., 2019), the range for linear detection of DON spanned from 5.31 to 99.37 ng/mL, exceeding that of the other three mycotoxins. Furthermore, the authors validated these findings using HPLC-MS/MS, confirming the method's reliability.

11.6 CONCLUSION AND FUTURE PROSPECTIVE

Surface-based methods hold great promise for assessing food safety and quality due to their precision, adaptability, speed, repeatability, and minimal sample preparation requirements. These attributes make these methods more reliable and convenient than traditional analytical methods. This chapter explores the progress of research on surface-based methods for detecting pesticides, antibiotic residues, food toxins, microorganisms, metal ions, heavy metals, and more. While techniques such as advanced Raman spectroscopy, SPR, LSPR, surface-based biosensors, and microfluidic detectors have been widely used in food detection, they still have some limitations. Efforts are ongoing to enhance the portability and accuracy of these diagnostic tools. Surface-based method technologies have been combined with other techniques in recent years to improve the precision of test results, accelerating the potential application of these methods in ensuring food safety and quality. However, further research is needed to address complex food interferences in order to achieve rapid, nondestructive, and accurate diagnostics. To establish surface-based methods as a standard tool for food safety and quality, future research should focus on reducing the technique's cost, standardizing test procedures, designing new platforms, and integrating surface-based methods with other approaches to overcome limitations. These advancements will ensure food safety and quality for local consumers and enhance other quality standards that meet global trade requirements. In the near future, we can expect these technologies to become more advanced with established diagnostic protocols and to have widespread applications in the food industry.

REFERENCES

Abedi-Firoozjah, R., Alizadeh-Sani, M., Zare, L., Rostami, O., Salim, S. A., Assadpour, E., . . . Jafari, S. M. (2024). State-of-the-art nanosensors and kits for the detection of antibiotic residues in milk and dairy products. *Advances in Colloid and Interface Science, 328*, 103164.

Abedi-Firoozjah, R., Ebdali, H., Soltani, M., Abdolahi-Fard, P., Heydari, M., Assadpour, E., . . . Jafari, S. M. (2024). Nanomaterial-based sensors for the detection of pathogens and microbial toxins in the food industry; a review on recent progress. *Coordination Chemistry Reviews, 500*, 215545.

Aghamirzaei, M., Khiabani, M. S., Hamishehkar, H., Mokarram, R. R., & Amjadi, M. (2021). Plasmonic sensor for detection of β-lactam antibiotics based on the conjugated antibody with gold nanoparticles. *Journal of Applied Spectroscopy, 88*, 233–241.

Akgönüllü, S., Armutcu, C., & Denizli, A. (2022). Molecularly imprinted polymer film based plasmonic sensors for detection of ochratoxin A in dried fig. *Polymer Bulletin, 79*(6), 4049–4067.

Alsammarraie, F. K., Lin, M., Mustapha, A., Lin, H., Chen, X., Chen, Y., . . . Huang, M. (2018). Rapid determination of thiabendazole in juice by SERS coupled with novel gold nanosubstrates. *Food Chemistry, 259*, 219–225.

Asgari, S., Wu, G., Aghvami, S. A., Zhang, Y., & Lin, M. (2021). Optimisation using the finite element method of a filter-based microfluidic SERS sensor for detection of multiple pesticides in strawberry. *Food Additives & Contaminants: Part A, 38*(4), 646–658.

Atalay, Y. T., Vermeir, S., Witters, D., Vergauwe, N., Verbruggen, B., Verboven, P., . . . Lammertyn, J. (2011). Microfluidic analytical systems for food analysis. *Trends in Food Science & Technology, 22*(7), 386–404.

Bhardwaj, H., Sumana, G., & Marquette, C. A. (2020). A label-free ultrasensitive microfluidic surface Plasmon resonance biosensor for Aflatoxin B1 detection using nanoparticles integrated gold chip. *Food Chemistry, 307*, 125530.

Bozkurt, A. G., Buyukgoz, G. G., Soforoglu, M., Tamer, U., Suludere, Z., & Boyaci, I. H. (2018). Alkaline phosphatase labeled SERS active sandwich immunoassay for detection of Escherichia coli. *Spectrochimica Acta Part A: Molecular and Biomolecular Spectroscopy, 194*, 8–13.

Cai, Y., He, X., Cui, P. L., Liu, J., Li, Z. B., Jia, B. J., . . . Yuan, W. Z. (2019). Preparation of a chemiluminescence sensor for multi-detection of benzimidazoles in meat based on molecularly imprinted polymer. *Food Chemistry, 280*, 103–109.

Chen, T., Cheng, G., Ahmed, S., Wang, Y., Wang, X., Hao, H., & Yuan, Z. (2017). New methodologies in screening of antibiotic residues in animal-derived foods: Biosensors. *Talanta, 175*, 435–442.

Chen, J., Dong, D., & Ye, S. (2018). Detection of pesticide residue distribution on fruit surfaces using surface-enhanced Raman spectroscopy imaging. *RSC Advances, 8*(9), 4726–4730.

Chen, X., Yao, C., & Li, Z. (2023). Microarray-based chemical sensors and biosensors: Fundamentals and food safety applications. *TrAC Trends in Analytical Chemistry, 158*, 116785. https://doi.org/10.1016/j.trac.2022.116785

Chen, Y., & Ming, H. (2012). Review of surface plasmon resonance and localized surface plasmon resonance sensor. *Photonic Sensors, 2*, 37–49.

Choi, S.-W., Chang, H.-J., Lee, N., Kim, J.-H., & Chun, H. S. (2009). Detection of mycoestrogen zearalenone by a molecularly imprinted polypyrrole-based surface plasmon resonance (SPR) sensor. *Journal of Agricultural and Food Chemistry, 57*(4), 1113–1118.

Crevillén, A. G., Ávila, M., Pumera, M., González, M. C., & Escarpa, A. (2007). Food analysis on microfluidic devices using ultrasensitive carbon nanotubes detectors. *Analytical Chemistry, 79*(19), 7408–7415.

Dale, A. M., Fischl, B., & Sereno, M. I. (1999). Cortical surface-based analysis: I. Segmentation and surface reconstruction. *Neuroimage*, *9*(2), 179–194.

Deng, D., Yang, H., Liu, C., Zhao, K., Li, J., & Deng, A. (2019). Ultrasensitive detection of Sudan I in food samples by a quantitative immunochromatographic assay. *Food Chemistry*, *277*, 595–603.

Dhakal, S., Chao, K., Huang, Q., Kim, M., Schmidt, W., Qin, J., & Broadhurst, C. L. (2018). A simple surface-enhanced Raman spectroscopic method for on-site screening of tetracycline residue in whole milk. *Sensors*, *18*(2), 424.

Dorokhin, D., Haasnoot, W., Franssen, M. C., Zuilhof, H., & Nielen, M. W. (2011). Imaging surface plasmon resonance for multiplex microassay sensing of mycotoxins. *Analytical and Bioanalytical Chemistry*, *400*, 3005–3011.

Dugandžić, V., Kupfer, S., Jahn, M., Henkel, T., Weber, K., Cialla-May, D., & Popp, J. (2019). A SERS-based molecular sensor for selective detection and quantification of copper (II) ions. *Sensors and Actuators B: Chemical*, *279*, 230–237.

Durucan, O., Rindzevicius, T., Schmidt, M. S., Matteucci, M., & Boisen, A. (2017). Nanopillar filters for surface-enhanced Raman spectroscopy. *ACS Sensors*, *2*(10), 1400–1404.

Dybas, J., Grosicki, M., Baranska, M., & Marzec, K. M. (2018). Raman imaging of heme metabolism in situ in macrophages and Kupffer cells. *Analyst*, *143*(14), 3489–3498.

Dzantiev, B. B., Byzova, N. A., Urusov, A. E., & Zherdev, A. V. (2014). Immunochromatographic methods in food analysis. *TrAC Trends in Analytical Chemistry*, *55*, 81–93.

Fu, X., Wang, Z., Li, J., Ma, S., Fu, G., Jin, W., . . . Dong, Y. (2021). Microcavity fiber SERS probe coated with Ag nanoparticles for detecting antibiotic in milk. *IEEE Photonics Journal*, *13*(2), 1–13.

Gao, D., Ma, Z., & Jiang, Y. (2022). Recent advances in microfluidic devices for foodborne pathogens detection. *TrAC Trends in Analytical Chemistry*, *157*, 116788. https://doi.org/10.1016/j.trac.2022.116788

Gervais, L., & Delamarche, E. (2009). Toward one-step point-of-care immunodiagnostics using capillary-driven microfluidics and PDMS substrates. *Lab on a Chip*, *9*(23), 3330–3337.

Gordon, A. (2017). Introduction: effective implementation of food safety and quality systems: Prerequisites and other considerations. In *Food Safety and Quality Systems in Developing Countries* (pp. 1–19). Elsevier.

Gui, R., Jin, H., Guo, H., & Wang, Z. (2018). Recent advances and future prospects in molecularly imprinted polymers-based electrochemical biosensors. *Biosensors and Bioelectronics*, *100*, 56–70.

Guo, Y., Girmatsion, M., Li, H.-W., Xie, Y., Yao, W., Qian, H., . . . Mahmud, A. (2021). Rapid and ultrasensitive detection of food contaminants using surface-enhanced Raman spectroscopy-based methods. *Critical Reviews in Food Science and Nutrition*, *61*(21), 3555–3568.

Harshita Wu, H. F., & Kailasa, S. K. (2023). Recent advances in nanomaterials-based optical sensors for detection of various biomarkers (inorganic species, organic and biomolecules). *Luminescence*, *38*(7), 954–998.

Hassan, M. M., Zareef, M., Xu, Y., Li, H., & Chen, Q. (2021). SERS based sensor for mycotoxins detection: Challenges and improvements. *Food Chemistry*, *344*, 128652.

Huang, J., He, Z., Cao, J., Hong, J., Wu, Z., Gao, H., & Liao, X. (2018). Electrochemical immunosensor detection for lactoferrin in milk powder. *International Journal of Electrochemical Science*, *13*, 7816–7826.

Huang, S., Xu, J., Zheng, J., Zhu, F., Xie, L., & Ouyang, G. (2018). Synthesis and application of magnetic molecularly imprinted polymers in sample preparation. *Analytical and Bioanalytical Chemistry*, *410*, 3991–4014.

Jackman, J. A., Ferhan, A. R., & Cho, N.-J. (2019). Surface-based nanoplasmonic sensors for biointerfacial science applications. *Bulletin of the Chemical Society of Japan*, *92*(8), 1404–1412. https://doi.org/10.1246/bcsj.20190112

Jiang, L., Hassan, M. M., Ali, S., Li, H., Sheng, R., & Chen, Q. (2021). Evolving trends in SERS-based techniques for food quality and safety: A review. *Trends in Food Science & Technology, 112*, 225–240.

Jiang, Y., & Wu, J. (2019). Recent development in chitosan nanocomposites for surface-based biosensor applications. *Electrophoresis, 40*(16–17), 2084–2097.

Jiao, A., Cui, Q., Li, S., Li, H., Xu, L., Tian, Y., . . . Chen, M. (2022). Aligned TiO2 nanorod arrays decorated with closely interconnected Au/Ag nanoparticles: Near-infrared SERS active sensor for monitoring of antibiotic molecules in water. *Sensors and Actuators B: Chemical, 350*, 130848. https://doi.org/10.1016/j.snb.2021.130848

Joshi, S., Annida, R. M., Zuilhof, H., van Beek, T. A., & Nielen, M. W. (2016a). Analysis of mycotoxins in beer using a portable nanostructured imaging surface plasmon resonance biosensor. *Journal of Agricultural and Food Chemistry, 64*(43), 8263–8271.

Joshi, S., Segarra-Fas, A., Peters, J., Zuilhof, H., Van Beek, T. A., & Nielen, M. W. (2016b). Multiplex surface plasmon resonance biosensing and its transferability towards imaging nanoplasmonics for detection of mycotoxins in barley. *Analyst, 141*(4), 1307–1318.

Kalachyova, Y., Erzina, M., Postnikov, P., Svorcik, V., & Lyutakov, O. (2018). Flexible SERS substrate for portable Raman analysis of biosamples. *Applied Surface Science, 458*, 95–99.

King, T., Cole, M., Farber, J. M., Eisenbrand, G., Zabaras, D., Fox, E. M., & Hill, J. P. (2017). Food safety for food security: Relationship between global megatrends and developments in food safety. *Trends in Food Science & Technology, 68*, 160–175.

Krafft, B., Panneerselvam, R., Geissler, D., & Belder, D. (2020). A microfluidic device enabling surface-enhanced Raman spectroscopy at chip-integrated multifunctional nanoporous membranes. *Analytical and Bioanalytical Chemistry, 412*, 267–277.

Lace, A., & Cleary, J. (2021). A review of microfluidic detection strategies for heavy metals in water. *Chemosensors, 9*(4), 60.

Lan, Y.-B., Wang, S.-Z., Yin, Y.-G., Hoffmann, W. C., & Zheng, X.-Z. (2008). Using a surface plasmon resonance biosensor for rapid detection of Salmonella typhimurium in chicken carcass. *Journal of Bionic Engineering, 5*(3), 239–246.

Li, J.-L., Sun, D.-W., Pu, H., & Jayas, D. S. (2017). Determination of trace thiophanate-methyl and its metabolite carbendazim with teratogenic risk in red bell pepper (Capsicumannuum L.) by surface-enhanced Raman imaging technique. *Food chemistry, 218*, 543–552.

Li, M., Wu, H., Wu, Y., Ying, Y., Wen, Y., Guo, X., & Yang, H. (2017). Heterostructured cube Au–Ag composites for rapid Raman detection of antibiotic ciprofloxacin. *Journal of Raman Spectroscopy, 48*(4), 525–529.

Li, C., Peng, Y., Wang, H., Liang, A., & Jiang, Z. (2019). A nanosol SERS method for quantitative analysis of trace potassium based on aptamer recognition and silver nanorod catalysis of Ag (I)-glucose reaction. *Sensors and Actuators B: Chemical, 281*, 53–59.

Li, H., Ahmad, W., Rong, Y., Chen, Q., Zuo, M., Ouyang, Q., & Guo, Z. (2020). Designing an aptamer based magnetic and upconversion nanoparticles conjugated fluorescence sensor for screening Escherichia coli in food. *Food Control, 107*, 106761.

Mamian-Lopez, M. B., Miguel, R. B., Araki, K., Temperini, M. L., & da Costa Ferreira, A. M. (2021). Multivariate probing of antitumor metal-based complexes damage on living cells through Raman imaging. *Spectrochimica Acta Part A: Molecular and Biomolecular Spectroscopy, 244*, 118838.

Mayer, K. M., & Hafner, J. H. (2011). Localized surface plasmon resonance sensors. *Chemical Reviews, 111*(6), 3828–3857.

Mogensen, K. B., & Kutter, J. P. (2009). Optical detection in microfluidic systems. *Electrophoresis, 30*(S1), S92–S100.

Moradi, M., Azizi-Lalabadi, M., Motamedi, P., & Sadeghi, E. (2021). Electrochemical determination of T2 toxin by graphite/polyacrylonitrile nanofiber electrode. *Food Science & Nutrition*, *9*(2), 1171–1179.

Moreno, V., Adnane, A., Salghi, R., Zougagh, M., & Ríos, Á. (2019). Nanostructured hybrid surface enhancement Raman scattering substrate for the rapid determination of sulfapyridine in milk samples. *Talanta*, *194*, 357–362.

Nie, Y., Teng, Y., Li, P., Liu, W., Shi, Q., & Zhang, Y. (2018). Label-free aptamer-based sensor for specific detection of malathion residues by surface-enhanced Raman scattering. *Spectrochimica Acta Part A: Molecular and Biomolecular Spectroscopy*, *191*, 271–276.

Nishi, H., Hiroya, S., & Tatsuma, T. (2015). Potential-scanning localized surface plasmon resonance sensor. *ACS Nano*, *9*(6), 6214–6221.

Paliwal, A., Sharma, A., Tomar, M., & Gupta, V. (2015). Room temperature detection of NO2 gas using optical sensor based on surface plasmon resonance technique. *Sensors and Actuators B: Chemical*, *216*, 497–503.

Panneerselvam, R., Liu, G.-K., Wang, Y.-H., Liu, J.-Y., Ding, S.-Y., Li, J.-F., . . . Tian, Z.-Q. (2018). Surface-enhanced Raman spectroscopy: Bottlenecks and future directions. *Chemical Communications*, *54*(1), 10–25.

Prabowo, B. A., Purwidyantri, A., & Liu, K.-C. (2018). Surface plasmon resonance optical sensor: A review on light source technology. *Biosensors*, *8*(3), 80.

Pu, H., Xiao, W., & Sun, D.-W. (2017). SERS-microfluidic systems: A potential platform for rapid analysis of food contaminants. *Trends in Food Science & Technology*, *70*, 114–126.

Pu, H., Xie, X., Sun, D.-W., Wei, Q., & Jiang, Y. (2019). Double strand DNA functionalized Au@ Ag Nps for ultrasensitive detection of 17β-estradiol using surface-enhanced raman spectroscopy. *Talanta*, *195*, 419–425.

Qiu, J., Qi, X., Li, X., Ma, Z., Tang, Y., Mi, X., . . . Zhang, R. (2018). Development of a spatial heterodyne Raman spectrometer with echelle-mirror structure. *Optics Express*, *26*(9), 11994–12006.

Sackmann, E. K., Fulton, A. L., & Beebe, D. J. (2014). The present and future role of microfluidics in biomedical research. *Nature*, *507*(7491), 181–189.

Sadeghi, E., Rahimi, F., Azizi, Z., Kaki, S., & Babakhanian, A. (2023). Fabrication of a sensitive electrochemical sensor based on hybrid polyamide/chromotropic acid nanofibers electrospun on glassy carbon electrode for Hg2+ sensing in drinking water and canned fish samples. *Food Chemistry*, *414*, 135467.

Safran, V., Göktürk, I., Bakhshpour, M., Yılmaz, F., & Denizli, A. (2021). Development of molecularly imprinted polymer-based optical sensor for the sensitive penicillin G detection in milk. *ChemistrySelect*, *6*(43), 11865–11875.

Sannomiya, T., Dermutz, H., Hafner, C., Vörös, J., & Dahlin, A. B. (2010). Electrochemistry on a localized surface plasmon resonance sensor. *Langmuir*, *26*(10), 7619–7626.

Saylan, Y., Yılmaz, F., Derazshamshir, A., Yılmaz, E., & Denizli, A. (2017). Synthesis of hydrophobic nanoparticles for real-time lysozyme detection using surface plasmon resonance sensor. *Journal of Molecular Recognition*, *30*(9), e2631.

Schasfoort, R. B. (2017). *Handbook of Surface Plasmon Resonance*. Royal Society of Chemistry.

Shankaran, D. R., Gobi, K. V., & Miura, N. (2007). Recent advancements in surface plasmon resonance immunosensors for detection of small molecules of biomedical, food and environmental interest. *Sensors and Actuators B: Chemical*, *121*(1), 158–177.

Shen, M., Duan, N., Wu, S., Zou, Y., & Wang, Z. (2019). Polydimethylsiloxane gold nanoparticle composite film as structure for aptamer-based detection of vibrio parahaemolyticus by surface-enhanced Raman spectroscopy. *Food Analytical Methods*, *12*, 595–603.

Shrivas, K., Nirmalkar, N., Deb, M. K., Dewangan, K., Nirmalkar, J., & Kumar, S. (2019). Application of functionalized silver nanoparticles as a biochemical sensor for selective detection of lysozyme protein in milk sample. *Spectrochimica Acta Part A: Molecular and Biomolecular Spectroscopy, 213*, 127–133.

Song, C., Li, J., Sun, Y., Jiang, X., Zhang, J., Dong, C., & Wang, L. (2020b). Colorimetric/SERS dual-mode detection of mercury ion via SERS-Active peroxidase-like Au@AgPt NPs. *Sensors and Actuators B: Chemical, 310*, 127849. https://doi.org/10.1016/j.snb.2020.127849

Stewart, S., Priore, R. J., Nelson, M. P., & Treado, P. J. (2012). Raman imaging. *Annual Review of Analytical Chemistry, 5*(1), 337–360.

Szlag, V. M., Rodriguez, R. S., He, J., Hudson-Smith, N., Kang, H., Le, N., . . . Haynes, C. L. (2018). Molecular affinity agents for intrinsic surface-enhanced Raman scattering (SERS) sensors. *ACS Applied Materials & Interfaces, 10*(38), 31825–31844.

Wackerlig, J., & Lieberzeit, P. A. (2015). Molecularly imprinted polymer nanoparticles in chemical sensing–Synthesis, characterisation and application. *Sensors and Actuators B: Chemical, 207*, 144–157.

Wang, C., Wang, J., Li, M., Qu, X., Zhang, K., Rong, Z., . . . Wang, S. (2016). A rapid SERS method for label-free bacteria detection using polyethylenimine-modified Au-coated magnetic microspheres and Au@ Ag nanoparticles. *Analyst, 141*(22), 6226–6238.

Wang, G., Wang, Y., Chen, L., & Choo, J. (2010). Nanomaterial-assisted aptamers for optical sensing. *Biosensors and Bioelectronics, 25*(8), 1859–1868.

Wang, K., Wang, Y., Li, Q., Liu, Z., & Liu, S. (2022a). A fluorescence and localized surface plasmon resonance dual-readout sensing strategy for detection of acetamiprid and organophosphorus pesticides. *Sensors and Actuators B: Chemical, 351*, 130977. https://doi.org/10.1016/j.snb.2021.130977

Wang, P., Ding, L., Zhang, Y., & Jiang, X. (2023). A novel aptamer biosensor based on a localized surface plasmon resonance sensing chip for high-sensitivity and rapid enrofloxacin detection. *Biosensors, 13*(12), 1027.

Wang, X., Liu, G., Hu, R., Cao, M., Yan, S., Bao, Y., & Ren, B. (2022b). Chapter 1—Principles of surface-enhanced Raman spectroscopy. In Y. Wang (Ed.), *Principles and Clinical Diagnostic Applications of Surface-Enhanced Raman Spectroscopy* (pp. 1–32). Elsevier.

Wang, Y., Chen, H., & Jiang, L. (2021). A highly reproducible SERS sensor based on an Au nanoparticles/graphene oxide hybrid nanocomposite for label-free quantitative detection of antibiotics. *Analyst, 146*(18), 5740–5746.

Wei, H., Abtahi, S. M. H., & Vikesland, P. J. (2015). Plasmonic colorimetric and SERS sensors for environmental analysis. *Environmental Science: Nano*, 2(2), 120–135.

Wei, T., Ren, P., Huang, L., Ouyang, Z., Wang, Z., Kong, X., . . . He, Q. (2019). Simultaneous detection of aflatoxin B1, ochratoxin A, zearalenone and deoxynivalenol in corn and wheat using surface plasmon resonance. *Food Chemistry, 300*, 125176.

Wei, X., Yin, M., Zhang, L., Lin, H., Wang, J., Xie, W., & Xu, D. (2022). Surface Plasmon Resonance (SPR) biosensor for detection of mycotoxins: A review. *Journal of Immunological Methods, 510*, 113349. doi:https://doi.org/10.1016/j.jim.2022.113349

Willets, K. A., & Van Duyne, R. P. (2007). Localized surface plasmon resonance spectroscopy and sensing. *Annu. Rev. Phys. Chem., 58*(1), 267–297.

Wu, J., & Gu, M. (2011). Microfluidic sensing: state of the art fabrication and detection techniques. *Journal of biomedical optics, 16*(8), 080901-080901-080912.

Xia, L., Chen, M., Zhao, X., Zhang, Z., Xia, J., Xu, H., & Sun, M. (2014). Visualized method of chemical enhancement mechanism on SERS and TERS. *Journal of Raman Spectroscopy, 45*(7), 533–540.

Xu, X., Ma, X., Wang, H., & Wang, Z. (2018). Aptamer based SERS detection of *Salmonella typhimurium* using DNA-assembled gold nanodimers. *Microchimica Acta*, *185*, 1–8.

Xu, J., Li, X., Wang, Y., Hua, T., Guo, R., Miao, D., & Jiang, S. (2020). Flexible, stable and sensitive surface-enhanced Raman scattering of graphite/titanium-cotton substrate for conformal rapid food safety detection. *Cellulose*, *27*, 941–954.

Yang, M., Liu, G., Mehedi, H. M., Ouyang, Q., & Chen, Q. (2017). A universal SERS aptasensor based on DTNB labeled GNTs/Ag core-shell nanotriangle and CS-Fe3O4 magnetic-bead trace detection of Aflatoxin B1. *Analytica Chimica Acta*, *986*, 122–130.

Yola, M. L., Atar, N., & Eren, T. (2014). Determination of amikacin in human plasma by molecular imprinted SPR nanosensor. *Sensors and Actuators B: Chemical*, *198*, 70–76.

Zhang, J., He, L., Zhang, X., Wang, J., Yang, L., Liu, B., . . . Zhang, Z. (2017). Colorimetric and SERS dual-readout for assaying alkaline phosphatase activity by ascorbic acid induced aggregation of Ag coated Au nanoparticles. *Sensors and Actuators B: Chemical*, *253*, 839–845.

Zhang, F., Liu, M., Liu, R., Li, J., Sang, Y., Tang, Y., . . . Wang, S. (2021). A broad-spectrum sensing strategy for the tetracycline family of antibiotics based on an ovalbumin-stabilized gold nanocluster and its application in a pump-free microfluidic sensing platform. *Biosensors and Bioelectronics*, *171*, 112701. doi:https://doi.org/10.1016/j.bios.2020.112701

Zhao, S., Huang, J., Lei, J., Huo, D., Huang, Q., Tan, J., . . . Tian, F. (2021). A portable and automatic dual-readout detector integrated with 3D-printed microfluidic nanosensors for rapid carbamate pesticides detection. *Sensors and Actuators B: Chemical*, *346*, 130454. doi:https://doi.org/10.1016/j.snb.2021.130454

Zhou, J., Qi, Q., Wang, C., Qian, Y., Liu, G., Wang, Y., & Fu, L. (2019). Surface plasmon resonance (SPR) biosensors for food allergen detection in food matrices. *Biosensors and Bioelectronics*, *142*, 111449.

Zhu, J., Chen, Q., Kutsanedzie, F. Y., Yang, M., Ouyang, Q., & Jiang, H. (2017). Highly sensitive and label-free determination of thiram residue using surface-enhanced Raman spectroscopy (SERS) coupled with paper-based microfluidics. *Analytical Methods*, *9*(43), 6186–6193.

Zhu, C., Hu, X., & Wang, X. (2019). Silver nanocubes/graphene oxide hybrid film on a hydrophobic surface for effective molecule concentration and sensitive SERS detection. *Applied Surface Science*, *470*, 423–429.

Zhu, A., Jiao, T., Ali, S., Xu, Y., Ouyang, Q., & Chen, Q. (2021). SERS sensors based on aptamer-gated mesoporous silica nanoparticles for quantitative detection of staphylococcus aureus with signal molecular release. *Analytical Chemistry*, *93*(28), 9788–9796.

Zhu, L., Wu, Q., Mei, X., Li, Y., & Yang, J. (2023). based microfluidic sensor array for tetracycline antibiotics discrimination using lanthanide metalcarbon quantum dots composite ink. *Advanced Composites and Hybrid Materials*, *6*(6), 221.

Zhu, L., Hu, Z., Shen, Y., & Wang, Y. (2023). Preparation and application of lysozyme molecularly imprinted surface plasmon resonance biosensors. *Microchemical Journal*, *190*, 108665

Zhang, Q., Ma, S., Meng, W., Zheng, Y., Yin, L., Wang, H., . . . Su, S. (2024). Smartphone-based plant-wearable microfluidic sensor with self driven electrolyte for in-situ detection of methyl parathion. *Sensors and Actuators B: Chemical*, *418*, 136254. doi:https://doi.org/10.1016/j.snb.2024.136254

Zou, Q., Li, X., Xue, T., Zheng, J., & Su, Q. (2019). SERS detection of mercury (II)/lead (II): A new class of DNA logic gates. *Talanta*, *195*, 497–505.

12 Principle, Fundamentals, and Application of Hyphenated Techniques in Food Analysis

Rhuthuparna Malayil, Amitha Thankachan, Sandeep Singh, and Anjana Munshi

12.1 INTRODUCTION

Food scarcity has led to safety concerns at the global level, although this issue is not being addressed aggressively (Grunert, 2005). Customers consistently demand high standards for food quality and safety, which increases the legality of food processing. A sophisticated quality inspection system guarantees secure food processing procedures that involve essential labeling providing the ingredient and nutritional information, for the quality and safety of the food (Wu and Sun, 2013).

The most extensively used and favored approach for ensuring quality is human visual inspection. However, the process is highly laborious and time-consuming. Mass spectrometry (MS) and high-performance liquid chromatography (HPLC) are two extensively used instruments for chemical analysis of the food. However, they still have several drawbacks such as laborious sample preparation processes and challenges in managing huge numbers of samples (Wu and Sun, 2013). Food analysis can be done quickly, nondestructively, affordably, and sustainably using spectroscopic techniques. NMR is a spectroscopic method for figuring out a substance's molecular structure and characteristics. Additionally, it is effectively used to check the quality of various food samples. High-field NMR has demonstrated its capacity to precisely determine critical edible oil or fatty acids' quality characteristics, particularly the ability to identify possible adulteration (Kharbach et al., 2023). Infrared spectroscopy (IR) is used to analyze food ingredients quantitatively and qualitatively, including moisture levels, proteins, lipids, and carbohydrates. It is also used to identify functional groups, carbon, and nitrogen (Alamprese et al., 2013, Kharbach et al., 2023). The development of effective and accurate alternative for the analysis of quality-related aspects of food products is indeed important.

DOI: 10.1201/9781003482611-12

12.2 PARAMETERS OF FOOD QUALITY AND SAFETY ANALYSIS

The biggest challenge facing the food production sector is creating high-quality products. Organic food is always constrained because of quality expectations and promises. Product-related aspects and the manufacturing process are assessed through several criteria to define these qualities. Evaluation of each criterion should be through proper parameters (Kahl et al., 2012). According to consumer requirements, the quality of food products is decided by their physical characteristics, chemical composition, level of contaminants, energy content, sensory attributes, shelf life, packaging, and labeling (Figure 12.1) (Molnár et al., 1979). Hierarchical and dynamic interactions among these qualities are also regarded as distinctive features of food quality. In addition to these, food safety has primary significance for quality of the food (Molnár et al., 1979). All these characteristics are measured independently to create an index of overall quality.

12.3 MAJOR TECHNIQUES APPLIED IN FOOD ANALYSIS

12.3.1 Genomics and Food Safety

Recently, researchers have given more attention to the contributions of 'genomics' technology in the food processing industry. DNA-based techniques are widely used for food analysis. Next-generation sequencing (NGS) and PCR are two well-accepted approaches being utilized for food authentication and the detection of even minute traces of biological adulterants in food products. Fraudulent practices like substitution of the ingredient by the cheaper ones are very common in the food industry. As the DNA is stable under a variety of food processing conditions, DNA-based methodologies are a promising

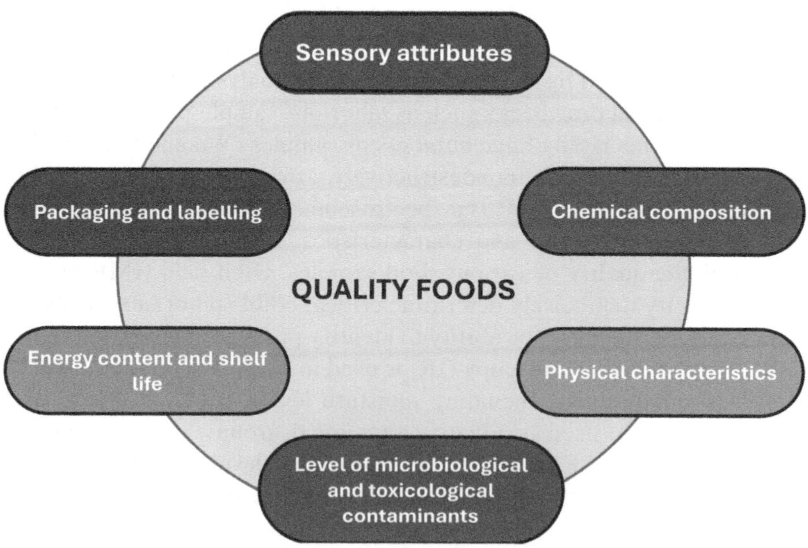

FIGURE 12.1 Different parameters to check the quality of food products.

way of addressing this issue (Figure 12.2) (Catalano et al., 2016, Haynes et al., 2019). A standard PCR reaction requires three steps (denaturation-annealing—extension) carried out for 20–50 times. PCR is extended to many other techniques like RT-PCR, AFLP or RAPD, RFLP, Multiplex PCR assay, and SSCP (Hill, 1996). A specific area of DNA amplified by PCR can be used as a molecular biomarker for qualitative species identification. Closely related species of animals and plants can be differentiated by the separate DNA regions within the mitochondria and chloroplast, respectively.

Another important application of PCR is the detection of genetically modified organisms (GMOs). PCR amplification and subsequent sequencing can detect the DNA sequence (transgenes) specific for the GMOs (Nielsen et al., 2012). Additionally biological identification is carried out through DNA barcoding and FINS. FINS utilizes sequencing of short, standardized gene fragments for distinguishing many products (Hebert et al., 2003, Bartlett and Davidson, 1992). Genomics plays a major role in the area of food microbiology. It helps to provide a molecular fingerprint of various microorganisms in food, which in turn predicts the possible preservation treatment for the effective and synergistic inhibition of unwanted microbial growth (Brul et al., 2006). NGS is being applied for the accurate identification of unculturable microbial taxa that are present in small numbers. Knowledge on the microbial diversity and dynamics could improve the monitoring and safety of food products (Mayo et al., 2014). Likewise, metagenomics also helps in the identification of non-culturable pathogens and characterizes the diversity in food systems (Bergholz et al., 2014). NGS enables the identification of plant, animal, fungal, and microalgae ingredients in food commodities. The quality of the DNA is important for the better analysis of the

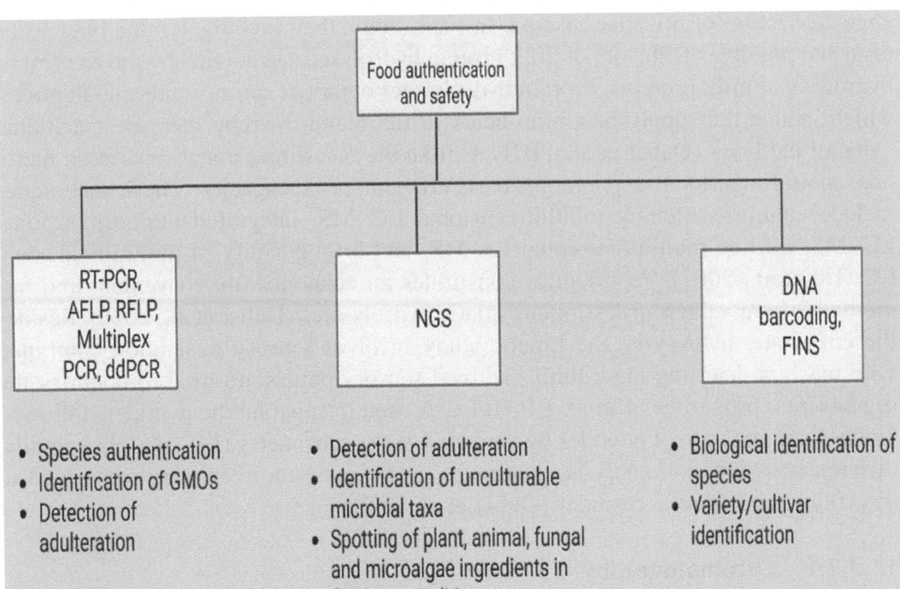

FIGURE 12.2 DNA-based methods for food authentication and safety analysis.

DNA sequences of interest present in a food sample. DNA can be extracted from any kind of materials including fresh, raw, dried, powdered, and highly processed materials. Each food matrix requires specific extraction procedures (Ripp et al., 2014).

12.3.2 Proteomics and Food Analysis

Proteomics is a discipline that deals with the large-scale analysis of the structural and functional aspects of proteins, protein quantification, and study of protein–protein interactions in a biological system. MS-based proteomics tool is a highly powerful strategy used for protein identification and quantification. Proteomic tools are widely used for food quality and safety studies. Food proteomics and targeted food proteomics are two proteomic approaches used in food analysis. Discovery of food proteomics follows a bottom-up proteomics strategy where potential food protein/peptide biomarkers are identified by proteome analysis. In the bottom-up proteomics strategy, the protein of interest is digested by proteases like trypsin and analyzed through MS. Targeted food proteomics is a sequential approach performed after the discovery phase, where the food products are precisely searched for the selected peptide biomarkers (Carrera, 2021). For the analysis of huge number of proteins encoded by the genome, high-throughput proteomic technologies like one-dimensional and multidimensional liquid chromatography–mass spectrometry (LC-MS) are available. Integration of ESI-MS with LC is the most preferred method for proteome analysis. MALDI is often advised instead of ESI-MS as they offer higher speed of analysis, lower sample consumption, and higher tolerance toward sample contaminants (such as buffers, salts, and surfactants) (Hu et al., 2007). In a recent study, Daher et al. performed LC-ESI-Q-TOF-MS/MS and enzymatic hydrolysis kinetics for the peptide characterization of micellar caseins. In their study, they investigated the unpleasant taste and possible off-flavors of milk protein hydrolysates generated by the enzymatic hydrolysis of milk proteins. Protein hydrolysates comprise amino acids and peptides, which induce the supply of amino acids in the blood, thereby increase nutritional value of the body (Daher et al., 2021). Unlike the one-dimensional separation methods, multidimensional separations have efficient peak capacity. These techniques include column-switching multidimensional LC–MS, integrated multidimensional LC–MS, off-line multidimensional LC-MS, and orthogonality of multidimensional LC (Hu et al., 2007). Peptidomics constitutes an economically convenient and fast method to assess the sourness of enzymatic hydrolysates (Daher et al., 2020). Besides the enzymatic hydrolysis, the kinetic study involves sensory evaluation combined with machine learning algorithms. Trained sensory panels are used to quantify the organoleptic properties of food. RP-HPLC is used to separate the peptides, followed by the fragmentation of peptides by tandem mass spectrometry (MS/MS). Eventually, differential expression analysis, heat maps, and regression trees were performed to determine the nature of peptides (Daher et al., 2021).

12.3.2.1 Chromatography

Chromatography is a biophysical method employed for the separation, purification, and identification of components of a mixture. The technique comprises mainly three components: stationary phase, mobile phase, and separated molecule. The efficiency of separation depends on properties like adsorption, partition, and affinity of the components to be separated. There are different chromatography methods available for separation. Some of them include column chromatography, thin-layer chromatography (TLC), paper chromatography, gas chromatography (GC), ion exchange chromatography, HPLC, etc. (Coskun, 2016).

GC is widely used for the analysis of food composition, food additives, natural products, flavor and aroma components, and packaging materials. In GC, the gas stream carry the sample into a column where the components are separated on the basis of adsorption or partition. Additionally, it helps to detect the trace concentrations of contaminants such as pesticides, fumigants, environmental pollutants, and natural toxins in complex food. GC typically has an interest in analyzing any volatile, semi-volatile, nonpolar, and semi-polar component in food. Independent of the polarity and volatility, HPLC can be used for separating all types of chemicals uniquely for drugs, amino acids/proteins, carbohydrates, and, lipids (Lehotay and Hajšlová, 2002). Even though GC is more advantageous, the biochemical applications of HPLC turned the interest of analysts to use HPLC primarily. GC is more efficient for the characterization of lipids, drugs, pesticides, and carbohydrates. Liquid chromatography is primarily used for the measurement of vitamin A, β-carotene, thiamine, and riboflavin in food products. Separation of highly complex samples is still challenging. To accommodate this problem, the research world is introduced with the concept of comprehensive 2D chromatography, in which all the components are subjected to two separations. As compared to other classic chromatographic methods, it is equipped with high-resolution power (Tranchida et al., 2004).

12.3.2.2 Mass Spectrometry (MS)

MS separates ionized molecules based on their specific charge/mass ratio. MS instruments comprise different modes of ionization such as Electrospray Ionization (ES),Electron Ionization (EI), Matrix-Assisted Laser Desorption/Ionization (MALDI), and Atmospheric Pressure Chemical Ionization (APCI). The role of MS in food research is well established. It is becoming an invaluable tool in food analysis and quality control. MS is extensively used for peptide and protein characterization. The possibilities of MS in food analysis are extended by coupling with various other technologies like LC–MS, APCI-MS, GC–MS, Matrix-assisted laser desorption/ionization time-of-flight mass spectrometry (MALDI-TOF-MS), High-Performance Liquid Chromatography–Mass Spectrometry (HPLC-MS), and Inductively coupled plasma–mass spectrometry (ICP-MS) (Careri et al., 2002). HPLC-MS is an ideal tool for precise and reproducible quantitative analysis. In food products, the presence of heat-labile compounds like proteins and vitamins can be easily identified by HPLC-MS. An interface connected between HPLC and MS transfers the separated compounds from liquid chromatograph to the MS ion source. Over the last few years, ICP-MS has been utilized to study the inorganic specifications in food analysis (Careri et al., 1998, Cramer and Humpf, 2012). A combination of Liquid Chromatography–Inductively Coupled Plasma–Mass Spectrometry (LC-ICP-MS) has been applied in the field of nutritional and toxicological research for the identification of the chemical nature of an element (Vela et al., 1993). High-Performance Liquid Chromatography–Inductively Coupled Plasma–Mass Spectrometry (HPLC-ICP-MS) is observed to be ideal for the elemental speciation analysis. Shona McSheehy et al. evaluated the naturally occurring selenium species in garlic extract by two-dimensional HPLC with parallel ICP-MS and Electrospray Ionization–Tandem Mass Spectrometry (ESI-MS-MS), because of their cancer-preventing activity. The peaks in the chromatographic profile provide information regarding the compounds under investigation. In another study, researchers separated eight selenium compounds from selenium nutritional supplements through a combination of HPLC–ICP-MS (McSheehy et al., 2000, Zheng et al., 2000). HPLC–ICP-MS was also used to quantify arsenic compounds in various baby foods and freeze-dried apple samples, in which total arsenic content was measured by ICP-MS analysis of the extracts followed by separation through anion-exchange HPLC column (Caruso et al., 2001). Furthermore, the identification of polyphenolic compounds and antioxidants in complex mixtures can be successfully targeted by HPLC-MS, GC–MS, and MALDI-TOF-MS (Careri et al., 2002). GC–MS, LC–MS, and LC–MS–MS aid in monitoring the xenobiotic toxic compounds in foods, thereby ensuring safety in food consumption. Recently, a new MS technique biomolecular interaction analysis mass spectrometry (BIA-MS) is introduced, which incorporates surface plasmon resonance-based biomolecular interaction analysis (SPR-BIA) and MALDI-TOF-MS (Nelson et al., 1997).

Lipid analysis using the MS technique is widely practiced in food-processing research. Lipid profiling for the detection of triglycerides, fatty acids, waxes, and sterols can be done using MS. Proteins are the major components that determine the biological functions of food products. MS is frequently used to check food proteins and peptides. MALDI-TOF-MS has been used to evaluate the authenticity of milk. It helps to analyze the presence of powdered milk in fresh raw milk. This is a highly optimized and time-saving method. The sample for MALDI-MS is coated with a energy-adsorbent matrix. Sample within the matrix then ionized with the laser beam. The generated protonated ions are separated on the basis of their mass-to-charge ratio (m/z) and measured using mass analyzers like time of flight (TOF) (Singhal et al., 2015). The instrument can separate the peak by a few Daltons. Hence, MALDI-MS has been used to detect the presence of small quantity of cow milk in either raw ewe or water buffalo milk samples based on the variants A and B of cow and ewe β-lactoglobulins. The protein patterns generated from the whey proteins, α-lactalbumin, and β-lactoglobulin define the presence of adulteration (Cozzolino et al., 2001). LC–MS analysis is used for the characterization and determination of fat-soluble and water-soluble vitamins in both food and pharmaceutical areas. Another MS technique for food quality screening is pyrolysis-mass spectrometry (Py-MS), which analyzes the food authenticity and determines the food and beverage adulteration. It has been used to characterize milk samples, vinegar, and wines; for the detection of caffeine content in various beverages; and for the classification of cocoa butter. Py-MS involves thermal degradation of the sample in an inert atmosphere followed by MS analysis (Guillou et al., 1999).

12.3.3 Enzymatic Protein Hydrolysis

Enzymatic protein hydrolysis is a highly sophisticated and complex novel methodology applied in the food industry for the characterization of foods and to produce edibles with extended functional properties. Food-derived bioactive peptides yielded by protein hydrolysis are functionally very important as they contribute to human health (Kristinsson and Rasco, 2000). In enzymatic protein hydrolysis, proteins are cleaved using specific proteases and form amino nitrogen, which are measured through titration (Ghosh et al., 2017). Microbial fermentation or chemical hydrolysis is another method used for bioactive peptide production. Nevertheless, enzymatic hydrolysis is more advantageous due to the selectivity of commercial enzymes, absence of secondary products, and use of mild temperature and pH conditions. Moreover, this method is more environmentally friendly. The following factors like multiple substrates and reactions, the undefined nature of substrates, the heterogeneous reaction system, and modulated thermal inactivation of the enzyme make it more intricate to develop theoretical models. However, the enzymatic breakdown of proteins using various enzymes and protein sources has been modeled using the traditional Michaelis-Menten kinetics (O'Meara and Munro, 1985).

12.3.4 Near Infrared (NIR) Spectroscopy

Spectroscopy is a technique is utilized for the measurement of spectra produced when matter interacts with electromagnetic radiation. NIR spectroscopy measures the absorption at a wavelength range of 780–2,500 nm. The main principle behind the spectroscopic analysis is Beer-Lambert law. When the light passes through the sample, a fraction of the light is reflected by the sample and recorded by the detector placed on the same side that translates changes in light intensity to clinically useful information. In NIR spectroscopy, a single polychromatic thermal source is generally used as a light source.

$$I_{trans} = I_{In} e^{-DC\varepsilon},$$

where I_{trans} = intensity of transmitted light, I_{In} = intensity of incident light, e = natural log base (2.71828), D = path length of light through the solution, C = concentration of the absorbing molecule, and ε = specific extinction coefficient.

In foods, the NIR spectra comprise overlapping absorptions corresponding to the fundamental vibrations of CH, OH, and NH chemical bonds, whereas a series of overtone and combination bands exist for each vibration. Previously classical absorption spectroscopy was used to analyze the water, protein, fat, and carbohydrate content in food samples. However, NIR is more advantageous than traditional spectroscopy in terms of accessibility and simplicity. Additionally, NIR doesn't need any sample preparation, and therefore it can analyze quickly. Although this method is exceptional, its dependence on less precise reference method constrained it from being the best. Different sample presentation methods make it more approachable in the field of food analysis. There are three different instruments feasible for NIR, and choosing the right one determines the successful application of NIR (Osborne and Fearn, 1986).

NIR has been widely used in food industry. In the wheat breeding program, it has been used to test the quality of the cross breed and helps the breeders to check the protein content and to speed up the analysis by eliminating the intermediate grinding procedure. In many countries, protein content of wheat determines the price value. Canada used NIR technology for testing of the wheat protein in the Canadian wheat segregation program. Through this, Canada established an eco-friendly large-scale protein testing of grain. Additionally, they predicted functional properties like water absorption, flour yield, dough development, damaged starch, extensibility, and loaf volume. These services finally ended up with the baking of a good product without any energy wastage. Optimum fertilizer requirements of cereal crops like rice and wheat can be determined using NIR. This is possible by analyzing the total nitrogen and carbohydrate content of the plant tissue samples (McGrath et al., 1997). Many biscuits and sweet bakeries utilize NIR to assess the protein and moisture content of several raw ingredients such as flour, sucrose, and chocolate. Norris (1989) proposed an NIR spectrum of ground grain as an international standard reference method for wheat hardness (Norris, 1989). In the dairy industry, various products such as liquid and dried whole milk, skim milk, cream, and traditional and processed cheese are being analyzed for the protein, moisture, fat, and lactose content via NIR.

12.3.5 Imaging Techniques

Imaging technologies have become an invaluable tool for nondestructive detection of food quality and safety. There are several imaging technologies that are particularly used for food analysis. They include soft X-ray imaging, ultrasound imaging, fluorescence imaging, magnetic resonance imaging, hyper spectral imaging, thermal imaging, and odor imaging. The efficiency of the systematic analysis by this method depends on the processing capacity of its computer hardware and the quality of camera attached (Chen et al., 2013a). X-ray is a form of electromagnetic radiation that comes within the wavelength range of 0.01–10 nm. Based on the photon energy, there are two categories of X-ray, namely, soft X-ray (10 keV) and hard X-ray (10–120 keV). As the high penetrating ability of hard X-rays impairs the food, the scientists showed their interest in soft X-ray imaging technique for food inspection. The soft XRI system generally comprises an X-ray source, image-acquisition card and computer, a sensor for X-ray detection, conveying belt, and a stepping motor. While the penetration of X-ray to the food product, some energy is lost, and the energy comes out detected by the sensor and converts into an image. As the image directly predicts the internal defects and quality of the agricultural products and animal products, it has been widely used in the food industry. Presence of foreign contaminants further reduces the energy and appears as a darker shade of gray (Bischof et al., 2007). X-ray micro-computed tomography has also been employed for the microstructure characterization of frozen products (Mousavi et al., 2005).

Compared to other imaging techniques, ultrasound imaging (UI) is easy and economic. Apart from the difference in the wave energy, the principle is like that of XRI. Ultrasound utilizes mechanical waves above 20 kHz. UI is majorly used for food packaging and foreign objects. Ultrasound wave scatters when it travels through a nonuniform medium. As compared to the host tissue, the amount of energy reflected and transmitted by the foreign objects is quite different. The difference in relative acoustic impedances, time-of-flight, and velocity changes the material characteristics (Povey and McClements, 1988).

12.3.6 X-Ray Diffraction (XRD)

There are several other top-level technologies practiced in the food industry. XRD is one of the core and coherent methods being used for quality checks such as screening of food impurities. XRD is also considered as an ideal method to assess the density of food. The X-rays used for this method were generated by bombarding tungsten, molybdenum, or copper, with high-energy electrons. The X-rays diffracted after hitting the sample are collected by the detector and analyzed. XRD follows the Bragg's law,

$$n\lambda = 2d \sin\theta$$

where n = positive integer, λ = wavelength of incident X-ray, d = lattice separation, and θ = scattering angle

Simulation of the diffraction data can be performed using various software applications like CrystalDiffract, MAUD, or X'Pert High Score Plus. The area under the peak or the position of the peak at a particular 2θ or the total diffractograph will define the crystallinity. For a long time, it has been widely used in the field of material science. In recent years, it also started gain recognition in the area of food analysis (Purohit et al., 2019). XRD is used to check the miscibility and morphology of antimicrobial films, which helps to extend the shelf life of fresh food products (Tripathi et al., 2009). Moreover, they can be utilized for structural analysis, understanding crystal polymorphism and amorphous content of food components like oil fat, cereal starch, sugar, chocolate, and cocoa butter, and to evaluate the powder quality of dehydrated fruit juices. The peaks in the diffractogram can be used to differentiate between parboiled sound kernels (SK) and fungal-infested rice kernels (FIK), in which diffraction angles at 15.3°, 18.4° and 20°, are underdeveloped in parboiled FIK compared to SK. The moisture content of the cooked rice can be examined by analyzing the reflection angle of 5.5 degrees. The difference in the organization of hydrocarbon chains of fatty acids in triglycerol of bovine milk can be identified with short and long spacings of XRD. In chocolate processing, the crystallinity of cocoa butter, sugars, and fat molecule can be identified by XRD (Purohit et al., 2019).

12.4 RECENT INNOVATION IN HYPHENATED TECHNIQUES FOR FOOD ANALYSIS

Electrochemical methods are emerging as a powerful tool for the determination and quantification of antioxidants present in food and beverages. These assays are quick, precise, simple, and environmentally friendly than the conventional spectrophotometric methods. Commonly used electrochemical methods include chronoamperometry, differential pulse voltammetry (DPV), cyclic voltammetry, and square wave voltammetry. DPV produces a voltammogram, in which the height of the peak is directly proportional to the concentration of the analyte measured. DPV enables improved resolution for species with identical redox potentials. It generally comprises a electrochemical cell, where chemical reactions will occur, and a supporting electrolyte to reduce the effect of electromigration. The pulse generates a current signal through the redox reaction. This signal is plotted against the applied potential to plot the voltammogram. DPV is widely used for the detection of vanillin in beverages and other food stuffs. Moreover, different classes of flavor enhancers, synthetic dyes, preservatives, and antioxidants in foodstuffs can be analyzed using DPV (Goncalves-Filho and De Souza, 2024). Chronoamperometry is a microfluidic-based noval analytical method to detect safe and high-quality products. This method is very cost-effective and requires only 10 μL analyte. Most recently, chronoamperometry has been applied to distinguish milk based on the quality or adulteration (Souri et al., 2019). Electrochemical methods are advantageous over several other methods and emerging as alternatives for quantifying the antioxidant capacity of food products (Hoyos-Arbelaez et al., 2017). Similarly colored samples like wine and fruit juices that are difficult to read through spectrophotometric techniques can be easily analyzed by electrochemical methods (Sousa et al., 2004; Lino et al., 2014). Electrochemical biosensors are analytical devices used for the screening of foreign particles and other toxic substances for food safety. Capillary electrophoresis (CE) and capillary electro chromatography (CEC) are two recently developed electromigration techniques in foodomics approach for the analysis of small organic ions to large macromolecules. In order to characterize compounds and determine the mass of analytes, CE and CEC have been combined with MS methods. These methods use electrophoretic and electroosmotic mobilities generated by applying high electric field. CEC is hybrid form of HPLC and CE. Instead of pressure-driven flow, CEC uses a strong EOF, which helps to transfer the analytes into the capillary column. These analytical techniques are characterized by high mass sensitivity and chromatographic efficiency (Della Posta et al., 2022). Recently, Sude Oflu and his colleagues developed a highly accurate and optimized approach, namely spray assisted droplet formation-liquid phase microextraction (SADF-LPME), for the derivative and preconcentrate target analyte. In their study, they combined SADF-LPME with ID[4] to determine methamphetamine at trace levels in urine and serum samples. ID is a calibration strategy that increases the precision of the method (Oflu et al., 2023). In recent years, X-ray micro-computed tomography (μCT) has been taking the lead for the investigation of food microstructure and quality, internal disorders, pore size distribution, and evaluation of specific components. μCT can be applied to various commodities including fruits and vegetables, confectionery and baked products, fish, dairy, and cereals. In agricultural field, μCT is applied to sort food by developing classification algorithms. A significant advantage of μCT over other methods is that it is noninvasive and nondestructive and requires no sample preparation. μCT generally follows the basic principles of absorption physics and reconstruction mathematics. By measuring changes in X-ray attenuation, which includes absorption and scattering, μCT assesses a sample's internal structure including the density and compositional differences. The rotating sample after X-ray illumination produces a series of 2-D radiographs, which are subsequently processed by a computer to create a 3D volume (Figure 12.3) (Schoeman et al., 2016).

A noncontact ultrasonic system is used to evaluate the properties of food in containers. Precise physiochemical changes in palm oil and milk-based drinks are measured using the noncontact method. Crystallization polymorphism of palm oil due to temperature variation can be evaluated by observing the attenuation of the ultrasonic signal (Gan et al., 2006) (Figure 12.4). Combined Mass Spectrometry-Nuclear Magnetic Resonance (MS-NMR) is a new technology that has been developed for the qualitative and quantitative determination of amino acids and other ingredients in dietary supplements. In MS-NMR, one part of the sample is directed to MS, while the remaining part

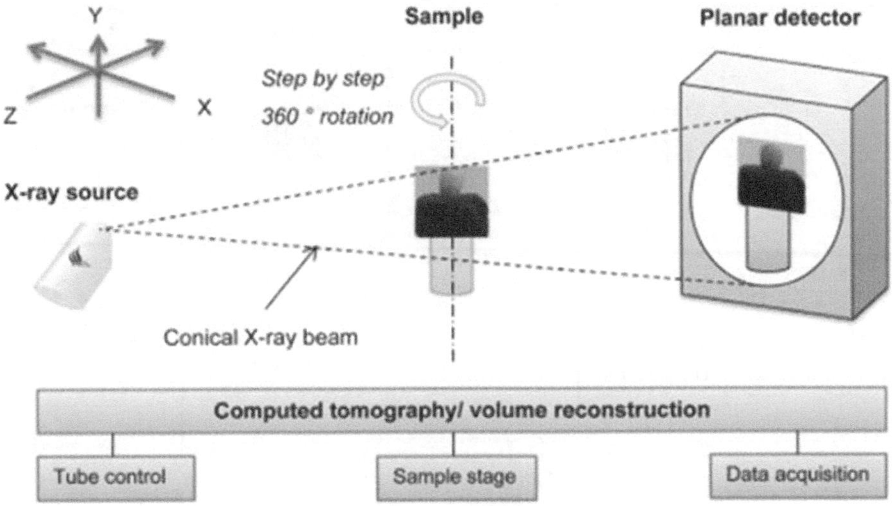

FIGURE 12.3 Schematic representation of the principle of μCT. The X-rays passing through the sample are collected by the detector, converting them into digital projection images/radiographs.

Source: Schoeman et al. (2016)

goes through the NMR. MS uses flow injection electrospray ionization (FIA-ESI-MS), where the sample is directly injected to the ionization source. The mass accuracy and the fragmentation information provide the information of AAs. The analysis involves the following steps: sample preparation, flow injection, ESI-MS acquisitions, selection of NMR experiments, and nonuniform sampling (NUS) technique for data acquisition. NMR confirms the MS data through 1D and 2D experiment along with NUS technique. Unlike the chromatography separation, this method is very fast and easy (Palaric et al., 2018).

A new technology, capillary zone electrophoresis (CZE), has been explored to analyze different chemical species together. As it requires a small amount of sample and reagent, shorter analysis time, and reduced waste generation, it is highly preferred over other techniques. A fused-silica capillary is used for the separation in a capillary electrophoresis system. The sample is injected hydrodynamically and detected at 210 nm. Once pKa, electrophoretic mobility, and other separation process components for each analyte are available through PeakMaster software, mobility versus pH plot can be created. This method has been widely used to evaluate the principle components (amino acids, proteins, peptides, saccharides, and metal ions), additives (sweeteners, cyclamate, aspartame, and saccharine), and other residues in foodstuffs. For example, in a recent study, CZE was used to determine the level of phosphate, citrate, and sulfite together in seafood (Goncalves et al., 2020). Supercritical fluid chromatography Mass spectroscopy (SFC-MS) is a powerful separation technique applied to many areas including food analysis. It can be considered as an alternative or extension to routine analysis techniques like GC–MS/LC–MS. SFC-MS is performed in packed columns with ESI and APCI as the ionization sources. Through a capillary

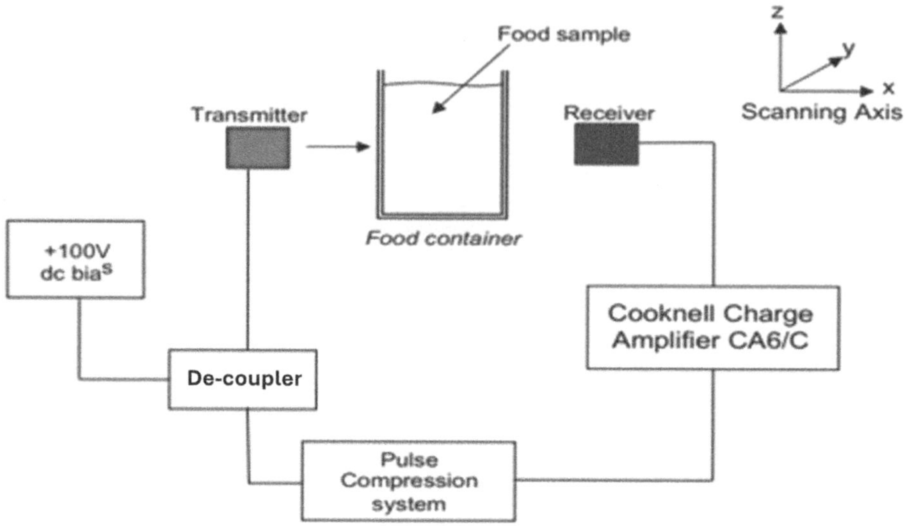

FIGURE 12.4 Experimental arrangement of noncontact ultrasound to measure food properties. *Source:* Gan et al., (2006)

tubing, SFC full-flow is introduced into the MS detector. Because the majority of the mobile phase (CO_2) is volatilized at the backpressure regulator's outlet, very little of the organic solvent enters the MS interface during the coupling of SFC to MS. There are two interfaces available for SFC-MS: one is pre-BPR splitter with sheath pump and the other is BPR and sheath pump with no splitter. Both the interfaces prevent solute precipitation when CO_2 volatilizes as the makeup fluid is introduced before the MS detector. SFC-MS has been applied for the identification of food contaminants (pesticide compounds, transformation product compounds, and polycyclic aromatic hydrocarbon compounds) and nutrients (lipids, carotenoids, vitamins, and free amino acids) (Figure 12.5) (Toribio et al., 2021). The basic working principle of ion mobility spectrometry is based on the electrophoretic mobility of ions present in gas and the separation of analytes in accordance with their properties like size, shape, and charge. In case of TIMS separation, the ions are pushed through TIMS funnel by movable gas flow. The localization ions within the TIMS funnel are passed through the electric field gradient (EFG) that acts as an opposing force. Later, analytes are eluted and separated in accordance with their characteristic mobility through the successive decline of EFG. The trapped ion mobility spectrometry (TIMS) techniques are being used considerably in lipidomics as they enable the lipid identification. The integration of SFC and TIMS enables high-speed and high-resolution lipidomic analysis. The selectivity, resolution, and quantification of the separation have increased with the coupling of TIMS as the post-ionization technique in SFC (Scholz et al., 2024).

Fourier Transform Ion Cyclotron Resonance Mass Spectrometry (FT-ICR-MS) is a sophisticated technique for ultra high-resolution MS. The process in FT-ICR-MS starts with the ionization of analytes that are produced from an ionization source, which are accumulated and transferred into ion cyclotron resonance (ICR) cell in packets. The ICR

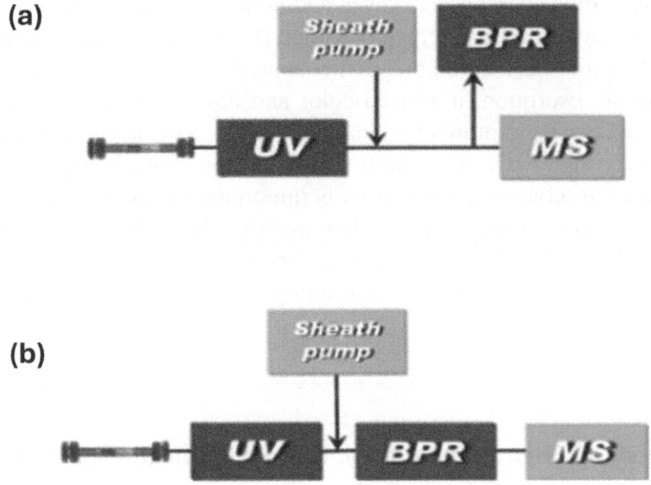

FIGURE 12.5 Schematic representation of commonly used SFC-MS interfaces: (a) pre-BPR splitter with sheath pump and (b) BPR and sheath pump with no splitter.

Source: Toribio et al. (2021)

cell consists of two trapping plates, two excitation plates, and two detection plates. The ions inside ICR cell move in small orbits in a noncoherent manner and excited by a radiofrequency (RF) pulse perpendicular to magnetic field. This cyclotron motion is considered as the basis of mass-to-charge measurement in FT-ICR-MS. It is widely used for the analysis of intact glucosinolates in Brassicaceae extracts, flavan-3-ols in extracts of seeds of grape pomace, and the profiling of glycoalkaloids (Gosset-Erard et al., 2023).

12.5 CHALLENGES AND LIMITATION OF SURFACE-BASED ANALYTICAL TECHNIQUES IN FOOD ANALYSIS

Although the analytical development in food analysis has progressed so far, there are still lots of issues to be addressed. Analysis of molecular-level functioning of the food compounds is the major challenge associated with food analysis. The information regarding the interaction of food constituents with genes and their subsequent influence on protein production and metabolites are inevitable for modern food science and nutrition. In this context, advanced food omics approach serves as a new discipline that addresses food and nutrition (Herrero et al., 2012). Currently, the detection-specific microbial pathogens have been made more simple and achievable by the use of genetic-based methods such as microarray based technologies (Lancova et al., 2011). Besides the detection methods, pre-analytical processing preceding detection is also important. In this context, fast release of results in compliance with official criteria and without disrupting the selectivity, accuracy, sensitivity, specificity, and recovery of methods is one of the analytical challenges existing in food science (LeDoux, 2011).

Food allergy is an important issue in food analysis, as it causes severe problems in sensitive individuals. LC and MS have been frequently used for the identification and quantification of food allergens. However, additional problems come across while food processing may interfere the allergen detection (Faeste et al., 2011). Incorporation of nanomaterials in food is a common practice nowadays, as they contribute to delivery system, nutrient absorption, improved color and flavor, contaminant detection, and food packaging. So, we require efficient technologies for characterizing the nanomaterials and study their toxic effect on food (Magnuson et al., 2011). Additionally, the identification of food-associated viruses is important for food safety. The currently used methods need to be improved further, as they rely on the type of virus and food matrix (Schrader et al., 2011). Above all, dedicated workers, laboratories, and equipment are necessary for the successful running of the multidimensional techniques.

TABLE 12.1
List of Frequently Used Techniques and Their Applications in Food Analysis

Sl No	Techniques	Principle	Application in Food Analysis	References
1	NGS and PCR	Amplification and sequencing of individual DNA or RNA	Identify small amount of biological adulterants and fungal or algal contents in in food products, accurate identification of unculturable microbial taxa, and the detection of GMOs.	Hill (1996), Ripp et al. (2014), Catalano et al. (2016), Haynes et al. (2019)
2	MS	Separation of ionized molecules based on charge/mass ratio	Protein identification and quantification, lipid analysis in food-processing research	Careri et al. (2002)
3	ICP-MS	Argon plasma is used to ionize elements and for subsequent detection by mass spectrometer	Inorganic specifications and element speciation in food analysis	Careri et al. (1998)
4	HPLC-ICP-MS, LC–MS, GC–MS, and HPLC-MS	An interface connected between HPLC and MS transfers the separated compounds from liquid chromatograph to the MS ion source	Used to quantify arsenic compounds in various baby foods and freeze-dried apple samples, detection of heat-labile compounds like proteins and vitamins	Caruso et al. (2001), Cramer and Humpf (2012)
5.	MALDI-TOF-MS	Sample entrapped within a matrix is ionized by laser beam, and the charged analytes are then measured using time of flight (TOF).	Identification of polyphenolic compounds and antioxidants in complex mixtures, monitoring the xenobiotic toxic compounds in foods, thereby ensuring safety in food consumption, evaluate the authenticity of milk, and identification of vitamins in both food and pharmaceutical areas	Cozzolino et al. (2001), Careri et al. (2002), Singhal et al. (2015)

TABLE 12.1 *(Continued)*
List of Frequently Used Techniques and Their Applications in Food Analysis

Sl No	Techniques	Principle	Application in Food Analysis	References
6.	Py-MS	Involve thermal degradation of the sample in an inert atmosphere followed by the chemical profiling of the sample by MS	Analyzes the food authenticity and determines the food and beverage adulteration, used to characterizemilk samples, vinegar, and wines, the detection of caffeine content in various beverages, and the classification of cocoa butter	Guillou et al. (1999)
7	GC	The gas stream carries the sample into a column where the components are separated on the basis of adsorption or partition.	Analysis of food additives, packaging materials, natural products, flavor and aroma components.	Lehotay and Hajšlová (2002)
8	HPLC	HPLC uses polar or nonpolar column as a stationary phase, where samples get separated based on polarity.	Separating all types of chemicals uniquely for drugs, amino acids/proteins, carbohydrates, and lipids	Lehotay and Hajšlová (2002)
9	Enzymatic protein hydrolysis	Proteins are cleaved using specific proteases and form amino nitrogen, which is measured through titration.	Characterization of foods and to produce edibles with extended functional properties	Kristinsson and Rasco (2000), Ghosh et al. (2017)
10	NIR spectroscopy	Uses electromagnetic radiation to analyze the molecular vibrational excitation of the sample.	Tests the quality of the cross breed, assesses the protein and moisture content of several raw ingredients such as flour, sucrose, and chocolate.	Norris (1989), McGrath et al. (1997)
11	NMR	NMR is a quantitative method that utilizes magnetic property of certain nuclei with an odd atomic number and a nuclear spin (S).	Quality control, molecular mechanisms as well as interactions of food components, compositional analysis, food authentication, structural characterization, sensory evaluation, etc.	Williamson and Hatzakis (2017), Hatzakis (2019)
12	X-ray imaging	While in the penetration of X-ray into the food product, some energy is lost, and the energy comes out detected by the sensor and converts into an image.	Predicts the internal defects and quality of the agricultural products and animal products, microstructure characterization of frozen products.	Mousavi et al. (2005), Bischof et al. (2007)

(Continued)

TABLE 12.1 *(Continued)*
List of Frequently Used Techniques and Their Applications in Food Analysis

Sl No	Techniques	Principle	Application in Food Analysis	References
13	UI	Uses soundwaves that interact with the host tissues to produce the structural and functional information.	Detects foreign matters and infirmity in cheese and poultry materials	Cho and Irudayaraj (2003)
15	XRD	High-energy electrons are bombarded with X-rays generated by the metals such as tungsten, and then directed toward the sample.	Assessment of quality checks for foreign particles and analyzing the density of food products	Purohit et al. (2019)
16	DPV	DPV produces a voltammogram, in which the height of the peak is directly proportional to the concentration of the analyte measured.	Identification of different classes of flavor enhancers, synthetic dyes, preservatives, and antioxidants in foodstuffs	Goncalves-Filho and De Souza (2024)
17	CE and CEC	This method uses electrophoretic and electroosmotic mobilities generated by applying high electric field.	Small organic ions to large macro molecules	Della Posta et al. (2022)
18	μCT	μCT generally follows the basic principles of absorption physics and reconstruction mathematics.	Various commodities including fruits and vegetables, confectionary and baked products, fish dairy, and cereals	Schoeman et al. (2016)
19	SFC-MS	SFC-MS is performed in packed columns with ESI and APCI as the ionization sources	Identification of food contaminants (pesticide and compounds), transformation product compounds, and nutrients	Toribio et al. (2021)

Moreover, a good knowledge about the principles and benefits of available methods is important for the selection of suitable ones (Table 12.1).

12.6 FUTURE SCOPES OF FOOD ANALYSIS

Food analysis will create more possibilities in the food technology field in several ways. As globalization has created complexity in the purity of the raw materials used in food products, the need for food analysis has heightened (García-Cañas et al., 2012).

Advancement in the hyphenated techniques has resulted in relevant development in the food analysis. Techniques like MS made possible to determine or confirm the presence of several compounds such as xenobiotic substances in the food products (Careri et al., 1996). Also, combining techniques helps to analyze samples more efficiently and in short time run. The selective and more precise food analysis will help to create global food security. As the quality of food products and the practices in food processing will be improved with the accurate food analysis. The determination of adulteration and other food additives will become more sophisticated and precise due to technical developments in the field of food analysis. The nutritive value, chemical composition, shelf-life, etc., of food products can be easily detected through food analysis (Trugo et al., 2003). Assessing the interaction of electromagnetic radiation with food can help with the detection and content of several dietary constituents in addition to the research of food qualities. It has been possible to create prediction models because of recent advancements in spectroscopic techniques in the food analysis (García-Cañas et al., 2012). Also, the advancements in spectroscopic techniques in food analysis can help to understand the complex structural peculiarities and chemical makeup of food matrix. Moreover, they facilitate classification and identification of different food types, contamination detection, and monitoring of quality changes (Kharbach et al., 2023). The consumers' concern about the safety of the food led to an increase in the development of advanced analytical methods and techniques in food science. Analytical chemistry is involved in large number of food properties like processing and preparation foods, detection of adulteration, inactivation of toxins, and characterization of chemical composition of foods (Senorans et al., 2003).

12.7 CONCLUSION

There is a saying that 'we are, what we eat'. Hence, we still need sophisticated analytical methods for more adequate quality analysis of food. A passable supply of food and nutrients is essential for human survival. The increasing consumer concerns on food safety have led to a strong drive for the better quality of products both in the fresh market and food industry. The quality measures include nutritive values, functional properties, sensory properties, chemical constituents, and so on. So the measurement of these food properties necessitates advanced analytical techniques. There is no absolute method for every purpose. Combinations of all parameters are applied for an accurate assessment of food products. The reliability of the quality analysis methods can be refined by coupling efficient technologies with advanced statistical methods. Development of such urbane methods can indeed improve the understanding of the determinants of food quality and safety analysis.

REFERENCES

ALAMPRESE, C., CASALE, M., SINELLI, N., LANTERI, S. & CASIRAGHI, E. 2013. Detection of minced beef adulteration with turkey meat by UV–vis, NIR and MIR spectroscopy. *Food Science and Technology*, 53, 225–232.

BARTLETT, S. E. & DAVIDSON, W. S. 1992. FINS (forensically informative nucleotide sequencing): A procedure for identifying the animal origin of biological specimens. *Biotechniques*, 12, 408–411.

BERGHOLZ, T. M., MORENO SWITT, A. I. & WIEDMANN, M. 2014. Omics approaches in food safety: Fulfilling the promise? *Trends Microbiol*, 22, 275–281.

BISCHOF, J. C., MAHR, B., CHOI, J. H., BEHLING, M. & MEWES, D. 2007. Use of X-ray tomography to map crystalline and amorphous phases in frozen biomaterials. *Annals of Biomedical Engineering*, 35, 292–304.

BRUL, S., SCHUREN, F., MONTIJN, R., KEIJSER, B. J., VAN DER SPEK, H. & OOMES, S. J. 2006. The impact of functional genomics on microbiological food quality and safety. *International Journal of Food Microbiology*, 112, 195–199.

CARERI, M., BIANCHI, F. & CORRADINI, C. 2002. Recent advances in the application of mass spectrometry in food-related analysis. *Journal of Chromatography A*, 970, 3–64.

CARERI, M., MANGIA, A. & MUSCI, M. 1998. Overview of the applications of liquid chromatography–mass spectrometry interfacing systems in food analysis: Naturally occurring substances in food. *Journal of Chromatography A*, 794, 263–297.

CARERI, M., MANGIA, A. & MUSCI, M. 1996. Applications of liquid chromatography-mass spectrometry interfacing systems in food analysis: Pesticide, drug and toxic substance residues. *Journal of Chromatography A*, 727, 153–184.

CARRERA, M. 2021. Proteomics and food analysis: Principles, techniques, and applications. *Foods*, 10.

CARUSO, J. A., HEITKEMPER, D. T. & B'HYMER, C. 2001. An evaluation of extraction techniques for arsenic species from freeze-dried apple samples. *Analyst*, 126, 136–140.

CATALANO, V., MORENO-SANZ, P., LORENZI, S. & GRANDO, M. S. 2016. Experimental review of DNA-based methods for wine traceability and development of a single-nucleotide polymorphism (SNP) genotyping assay for quantitative varietal authentication. *Journal of Agricultural and Food Chemistry*, 64, 6969–6984.

CHEN, Q., ZHANG, C., ZHAO, J. & OUYANG, Q. 2013a. Recent advances in emerging imaging techniques for non-destructive detection of food quality and safety. *TrAC Trends in Analytical Chemistry*, 52, 261–274.

COSKUN, O. 2016. Separation techniques: Chromatography. *Northern Clinics of Istanbul*, 3, 156.

COZZOLINO, R., PASSALACQUA, S., SALEMI, S., MALVAGNA, P., SPINA, E. & GAROZZO, D. 2001. Identification of adulteration in milk by matrix-assisted laser desorption/ionization time-of-flight mass spectrometry. *Journal of Mass Spectrometry*, 36, 1031–1037.

CRAMER, B. & HUMPF, H. U. 2012. Applications of HPLC-MS techniques for the analysis of chemical contaminants and residues in food. In *Chemical Contaminants and Residues in Food*. Elsevier.

DAHER, D., DERACINOIS, B., BANIEL, A., WATTEZ, E., DANTIN, J., FROIDEVAUX, R., CHOLLET, S. & FLAHAUT, C. 2020. Principal component analysis from mass spectrometry data combined to a sensory evaluation as a suitable method for assessing bitterness of enzymatic hydrolysates produced from micellar casein proteins. *Foods*, 9.

DAHER, D., DERACINOIS, B., COURCOUX, P., BANIEL, A., CHOLLET, S., FROIDEVAUX, R. & FLAHAUT, C. 2021. Sensopeptidomic kinetic approach combined with decision trees and random forests to study the bitterness during enzymatic hydrolysis kinetics of micellar caseins. *Foods*, 10.

DELLA POSTA, S., FANALI, C., GALLO, V. & FANALI, S. 2022. Recent advances in the hyphenation of electromigration techniques with mass spectrometry. *TrAC Trends in Analytical Chemistry*, 116800.

FAESTE, C. K., RONNING, H. T., CHRISTIANS, U. & GRANUM, P. E. 2011. Liquid chromatography and mass spectrometry in food allergen detection. *Journal of Food Protection*, 74, 316–345.

GAN, T. H., PALLAV, P. & HUTCHINS, D. A. 2006. Non-contact ultrasonic quality measurements of food products. *Journal of Food Engineering*, 77, 239–247.

GARCÍA-CAÑAS, V., SIMÓ, C., HERRERO, M., IBÁÑEZ, E. & CIFUENTES, A. 2012. *Present and Future Challenges in Food Analysis: Foodomics*. ACS Publications.

GHOSH, B. C., PRASAD, L. N. & SAHA, N. P. 2017. Enzymatic hydrolysis of whey and its analysis. *Journal of Food Science and Technology*, 54, 1476–1483.

GONCALVES-FILHO, D. & DE SOUZA, D. 2024. Trends in pulse voltammetric techniques applied to foodstuffs analysis: The food additives detection. *Food Chemistry*, 454, 139710.

GOSSET-ERARD, C., AUBRIET, F., LEIZE-WAGNER, E., FRANCOIS, Y. N. & CHAIMBAULT, P. 2023. Hyphenation of Fourier transform ion cyclotron resonance mass spectrometry (FT-ICR MS) with separation methods: The art of compromises and the possible—A review. *Talanta*, 257, 124324.

GRUNERT, K. G. 2005. Food quality and safety: Consumer perception and demand. *European Review of Agricultural Economics*, 32, 369–391.

GUILLOU, C., LIPP, M., RADOVIC, B., RENIERO, F., SCHMIDT, M. & ANKLAM, E. 1999. Use of pyrolysis–mass spectrometry in food analysis: Applications in the food analysis laboratory of the European Commissions' Joint Research Centre. *Journal of Analytical and Applied Pyrolysis*, 49, 329–335.

HATZAKIS, E. 2019. Nuclear magnetic resonance (NMR) spectroscopy in food science: A comprehensive review. *Comprehensive Reviews in Food Science and Food Safety*, 18, 189–220.

HAYNES, E., JIMENEZ, E., PARDO, M. A. & HELYAR, S. J. 2019. The future of NGS (Next Generation Sequencing) analysis in testing food authenticity. *Food Control*, 101, 134–143.

HEBERT, P. D., CYWINSKA, A., BALL, S. L. & DEWAARD, J. R. 2003. Biological identifications through DNA barcodes. *Proceedings of the Royal Society B: Biological Sciences*, 270, 313–321.

HERRERO, M., SIMO, C., GARCIA-CANAS, V., IBANEZ, E. & CIFUENTES, A. 2012. Foodomics: MS-based strategies in modern food science and nutrition. *Mass Spectrometry Reviews*, 31, 49–69.

HILL, W. E. 1996. The polymerase chain reaction: Applications for the detection of foodborne pathogens. *Critical Reviews in Food Science and Nutrition*, 36, 123–173.

HOYOS-ARBELAEZ, J., VAZQUEZ, M. & CONTRERAS-CALDERON, J. 2017. Electrochemical methods as a tool for determining the antioxidant capacity of food and beverages: A review. *Food Chemistry*, 221, 1371–1381.

HU, L., YE, M., JIANG, X., FENG, S. & ZOU, H. 2007. Advances in hyphenated analytical techniques for shotgun proteome and peptidome analysis—a review. *Analytica Chimica Acta*, 598, 193–204.

KAHL, J., BAARS, T., BÜGEL, S., BUSSCHER, N., HUBER, M., KUSCHE, D., REMBIAŁKOWSKA, E., SCHMID, O., SEIDEL, K. & TAUPIER-LETAGE, B. 2012. Organic food quality: A framework for concept, definition and evaluation from the European perspective. *Journal of the Science of Food and Agriculture*, 92, 2760–2765.

KHARBACH, M., ALAOUI MANSOURI, M., TAABOUZ, M. & YU, H. 2023. Current application of advancing spectroscopy techniques in food analysis: Data handling with chemometric approaches. *Foods*, 12.

KRISTINSSON, H. G. & RASCO, B. A. 2000. Fish protein hydrolysates: Production, biochemical, and functional properties. *Critical Reviews in Food Science and Nutrition*, 40, 43–81.

LANCOVA, K., DIP, R., ANTIGNAC, J.-P., LE BIZEC, B., ELLIOTT, C. T. & NAEGELI, H. 2011. Detection of hazardous food contaminants by transcriptomics fingerprinting. *TrAC Trends in Analytical Chemistry*, 30, 181–191.

LEDOUX, M. 2011. Analytical methods applied to the determination of pesticide residues in foods of animal origin. A review of the past two decades. *Journal of Chromatography A*, 1218, 1021–1036.

LEHOTAY, S. J. & HAJŠLOVÁ, J. 2002. Application of gas chromatography in food analysis. *TrAC Trends in Analytical Chemistry*, 21, 686–697.

LINO, F. M. A., DE SÁ, L. Z., TORRES, I. M. S., ROCHA, M. L., DINIS, T. C. P., GHEDINI, P. C., SOMERSET, V. S. & GIL, E. S. 2014. Voltammetric and spectrometric determination of antioxidant capacity of selected wines. *Electrochimica Acta*, 128, 25–31.

MAGNUSON, B. A., JONAITIS, T. S. & CARD, J. W. 2011. A brief review of the occurrence, use, and safety of food-related nanomaterials. *Journal of Food Science*, 76, R126–R133.

MAYO, B., RACHID, C. T., ALEGRIA, A., LEITE, A. M., PEIXOTO, R. S. & DELGADO, S. 2014. Impact of next generation sequencing techniques in food microbiology. *Current Genomics*, 15, 293–309.

MCGRATH, V. B., BLAKENEY, A. B. & BATTEN, G. D. 1997. Fructan to nitrogen ratio as an indicator of nutrient stress in wheat crops. *The New Phytologist*, 136, 145–152.

MCSHEEHY, S., YANG, W., PANNIER, F., SZPUNAR, J., ŁOBIŃSKI, R., AUGER, J. & POTIN-GAUTIER, M. 2000. Speciation analysis of selenium in garlic by two-dimensional high-performance liquid chromatography with parallel inductively coupled plasma mass spectrometric and electrospray tandem mass spectrometric detection. *Analytica Chimica Acta*, 421, 147–153.

MOLNÁR, P., VAJDA, Ö. & DZUR, K. 1979. Comparative study of food evaluation methods. In *Proceedings of the 23rd Conference of the European Organization for Quality Control* (pp. 17–30). Food Seminar.

MOUSAVI, R., MIRI, T., COX, P. W. & FRYER, P. J. 2005. A novel technique for ice crystal visualization in frozen solids using X-ray micro-computed tomography. *Journal of Food Science*, 70, e437–e442.

NELSON, R. W., KRONE, J. R. & JANSSON, O. 1997. Surface plasmon resonance biomolecular interaction analysis mass spectrometry. 2. Fiber optic-based analysis. *Analytical Chemistry*, 69, 4369–4374.

NIELSEN, E. E., CARIANI, A., MAC AOIDH, E., MAES, G. E., MILANO, I., OGDEN, R., TAYLOR, M., HEMMER-HANSEN, J., BABBUCCI, M., BARGELLONI, L., BEKKEVOLD, D., DIOPERE, E., GRENFELL, L., HELYAR, S., LIMBORG, M. T., MARTINSOHN, J. T., MCEWING, R., PANITZ, F., PATARNELLO, T., TINTI, F., VAN HOUDT, J. K., VOLCKAERT, F. A., WAPLES, R. S., FISHPOPTRACE, C., ALBIN, J. E., VIEITES BAPTISTA, J. M., BARMINTSEV, V., BAUTISTA, J. M., BENDIXEN, C., BERGE, J. P., BLOHM, D., CARDAZZO, B., DIEZ, A., ESPINEIRA, M., GEFFEN, A. J., GONZALEZ, E., GONZALEZ-LAVIN, N., GUARNIERO, I., JERAME, M., KOCHZIUS, M., KREY, G., MOUCHEL, O., NEGRISOLO, E., PICCINETTI, C., PUYET, A., RASTORGUEV, S., SMITH, J. P., TRENTINI, M., VERREZ-BAGNIS, V., VOLKOV, A., ZANZI, A. & CARVALHO, G. R. 2012. Gene-associated markers provide tools for tackling illegal fishing and false eco-certification. *Nature Communications*, 3, 851.

NORRIS, H. K. 1989. A definition of wheat hardness using near infrared reflection spectroscopy. *Cereal Foods World*, 34, 696–705.

OFLU, S., ERARPAT, S., ZAMAN, B. T., GUNKARA, O. T., BAKIRDERE, S. & TURAK, F. 2023. Combination of quadrupole isotope dilution mass spectrometry with simultaneous derivatization and spray assisted droplet formation-liquid phase microextraction for the determination of methamphetamine in human urine and serum samples by gas chromatography mass spectrometry. *Journal of Pharmacological and Toxicological Methods*, 119, 107207.

OSBORNE, B. G. & FEARN, T. 1986. *Near Infrared Spectroscopy in Food Analysis*, Longman Scientific & Technical New York.

O'MEARA, G. M. & MUNRO, P. A. 1985. Kinetics of the hydrolysis of lean meat protein by alcalase: Derivation of two alternative rate equations and their fit to experimental data. *Biotechnol Bioeng*, 27, 861–869.

PALARIC, C., PILARD, S., FONTAINE, J.-X., CAILLEU, D., MESNARD, F., RENAUD, T., PETIT, A. & MOLINIE, R. 2018. Combined MS-NMR approach for the quality control of food supplements containing amino acids. *Food Control*, 88, 217–228.

POVEY, M. J. W. & MCCLEMENTS, D. J. 1988. Ultrasonics in food engineering. Part I: Introduction and experimental methods. *Journal of Food Engineering*, 8, 217–245.

PUROHIT, S. R., JAYACHANDRAN, L. E., RAJ, A. S., NAYAK, D. & RAO, P. S. 2019. X-ray diffraction for food quality evaluation. In *Evaluation Technologies for Food Quality*. Elsevier.

RIPP, F., KROMBHOLZ, C. F., LIU, Y., WEBER, M., SCHÄFER, A., SCHMIDT, B., KÖPPEL, R. & HANKELN, T. 2014. All-Food-Seq (AFS): A quantifiable screen for species in biological samples by deep DNA sequencing. *BMC Genomics*, 15, 1–11.

SCHOEMAN, L., WILLIAMS, P., DU PLESSIS, A. & MANLEY, M. 2016. X-ray micro-computed tomography (μCT) for non-destructive characterisation of food microstructure. *Trends in Food Science & Technology*, 47, 10–24.

SCHOLZ, J., RUDT, E., GREMME, A., GASSMOLLER NEE WIENKEN, C. M., BORNHORST, J. & HAYEN, H. 2024. Hyphenation of supercritical fluid chromatography and trapped ion mobility-mass spectrometry for quantitative lipidomics. *Analytica Chimica Acta*, 1317, 342913.

SCHRADER, C., JOHNE, R., SCHIELKE, A. & ELLERBROEK, L. 2011. Food associated viruses and their detection-a review. *Cabidigitallibrary.*

SENORANS, J., IBANEZ, E. & CIFUENTES, A. 2003. New trends in food processing. *Critical Reviews in Food Science and Nutrition*, 43, 507–526.

SINGHAL, N., KUMAR, M., KANAUJIA, P. K. & VIRDI, J. S. 2015. MALDI-TOF mass spectrometry: An emerging technology for microbial identification and diagnosis. *Frontiers in Microbiology*, 6, 791.

SOURI, A., MEHRABADI, H. S. & HOSSEIN-BABAEI, F. (2019). *Classification of Dairy Products Using Chronoamperometry Performed in a Microfluidic Channel.* IEEE, 380–383.

SOUSA, W. R., DA ROCHA, C., CARDOSO, C. L., SILVA, D. H. S. & ZANONI, M. V. B. 2004. Determination of the relative contribution of phenolic antioxidants in orange juice by voltammetric methods. *Journal of Food Composition and Analysis,* 17, 619–633.

TORIBIO, L., BERNAL, J., MARTÍN, M. T. & ARES, A. M. 2021. Supercritical fluid chromatography coupled to mass spectrometry: A valuable tool in food analysis. *TrAC Trends in Analytical Chemistry*, 143, 116350.

TRANCHIDA, P. Q., DUGO, P., DUGO, G. & MONDELLO, L. 2004. Comprehensive two-dimensional chromatography in food analysis. *Journal of Chromatography A*, 1054, 3–16.

TRIPATHI, S., MEHROTRA, G. K. & DUTTA, P. K. 2009. Physicochemical and bioactivity of cross-linked chitosan-PVA film for food packaging applications. *International Journal of Biological Macromolecules*, 45, 372–376.

TRUGO, L. C., FINGLAS, P. M. J. L. C. T. & PM FINGLAS, E. 2003. Encyclopedia of food sciences and nutrition 1498–1506. *Cabidigitallibrary*, 1–10.

VELA, N. P., OLSON, L. K. & CARUSO, J. A. 1993. Elemental speciation with plasma mass spectrometry. *Analytical Chemistry,* 65, 585A–597A.

WU, D. & SUN, D.-W. 2013. Advanced applications of hyperspectral imaging technology for food quality and safety analysis and assessment: A review—Part II: Applications. *Innovative Food Science & Emerging Technologies*, 19, 15–28.

ZHENG, J., OHATA, M., FURUTA, N. & KOSMUS, W. 2000. Speciation of selenium compounds with ion-pair reversed-phase liquid chromatography using inductively coupled plasma mass spectrometry as element-specific detection. *Journal of Chromatography A*, 874, 55–64.

13 Assessing the Feasibility of Emerging Techniques in Food Industries

Priyanka Singh

13.1 INTRODUCTION

The global food industry is experiencing a paradigm shift driven by heightened concerns over food safety, authenticity, and quality assurance. Novel technologies used for food analysis include C-dots, smartphone-based analytical techniques, biosensors, laser and laser optics, aerogel, rapid microbial techniques (RMTs), spectral techniques and surface-based analysis. *C-dots* are fluorescent nanoparticles used for detecting contaminants like heavy metals and pesticides due to their strong luminescence and low toxicity. *Smartphone-based analytical techniques* use mobile cameras and apps to perform rapid, on-site tests for food adulterants and spoilage indicators, making diagnostics more accessible. *Biosensors*, which combine biological elements (like enzymes or antibodies) with detectors, offer highly specific, quick detection of pathogens, toxins, and allergens in food samples.

Laser and laser optics, such as laser-induced breakdown spectroscopy (LIBS), enable nondestructive elemental analysis of food products in real time. Aerogels, ultralight and porous materials, are used to extract and concentrate trace contaminants for better detection sensitivity. RMTs, including PCR and ATP bioluminescence, drastically cut down the time needed to detect microbial contamination compared to traditional culturing methods. Spectral techniques—like infrared (IR), Raman, and nuclear magnetic resonance (NMR) spectroscopy—offer detailed molecular analysis for identifying food composition and adulteration. Lastly, surface-based analysis methods such as SEM and XPS allow high-resolution imaging and chemical analysis of food surfaces and packaging materials.

This chapter aims to assess the commercial viability and feasibility of these emerging analytical techniques within the food sector. By examining their technological foundations, practical applications, and inherent limitations, chapter provide insights into how these tools can be integrated into the existing food quality assurance framework. The discussion also considers the regular economic and operational barriers that must be addressed for widespread industry adoption. Comparative study of all techniques is shown in Table 13.1.

DOI: 10.1201/9781003482611-13

TABLE 13.1
Comparative Study of Novel Analytical Techniques

Technology	Sensitivity and Specificity	Authenticity Applications	Reproducibility	Cost/Economic Viability	Limitations
Carbon Dots (C-dots)	Detect heavy metals, pesticides, dyes at nM–pM levels; fluorescence influenced by pH/ionic strength	Fluorescence emission used to identify pathogens and contaminants	Consistent under controlled conditions; synthesis variations affect performance; intra-assay CV <10% reported.	Cost-effective, sustainable, easy synthesis; portable devices for on-site testing.	Interference from proteins/fats in food matrices; lack of synthesis standardization; regulatory approval pending
Smartphone-Based Techniques	Detect pathogens, allergens, residues; e.g., *E. coli*, peanut allergens within minutes	Detect fraud/adulteration (olive oil, honey, wine)	High repeatability; immunosensor results comparable to PCR	Low-cost (~$140–250 setups), portable, useful for farmers/consumers	Require accessories (kits/sensors), accuracy depends on phone hardware/software, standardization and regulation needed
Biosensors	Highly sensitive to pesticides, pathogens, allergens; aptamers and antibodies improve specificity.	Identify biochemical markers (e.g., aflatoxin, Salmonella, wine quality).	RSD ~5–6%; reproducibility improving with nano-integration	Low per-test cost (paper/electrochemical formats); high initial R&D costs.	Shelf-life issues, batch-to-batch reproducibility, require skilled personnel for complex types
Laser & Laser Optics (LIBS, Raman, IR)	Rapid, nondestructive, sensitive to light elements; real-time elemental/molecular profiling	Detect adulterants, classify oils, verify origin, isotopic/chemical profiling	Generally high reproducibility, variability due to sample positioning ad environment	Faster and cheaper than AAS/ICP; commercially viable with ML integration	Sensitive to surface texture/environmental factors, requires calibration, scalability for industrial use limited

13.2 COMMERCIAL VIABILITY OF EMERGING TECHNIQUES

13.2.1 C-Dots

Sensitivity and Specificity: C-dots effectively detect low concentrations of contaminants such as pesticides and heavy metals (Zhang et al., 2022 and Biranje et al., 2021). However, environmental factors like pH and ionic strength can influence their fluorescence response, necessitating careful control (Mao et al., 2010). Kaur et al. (2020) developed glutathione-capped C-dots for the selective detection of mercury, achieving a limit of detection of 0.54 nM. However, interference of copper and chromium ions limited the practical application of detection of mercury in the field.

Hu et al. (2021) developed a high-efficiency spectrophotometric method using nitrogen-doped fluorescent C-dots (N-FCDs) for the ultrasensitive determination of carmine in food products. Their method demonstrated ultra-high sensitivity, with a detection limit of 11.2 nM, which is significantly lower than previously reported systems (0.28 µM) (Su et al., 2018) and one order of magnitude lower than high-performance liquid chromatography (HPLC) analysis (0.72 µM) (Lim et al., 2014). Additionally, Xu et al. (2018) reported CDs with flavonoid moieties conjugated with agarose hydrogel for the ultrasensitive detection of lead ions, achieving a detection limit as low as 55 pM. Another system reported by Bhamore et al. (2020) utilized glutathione-capped C-dots conjugated with agarose gel for visual detection by the naked eye. This system showed a good linear correlation with lead concentration and does not require heavy instrumentation, making it suitable for field applications.

Authenticity: The ability of C-dots to emit fluorescence at specific wavelengths allows for the differentiation of various foodborne pathogens and contaminants. This can aid in authenticating food products by detecting specific markers associated with food (Manzoor et al., 2023). However, complex food matrices can present challenges due to interference from other compounds, highlighting the need for more robust detection methods.

Cayuela et al. (2015) investigated the synthesis of highly photoluminescent CDs from different organic precursors using a bottom-up method. CDs derived from microcrystalline cellulose showed superior fluorescence intensity and stability against pH variations. Nanohybrids formed between c-MWCNTs and CDs enabled the detection of water-soluble c-MWCNTs in aqueous media. Thus, these CDs show potential as photoluminescent sensors for c-MWCNTs in environmental water samples impacted by industrial waste and diesel emissions.

13.2.1.1 Reproducibility of Results

Achieving reproducibility is critical for any analytical method. While C dots can yield consistent results under controlled conditions, variations in synthesis methods can lead to differences in properties. Research emphasizes the importance of standardized synthesis protocols to ensure consistent performance. Variations in precursor materials or reaction conditions can affect size, surface charge, and fluorescence characteristics. Yao et al. (2020) developed a method using magnetic nanoparticles decorated with CDs, coupled with silver nanoclusters as artificial enzymes for visual detection of *S. aureus*. This method could visually detect *S. aureus* ranging from 10 to 10^6 /CFU mL with a limit of detection (LOD) of 4.9 CFU/mL and 85.6% to 103.7% of recovery in food samples. Zhao et al. (2021) developed a highly sensitive fluorescent immunosensor to detect *Escherichia coli* O157:H7 in milk using microspheres labeled with CDs. The method showed an excellent repeatability with intra-assay variation (coefficient of variation) fewer than 10%. It was also sensitive enough to detect trace amounts of *E. coli* O157:H7 (fewer than 10 CFU) in real milk samples, achieving 100% accuracy after bacterial enrichment.

Cost of Analysis/Economic Viability: Despite the promising capabilities of C-dots, the complex nature of food matrices, such as proteins and fats, can interfere with fluorescence, reducing detection sensitivity. This challenge requires further research into more robust and specific methods. On the practical side, C-dots are being made cost-effective, sustainable, and nontoxic, making them suitable for large-scale production and use in food analysis (Huang and Ren, 2025). C-dots can be synthesized using relatively inexpensive materials and straightforward procedures, making them cost-effective compared to traditional detection methods like chromatography or mass spectrometry (MS) (Hu et al., 2021). The development of portable devices equipped with C dots facilitates on-site testing, significantly reducing the time and labor costs associated with laboratory analyses. For example, handheld fluorescence-based sensors can provide rapid results within minutes, enhancing efficiency in food safety monitoring (Shin et al., 2021). Hu et al. (2021) found that N-FCDs were quick, inexpensive, and easy to detect carmine in food. However, for widespread adoption, C-dots must meet regulatory standards for food safety and gain approval for commercial use. Overall, while challenges remain, C-dots offer significant promise in enhancing food safety and authenticity detection, with further development needed for broader implementation.

13.2.2 Smartphone-Based Rapid Analytical Techniques

The fusion of smartphone technology with advanced sensing systems has led to the development of compact, portable, and affordable analytical platforms—often referred to as lab-on-smartphone systems. These platforms utilize the built-in sensors, connectivity, and processing capabilities of smartphones, making them powerful tools for real-time monitoring and diagnostics in various sectors, particularly in resource-limited environments.

13.2.2.1 Sensitivity and Specificity

Smartphone-based systems must accurately detect minute quantities of harmful substances such as pathogens or chemical residues in food samples. Zhu et al. (2012)

introduced a smartphone-based platform capable of detecting *Escherichia coli* (*E. coli*) in liquid samples. Du Vall et al. (2015) highlighted how mobile phones integrated with custom apps can serve as point-of-care diagnostic tools. Coskun et al. (2013) developed an iTube platform that successfully detected peanut allergens in commercial cookies within approximately 20 minutes. This system also enabled data uploading for public health monitoring. Zhu et al. (2012) validated their *E. coli* detection system by ensuring minimal cross-reactivity with other bacterial species like *Salmonella*.

Authenticity Verification: Smartphone diagnostics are also applied in food fraud detection such as verifying whether products are truly organic or identifying ingredient adulteration. Casadei et al. (2021) demonstrated their utility in fraud detection scenarios like olive oil adulteration. Giordano et al. (2016) developed a portable electrochemical system integrated with a smartphone to analyze honey quality using voltammetric fingerprints. Park et al. (2014) developed a paper-based microfluidic chip and utilized to assess the flavor profile of ten red wines through a panel of chemical dyes. Images were captured using a smartphone camera, and the red–green–blue (RGB) pixel intensities were processed via principal component analysis (PCA) with pronounced accuracy and specificity. By employing eight dyes and analyzing two principal components, the wines could be differentiated according to grape variety and oxidation state.

Cost of analysis/economic viability: Smartphone-based methods significantly reduce analytical costs by eliminating the need for expensive lab infrastructure and skilled technicians. Ludwig et al. (2014) developed a smartphone-based fluorescence microscope for approximately $140, compared to conventional microscopes costing thousands. Das et al. (2016) built a smartphone spectrophotometer setup for under $250, offering a much cheaper alternative to standard lab devices. Although generally affordable, these systems may require purchasable accessories (e.g., sensors, test kits, software) that add to the overall cost—though still lower than traditional lab methods. Farmers, vendors, and small-scale producers can use these tools for rapid, low-cost food safety tests that align with regulations before distribution. Consumers can test food quality directly before purchase.

For widespread adoption, these systems must be affordable, user-friendly, and regulated to ensure accuracy. Industry acceptance and consumer trust also play a crucial role in market success.

13.2.3 Biosensors for Food Analysis

Sensitivity and Specificity: Modern biosensors offer high sensitivity, making them suitable for detecting low concentrations of food contaminants such as pesticides, pathogens, and allergens. Their performance is comparable to that of conventional methods like HPLC and ELISA, especially in time-sensitive applications like spoilage detection (Das and Mishra, 2022; Curulli, 2021).

The specificity of biosensors has been significantly enhanced through innovations in molecular recognition elements such as aptamers, antibodies, and molecularly imprinted polymers. These components allow for precise binding to target analytes, reducing false positives and supporting regulatory compliance—an important factor for commercial deployment. A representative application is a DNA-based piezoelectric biosensor targeting the *aflD* gene, which accurately detects aflatoxin-producing *Aspergillus* species in contaminated crops. The sensor exhibited performance in agreement with traditional gel electrophoresis (Tombelli et al., 2009).

Authenticity: Biosensors can be engineered to identify specific biochemical markers that indicate a product's origin, composition, or potential adulteration. For example, antibody-functionalized probes have been used to detect *Salmonella* in milk with a sensitivity of 2.3×10^3 CFU/mL (Jin et al., 2020). Similarly, Lanzellotto et al. (2014) utilized gold nanoparticles on carbon electrodes to quantify gallic acid in wine samples, achieving detection limits ranging from 0.03 to 0.30 mmol/L. Bhardwaj et al. (2020) developed a label-free, SPR-based immunosensor that sensitively detected aflatoxin B1 in wheat, demonstrating the commercial potential of biosensors in authenticity.

Commercial relevance: Real-time authentication of food products using biosensors can mitigate losses due to fraud and enhance consumer confidence, making them a compelling tool for producers and retailers (Bahadır and Sezgintürk, 2015).

Reproducibility of results: Shavronskaya et al. (2023) reported a TiO_2-based optical biosensor capable of accurately quantifying lactose in milk without any complex sample preparation. The system demonstrated consistent performance, with relative standard deviations (RSDs) of 5–6% across 20 trials, aligning with reproducibility metrics seen in photoelectron chemical and nanoparticle-based biosensors (Çakıroğlu et al., 2019; Nasiri et al., 2022).

Economic Viability: Biosensors represent a cost-effective alternative to conventional laboratory assays, especially when scaled for commercial production. Paper-based and electrochemical formats are particularly advantageous, offering low per-test costs with minimal training requirements (Mustafa et al., 2020). However, the initial expenses associated with R&D and integration into market-ready devices may be high (Sadana, 2005). Embedding biosensors into broader systems, such as smart packaging or portable readers, enhances utility and long-term cost-effectiveness.

13.2.4 Laser and Laser Optics in Food Analysis

Sensitivity and specificity: Laser-Induced Breakdown Spectroscopy (LIBS) offers several advantages over conventional techniques such as AAS, ICP-AES, and XRF, as it provides analysis free from extensive sample preparation. LIBS measurements can be performed without any pretreatment, utilizing only a small amount of material, which makes the method minimally destructive. It is an extremely fast analytical technique, capable of delivering results within a fraction of a second. Moreover, LIBS is versatile, being applicable to all states of matter, and is particularly sensitive to light elements that are often difficult to detect with other techniques. Another advantage is that direct access to the sample is not required, allowing for flexible applications. Additionally, LIBS can be combined with other analytical methods, such as Raman spectroscopy, to enable comprehensive multielement and molecular surface analysis (Shah et al., 2020)

Liu et al. (2019) effectively combined LIBS with PCA and PLS-DA to distinguish between genetically modified and non-modified maize samples, achieving 100% classification accuracy across 240 samples. High specificity arises from distinct molecular spectral signatures. Raman and mid-infrared lasers excel at distinguishing structurally similar compounds. In the context of food adulteration, Peng et al. (2020) applied LIBS coupled with feature selection algorithms (GA, VIP, SR) to quantify adulterants in acacia honey with high precision.

Authenticity: Laser techniques are widely used for verifying food authenticity, detecting adulterants (e.g., diluted milk, synthetic additives), and identifying origin markers through isotopic and chemical profiling. Caceres et al. (2013) used LIBS to classify edible oils from various countries, achieving 95% accuracy via neural network algorithms. Cama-Moncunill et al. (2017) employed LIBS to quantify trace metals in infant formula premixes, validating results through atomic absorption spectroscopy. The nondestructive nature of these techniques makes them particularly suitable for high-value products like wines and oils.

Reproducibility of Results: Laser-based systems are generally highly reproducible due to automated, calibrated optics. For example, Gazeli et al. (2020) demonstrated the classification of olive oils using LIBS in conjunction with machine learning methods like LDA and PCA, attaining ~99.2% accuracy. Despite this, variables such as sample positioning, surface texture, and environmental conditions (e.g., ambient light, temperature) may introduce variability. Using automated alignment and standardized software improves reproducibility across different instruments and laboratories.

Cost Analysis and Economic Feasibility: LIBS provide a faster, cheaper alternative for food quality assessments. According to Stefas et al. (2021), LIBS is especially promising in classifying olive oil by origin, variety, and purity. Its integration with machine learning tools has made it a viable technique for commercial food quality assurance.

13.2.5 Nanocomposites

Sensitivity and Specificity: Nanocomposites enhance the performance of nanosensors by providing large surface areas and functionalizable surfaces. These features enable the detection of extremely low levels of analytes such as toxins, allergens, and chemical residues in complex food matrices (D'Souza et al., 2017).

Authenticity: Nanocomposites are instrumental in verifying the composition and purity of food products. Surface-enhanced Raman scattering (SERS)-based sensors using gold or silver nanorods can rapidly differentiate bacterial strains such as *E. coli*, *Listeria monocytogenes*, and *Salmonella typhimurium* in less than 10 seconds (He et al., 2008; Chu et al., 2008). These materials can also aid in tracing geographic origin or identifying genetic modifications. For instance, SDI Europe Ltd. developed immunoassays for genetically modified (GM) ingredient detection using dipsticks and DNA biosensors (Tothill, 2001). SERS-barcoded nanosensors encapsulating AuNPs and DNA probes successfully identified Bt gene-transformed rice at concentrations as low as 0.1 pg/mL (Chen et al., 2012). Furthermore, benzoic acid in beverages was detected using a carbon nanotube-modified biosensor, offering better sensitivity (1.06 µA/µM) than HPLC methods (Kochana et al., 2012).

Reproducibility of Results: Nanocomposites must consistently perform across multiple batches and experimental setups. Challenges include variations in nanoparticle synthesis and sensor fabrication, which can affect sensitivity and stability. Mo et al. (2010) fabricated a novel modified electrode which comprised hydrophobic ionic liquid (i.e., trihexyltetradecyl phosphonium bis(trifluoromethylsulfonyl) imide, [P6,6,6,14][NTf2]), multiwalled carbon nanotubes (MWNTs), and cationic gemini surfactant (i.e., $C_{12}H_{25}N(CH_3)_2–C_4H_8–N(CH_3)_2C_{12}H_{25}Br_2$, $C_{12}–C_4–C_{12}$). Cyclic voltammetry and linear sweep voltammetry were used to investigate the electrochemical behavior of Sudan I. The components showed good synergic interaction in sensing Sudan I; thus, the modified electrode presented higher sensitivity. The modified electrode had good stability and repeatability. It was applied to the detection of Sudan I in hot chili powder and ketchup samples, and the recovery was acceptable.

Cost Analysis and Feasibility: While nanocomposites made from materials like gold or graphene are costly, alternatives such as carbon nanotubes or magnetic nanoparticles provide more economical options. These materials support high-performance sensors that are both accurate and affordable (Nath et al., 2008). Feasibility in the food sector depends on the ease of integrating these sensors into workflows, as well as sensor shelf-life and operational simplicity. Glucose-sensitive nanocomposite sensors, such as those using AuNPs or immobilized *Pseudomonas fluorescens*, have been successfully used to analyze commercial beverages (Ozdemir et al., 2010; Kirgoz et al., 2007).

Commercial Relevance: Real-time authentication of food products using biosensors can mitigate losses due to fraud and enhance consumer confidence, making them a compelling tool for producers and retailers (Bahadır and Sezgintürk, 2015).

Reproducibility of Results: Shavronskaya et al. (2023) reported a TiO_2-based optical biosensor capable of accurately quantifying lactose in milk without the need for complex sample preparation. The system demonstrated consistent performance, with RSDs of 5–6% across 20 trials, aligning with reproducibility metrics seen in photoelectrochemical and nanoparticle-based biosensors (Çakıroğlu et al., 2019; Nasiri et al., 2022).

Economic viability: Biosensors represent a cost-effective alternative to conventional laboratory assays, especially when scaled for commercial production. Paper-based and electrochemical formats are particularly advantageous, offering low per-test costs and minimal training requirements (Mustafa et al., 2020). However, the initial expenses associated with R&D and integration into market-ready devices can be high (Sadana, 2005). Embedding biosensors into broader systems, such as smart packaging or portable readers, enhances utility and long-term cost-effectiveness.

13.2.6 Aerogels in Food Analysis

13.2.6.1 Sensitivity, Specificity, and Authenticity

Aerogels are gaining traction in food analysis due to their tunable porosity, responsiveness to environmental changes, and suitability for sensor applications. Sohrabi et al. (2024) developed a pH-sensitive colorimetric aerogel composed of chitosan nanofibers (CHNF), ixiolirion tataricum anthocyanins (ITA), and starch nanoparticles (SNPs), which exhibited a distinct color transition from red to green across a pH range of 2–12. Similarly, Li et al. (2024) designed a smart freshness indicator for seafood using a gelatin/gum arabic (Gel/GA)-based aerogel label incorporating betanin as a colorimetric indicator.

13.2.6.2 Reproducibility of Results

Aerogels exhibit high reproducibility due to standardized synthesis protocols. Zhang et al. (2024) demonstrated that an anthocyanin-based aerogel sensor maintained consistent color transitions ($\Delta E >$ threshold for visible detection) from violet to blue and back over 12 cycles of ammonia exposure, underscoring its potential for reusable, colorimetric sensing applications.

13.2.6.3 Cost Analysis and Feasibility

While traditional silica-based aerogels are expensive due to complex synthesis processes, advancements in fabrication techniques-such as sol-gel methods and use of low-cost precursors-are improving the cost efficiency. Although aerogels currently involve high solvent consumption (e.g., supercritical CO_2 drying), innovations like continuous CO_2 extraction and reduced solvent usage could improve scalability (Manzocco et al., 2021). In high-value food sectors (e.g., truffles, wines, olive oils), the benefits of aerogel-based sensors may justify their initial cost. However, for routine testing in lower-value commodities, affordability remains a barrier. According to market data, the global aerogel industry was valued at US$1.25 billion in 2023 and is projected to grow at a CAGR of 16.8% from 2024 to 2030 (https://www.precisionbusinessinsights.com/market-reports/global-thermal-insulation-coating-industry-market).

13.2.7 Rapid Microbial Techniques

Sensitivity and Specificity: RMTs aim to detect low microbial loads with high accuracy. Sensitivity ensures the detection of minimal microbial presence, while specificity prevents cross-reactivity with non-target organisms. Yuan et al. (2023) developed a colorimetric immunoassay using a Pd/

Pt-decorated metal–organic framework hybrid nanozyme (MIL-88@Pd/Pt) with immunomagnetic nanobeads for *Salmonella* detection. The system identified *Salmonella typhimurium* at 32 CFU/mL within 60 minutes.

Authenticity: Microbial authenticity testing ensures the microbiological safety and label compliance of food products. Molecular techniques like PCR and DNA barcoding are used to confirm the absence of pathogens and verify microbial profiles. Fu et al. (2019) introduced a smartphone-readable dipstick assay using ambient light sensors to monitor organophosphate residues in vegetables. The results were comparable to GC–MS and Ellman assay benchmarks, validating its accuracy and field applicability.

Reproducibility of Results: Modern rapid testing platforms show consistent performance across multiple food matrices: Colorimetric methods can detect *Listeria*, *Salmonella*, and aflatoxins (Bai et al., 2023; Yuan et al., 2023; Xiong et al., 2018). Fluorescence-based systems detect as low as 1 CFU/mL of *S. aureus* (Chen et al., 2022). Photoelectrochemical and electrochemical sensors quantify contaminants with nanomolar sensitivity (He et al., 2021; Zhu et al., 2021). Molecular amplification (e.g., loop-mediated isothermal amplification (LAMP), PCR) enables detection limits as low as 10^{-1} CFU/mL (Lee et al., 2020) for pathogens like *Salmonella* and *E. coli.*

Economic viability: Rapid microbial detection platforms offer a cost-effective alternative by reducing testing time, labor, and sample holding durations. Lee et al. (2020) presented a colorimetric LAMP assay for *Vibrio parahaemolyticus*, with a detection time under 1 hour and minimal equipment requirements, making it suitable for widespread adoption in seafood monitoring programs.

13.3 LIMITATIONS OF EMERGING ANALYTICAL TECHNIQUES

Despite the promise of emerging techniques, several technical and practical limitations hinder their large-scale adoption:

Scalability: Most techniques show high performance in controlled lab settings but struggle with industrial-scale implementation. For example, mass production of C-dots or aerogels often lacks process consistency. Instruments like LIBS or biosensors may not yet support high-throughput environments.

Standardization and Regulation: The absence of harmonized protocols across institutions complicates method validation. Regulatory agencies require cross-laboratory reproducibility and standardized datasets, which emerging tools often lack.

Matrix interference: Complex food matrices containing proteins, fats, and polysaccharides can interfere with sensor signals. For instance, fluorescence of C-dots may be quenched by milk proteins, and electrochemical signals may be masked by lipids.

Reproducibility: The dependence on custom-synthesized materials and prototypes means that slight alterations in synthesis (e.g., pH, temperature) can drastically affect results, limiting reproducibility between batches or laboratories.

Initial development cost: Although the operational cost of biosensors or smartphone diagnostics is low, development costs—including design, optimization, and validation—are substantial. This can hinder adoption by small enterprises.

Technical expertise requirement: Advanced platforms such as SPR, Raman spectroscopy, or nucleic acid-based sensors require skilled personnel for accurate operation and interpretation, limiting accessibility in low-resource settings.

Data interpretation complexity: Techniques often generate complex data sets (e.g., spectral or molecular profiles) that require chemometric analysis or machine learning tools. Without integrated user-friendly interfaces, valuable data may be underutilized.

Shelf-life and stability: Sensor performance may degrade due to environmental exposure (e.g., humidity, temperature). Ensuring stability over time often demands additional protective packaging, increasing cost and complexity.

Integration with existing systems: Emerging tools may not be compatible with current food safety systems like HACCP or ISO 22000. Many are standalone, lacking digital interfaces or cloud-based data logging, thereby impeding seamless integration into regulatory workflows.

REFERENCES

Bahadır, E. B., & Sezgintürk, M. K. (2015). Applications of commercial biosensors in clinical, food, environmental, and biothreat/biowarfare analyses. *Analytical Biochemistry*, 478, 107–120.

Bai, X., Huang, J., Li, W., Song, Y., Xiao, F., Xu, Q., & Xu, H. (2023). Portabledual-mode biosensor based on smartphone and glucometer for onsitesensitive detection of Listeria monocytogenes. *The Science of the Total Environment*, 874, 162450. https://doi.org/10.1016/j.scitotenv.2023.162450

Bhamore, J. R., Park, T. J., & Kailasa, S. K. (2020). Glutathione-capped Syzygiumcumini carbon dot-amalgamated agarose hydrogel film for naked-eye detection of heavy metal ions. *Journal of Analytical Science and Technology*, 11(1). https://doi.org/10.1186/s40543-020-00208-8

Bhardwaj, H., Sumana, G., & Marquette, C. A. (2020). A label-free ultrasensitive microfluidic surface Plasmon resonance biosensor for Aflatoxin B1 detection using nanoparticles integrated gold chip. *Food Chemistry*, 307, 125530.

Biranje, A., Azmi, N., Tiwari, A., et al. (2021). Quantum dots based fluorescent probe for the selective detection of heavy metal ions. *Journal of Fluorescence*, 31, 1241–1250. https://doi.org/10.1007/s10895-021-02755-8

Caceres, J. O., Moncayo, S., Rosales, J. D., de Villena, F. J. M., Alvira, F. C., & Bilmes, G. M. (2013). Application of laser-induced breakdown spectroscopy (LIBS) and neural networks to olive oils analysis. *Applied Spectroscopy*, 67(9), 1064–1072.

Çakıroğlu, B., Demirci, Y. C., Gökgöz, E., & Özacar, M. (2019). A photoelectrochemical glucose and lactose biosensor consisting of gold nanoparticles, MnO_2 and $g\text{-}C_3N_4$ decorated TiO_2. *Sensors and Actuators B: Chemical*, 282, 282–289.

Cama-Moncunill, R., Casado-Gavalda, M. P., Cama-Moncunill, X., Markiewicz-Keszycka, M., Dixit, Y., Cullen, P. J., & Sullivan, C. (2017). Quantification of trace metals in infant formula premixes using laser-induced breakdown spectroscopy. *Spectrochimica Acta Part B: Atomic Spectroscopy*, 135, 6–14.

Casadei, E., Valli, E., Panni, F., Donarski, J., Gubern, J. F., Lucci, P., . . . & Toschi, T. G. (2021). Emerging trends in olive oil fraud and possible countermeasures. *Food Control*, 124, 107902.

Cayuela, A., Soriano, M. L., & Valcárcel, M. (2015). Photoluminescent carbon dot sensor for carboxylated multiwalled carbon nanotube detection in river water. *Sensors and Actuators B: Chemical*, 207, 596–601.

Chen, G., Wang, Z., Deng, M., Xiao, F., Bai, X., & Xu, H. (2022). Biomimetic dandelion-like magnetic nanoparticles for capture and detection of S. aureus and L. monocytogenes. *Sensors and Actuators B: Chemical*, 355, 131289. https://doi.org/10.1016/j.snb.2021.131289

Chen, K., Han, H., Luo, Z., Wang, Y., & Wang, X. (2012). A practicable detection system for genetically modified rice by SERS-barcoded nanosensors. *Biosensors and Bioelectronics*, 34(1), 118–124.

Cho, B. K., & Irudayaraj, J. M. K. (2003). Foreign object and internal disorder detection in food materials using noncontact ultrasound imaging. *Journal of Food Science*, *68*(3), 967–974.

Chu, H., Huang, Y., & Zhao, Y. (2008). Silver nanorod arrays as a surface-enhanced Raman scattering substrate for foodborne pathogenic bacteria detection. *Applied Spectroscopy*, 62(8), 922–931.

Coskun, A. F., Wong, J., Khodadadi, D., Nagi, R., Tey, A., & Ozcan, A. (2013). A personalized food allergen testing platform on a cellphone. *Lab on a Chip*, 13, 636–640.

Curulli, A. (2021). Electrochemical biosensors in food safety: Challenges and perspectives. *Molecules*, 26(10), 2940. https://doi.org/10.3390/molecules26102940

Das, A. J., Wahi, A., Kothari, I., & Raskar, R. (2016). Ultra-portable, wireless smartphone spectrometer for rapid, non-destructive testing of fruit ripeness. *Scientific Reports*, 6(1), 32504.

Das, J., & Mishra, H. N. (2022). Recent advances in sensors for detecting food pathogens, contaminants, and toxins: A review. *European Food Research and Technology*, 248(4), 1125–1148.

D'Souza, A. A., Kumari, D., & Banerjee, R. (2017). Nanocomposite biosensors for point-of-care—evaluation of food quality and safety. *Nanobiosensors*, 629–676. https://doi.org/10.1016/b978-0-12-804301-1.00015-1

DuVall, J. A., Borba, J. C., Shafagati, N., Luzader, D., Shukla, N., Li, J., Kehn-Hall, K., Kendall, M. M., Feldman, S. H., & Landers, J. P. (2015). Optical imaging of paramagnetic bead-DNA aggregation inhibition allows for low copy number detection of infectious pathogens. *PLoS ONE*, 10, e0129830.

Fu, Q., Zhang, C., Xie, J., Li, Z., Qu, L., Cai, X., Ouyang, H., Song, Y., Du, D., Lin, Y., & Tang, Y. (2019). Ambient light sensor based colorimetric dipstick reader for rapid monitoring organophosphate pesticides on a smart phone. *Analytica Chimica Acta*, 1092, 126–131.

Gazeli, O., Bellou, E., Stefas, D., & Couris, S. (2020). Laser-based classification of olive oils assisted by machine learning. *Food Chemistry*, 302, 125329.

Giordano, G. F., Vicentini, M. B. R., Murer, R. C., Augusto, F., Ferrão, M. F., Helfer, G. A., . . . Lima, R. S. (2016). Point-of-use electroanalytical platform based on homemade potentiostat and smartphone for multivariate data processing. *ElectrochimicaActa*, 219, 170–177. https://doi.org/10.1016/j.electacta.2016.09.157

He, L., Kim, N. J., Li, H., Hu, Z., & Lin, M. (2008). Use of a fractal-like gold nanostructure in surface-enhanced Raman spectroscopy for detection of selected food contaminants. *Journal of Agricultural and Food Chemistry*, 56(21), 9843–9847.

He, X., Ying, Y., Zhao, X., Deng, W., Tan, Y., & Xie, Q. (2021). Cobalt-doped tungsten trioxide nanorods decorated with Au nanoparticles for ultrasensitive photoelectrochemical detection of aflatoxin B1 based on aptamer structure switch. *Sensors and Actuators B: Chemical*, 332, 129528. https://doi.org/10.1016/j.snb.2021.129528

Hu, Q., Sun, H., Liu, L., Xiao, L., Yang, Z. Q., Rao, S., . . . & Han, J. (2021). Development of an ultrasensitive spectrophotometric method for carmine determination based on fluorescent carbon dots. *Food Additives & Contaminants: Part A*, 38(5), 731–740.

Huang, Z., & Ren, L. (2025). Large scale synthesis of carbon dots and their applications: A review. *Molecules*, 30(4), 774.

Jin, L., Li, T., Wu, B., Yang, T., Zou, D., Liang, X., Hu, L., Huang, G., & Zhang, J. (2020). Rapid detection of Salmonella in milk by nuclear magnetic resonance based on membrane integrated gold chip. *Food Chemistry*, 307, 125530.

Kaur, J., Komal, Renu, Kumar, V., Tikoo, K. B., Bansal, S., Kaushik, A., & Singhal, S. (2020) Glutathione modified fluorescent CdS QDs synthesized using environmentally benign pathway for detection of mercury ions in aqueous phase. *Journal of Fluorescence*, 30(4), 773–785. https://doi.org/10.1007/s10895-020-02545-8

Kirgoz, Ü. A., Timur, S., Odaci, D., Pérez, B., Alegret, S., & Merkoçi, A. (2007). Carbon nanotube composite as novel platform for microbial biosensor. *Electroanalysis: An International Journal Devoted to Fundamental and Practical Aspects of Electroanalysis*, 19(7–8), 893–898.

Kochana, J., Kozak, J., Skrobisz, A., & Woźniakiewicz, M. (2012). Tyrosinase biosensor for benzoic acid inhibition-based determination with the use of a flow-batch monosegmented sequential injection system. *Talanta*, 96, 147–152.

Lanzellotto, C., Favero, G., Antonelli, M. L., Tortolini, C., Cannistraro, S., Coppari, E., & Mazzei, F. (2014). Nanostructured enzymatic biosensor based on fullerene and gold nanoparticles: Preparation, characterization and analytical applications. *Biosensors and Bioelectronics*, 55, 430–437.

Lee, J.-E., Mun, H., Kim, S.-R., Kim, M.-G., Chang, J.-Y., & Shim, W.-B. (2020). A colorimetric Loop-mediated isothermal amplification (LAMP) assay based on HRP-mimicking molecular beacon for the rapid detection of *Vibrio parahaemolyticus*. *Biosensors and Bioelectronics*, 151, 111968. https://doi.org/10.1016/j.bios.2019.111968

Li, B., Chen, M., Wu, Y., Liu, B., & Huang, G. (2024). Effects of glycerol and PVA on gelatin/gum Arabic aerogels: Microstructure, mechanical properties, and applications for food monitoring. *Food Hydrocolloids*, 149, 109515.

Lim, H. S., Choi, J. C., Song, S. B., & Kim, M. (2014). Quantitative determination of carmine in foods by high-performance liquid chromatography. *Food Chemistry*, 158, 521–526.

Liu, X., Feng, X., Liu, F., Peng, J., & He, Y. (2019). Rapid identification of genetically modified maize using laser-induced breakdown spectroscopy. *Food and Bioprocess Technology*, 12(2), 347–357.

Ludwig, S. K., Zhu, H., Phillips, S., Shiledar, A., Feng, S., Tseng, D., . . . & Ozcan, A. (2014). Cellphone-based detection platform for rbST biomarker analysis in milk extracts using a microsphere fluorescence immunoassay. *Analytical and Bioanalytical Chemistry*, 406(27), 6857–6866.

Manzocco, L., Mikkonen, K. S., & García-González, C. A. (2021). Aerogels as porous structures for food applications: Smart ingredients and novel packaging materials. *Food Structure*, 28, 100188. https://doi.org/10.1016/j.foostr.2021.100188

Manzoor, S., Dar, A. H., Dash, K. K., Pandey, V. K., Srivastava, S., Bashir, I., & Khan, S. A. (2023). Carbon dots applications for development of sustainable technologies for food safety: A comprehensive review. *Applied Food Research*, 3(1), 100263.

Mao, X. J., Zheng, H. Z., Long, Y. J., Du, J., Hao, J. Y., Wang, L. L., & Zhou, D. B. (2010). Study on the fluorescence characteristics of carbon dots. *Spectrochimica Acta Part A: Molecular and Biomolecular Spectroscopy*, 75(2), 553–557.

Mo, Z., Zhang, Y., Zhao, F., Xiao, F., Guo, G., & Zeng, B. (2010). Sensitive voltammetric determination of Sudan I in food samples by using gemini surfactant–ionic liquid–multiwalled carbon nanotube composite film modified glassy carbon electrodes. *Food Chemistry*, 121(1), 233–237.

Mustafa, F., Finny, A. S., Kirk, K. A., & Andreescu, S. (2020). Printed paper-based (bio) sensors: Design, fabrication and applications. *Comprehensive Analytical Chemistry*. https://doi.org/10.1016/bs.coac.2020.02.002

Nasiri, H., Baghban, H., Teimuri-Mofrad, R., & Mokhtarzadeh, A. (2022). Graphitic carbon nitride/magnetic chitosan composite for rapid electrochemical detection of lactose. *International Dairy Journal*, 136, 105489.

Nath, S., Kaittanis, C., Tinkham, A., & Perez, J. M. (2008). Dextran-coated gold nanoparticles for the assessment of antimicrobial susceptibility. *Analytical Chemistry*, 80(4), 1033–1038.

Ozdemir, C., Yeni, F., Odaci, D., & Timur, S. (2010). Electrochemical glucose biosensing by pyranose oxidase immobilized in gold nanoparticle-polyaniline/AgCl/gelatin nanocomposite matrix. *Food Chemistry*, 119(1), 380–385.

Park S., T., Baynes, C., Cho, S. I., & Yoon, J. Y. (2014). Paper microfluidics for red wine tasting. *Rsc Advances*, 4(46), 24356–24362.

Peng, J., Xie, W., Jiang, J., Zhao, Z., Zhou, F., & Liu, F. (2020). Fast quantification of honey adulteration with laser-induced breakdown spectroscopy and chemometric methods. *Foods*, 9(3), 341.

Sadana, A., & Sadana, N. (2008). Market Size and Economics for Biosensors. *Fractal Analysis of the Binding and Dissociation Kinetics for Different Analytes on Biosensor Surfaces*, 317–334. https://doi.org/10.1016/B978-044453010-3.50014-1

Shah, S. K. H., Iqbal, J., Ahmad, P., Khandaker, M. U., Haq, S., & Naeem, M. (2020). Laser induced breakdown spectroscopy methods and applications: A comprehensive review. *Radiation Physics and Chemistry*, 170, 108666.

Shavronskaya, D. O., Nazarova, E. A., & Krivoshapkina, E. F. (2023). Optical bi-enzyme-titania biosensor system: A new way to detect lactose. *Biosensors and Bioelectronics: X*, 14, 100347.

Shin, Y. H., Gutierrez-Wing, M. T., & Choi, J. W. (2021). Recent progress in portable fluorescence sensors. *Journal of The Electrochemical Society*, 168(1), 017502.

Stefas, D., Gyftokostas, N., Nanou, E., Kourelias, P., & Couris, S. (2021). Laser-induced breakdown spectroscopy: An efficient tool for food science and technology (from the analysis of Martian rocks to the analysis of olive oil, honey, milk, and other natural earth products). *Molecules*, 26(16), 4981.

Sohrabi, N., Almasi, H., & Moradi, M. (2024). Controlled porosity intelligent aerogel based on chitosan nanofiber, anthocyanins and starch nanoparticles: Preparation, characterization and application in monitoring food spoilage. *Food Hydrocolloids*, 156, 110261.

Su, A., Wang, D., Shu, X., Zhong, Q., Chen, Y., Liu, J., & Wang, Y. (2018). Synthesis of fluorescent carbon quantum dots from dried lemon peel for determination of carmine in drinks. *Chemical Research in Chinese Universities*, 34(2), 164–168.

Tombelli, S., Mascini, M., Scherm, B., Battacone, G., & Migheli, Q. (2009). DNA biosensors for the detection of aflatoxin producing Aspergillus flavus and A. parasiticus. *Monatsh Chemistry*, 40, 901–907. https://doi.org/10.1007/s00706-009-0137-3.

Tothill, I. E. (2001). Biosensors developments and potential applications in the agricultural diagnosis sector. *Computers and Electronics in Agriculture*, 30(1–3), 205–218.

Williamson, K., & Hatzakis, E. (2017). NMR spectroscopy as a robust tool for the rapid evaluation of the lipid profile of fish oil supplements. *Journal of Visualized Experiments (JoVE)*, (123), e55547.

Xiong, Y., Pei, K., Wu, Y., Duan, H., Lai, W., & Xiong, Y. (2018). Plasmonic ELISA based on enzyme-assisted etching of Au nanorods for the highly sensitive detection of aflatoxin B1 in corn samples. *Sensors and Actuators B: Chemical*, 267, 320–327. https://doi.org/10.1016/j. snb.2018.04.027

Xu, J., Jie, X., Xie, F., Yang, H., Wei, W., & Xia, Z. (2018). Flavonoid moiety-incorporated carbon dots for ultrasensitive and highly selective fluorescence detection and removal of Pb2+. *Nano Research*, 11(7), 3648–3657. https://doi.org/10.1007/s12274-017-1931-6

Yao, S., Zhao, C., Liu, Y., Nie, H., Xi, G., Cao, X., . . . Wang, J. (2020). Colorimetric immunoassay for the detection of staphylococcus aureus by using magnetic carbon dots and sliver nanoclusters as o-phenylenediamine-oxidase mimetics. *Food Analytical Methods*. https://doi.org/10.1007/s12161-019-01683-5

Yuan, J., Duan, H., Wang, L., Wang, S., Li, Y., & Lin, J. (2023). A three-in-one hybrid nanozyme for sensitive colorimetric biosensing of pathogens. *Food Chemistry*, 408, 135212. https://doi.org/10.1016/j.foodchem. 2022.135212

Zhang, K., Li, Z., Zhao, W., Guo, J., Hashim, S. B., Khan, S., . . . & Zou, X. (2024). Aerogel colorimetric label sensors based on carboxymethyl cellulose/sodium alginate with black goji anthocyanin for monitoring fish freshness. *International Journal of Biological Macromolecules*, 265, 130466.

Zhang, X., Liao, X., Hou, Y., Jia, B., Fu, L., Jia, M., . . . & Kong, W. (2022). Recent advances in synthesis and modification of carbon dots for optical sensing of pesticides. *Journal of Hazardous Materials*, 422, 126881. https://doi.org/10.1016/j.jhazmat.2021.126881

Zhao, Y., Li, Y., Zhang, P., Yan, Z., Zhou, Y., Du, Y., . . . Yang, R. (2021). Cell-based fluorescent microsphere incorporated with carbon dots as a sensitive immunosensor for the rapid detection of Escherichia coli O157 in milk. *Biosensors and Bioelectronics*, 179, 113057. https://doi.org/10.1016/j.bios.2021.113057

Zhu, H., Sikora, U., & Ozcan, A. (2012). Quantum dot enabled detection of Escherichia coli using a cell-phone. *Analyst*, 137(11), 2541–2544.

Zhu, Y., Liu, P., Xue, T., Xu, J., Qiu, D., Sheng, Y., Li, W., Lu, X., Ge, Y., & Wen, Y. (2021). Facile and rapid one-step mass production of flexible 3D porous graphene nanozyme electrode via direct laser-writing for intelligent evaluation of fish freshness. *Microchemical Journal*, 162, 105855. https://doi.org/10.1016/j.microc.2020.105855

Index

Note: Page numbers in *italics* indicate a figure, and page numbers in **bold** indicate a table on the corresponding page.

A

aerogels, 4–5, 287, 295
- antibacterial properties, *133*, 133–134
- applications, 130–131, *131*
- challenges, 141–142
- classification, 126
- definition, 126
- drying technique, 127
- fabrication
  - atmospheric drying, 129
  - drying of wet gel, 129
  - freeze drying, 129
  - functional-material-loaded, 130
  - gel preparation, 128
  - solvent exchange, 128–129
  - supercritical drying, 129–130
- functional
  - inorganic, 127
  - organic, 127–128
- future scope, 145–146
- mechanical and insulation, 137–138, *138–140*
- moisture control, 131–133, *132*
- packaging, 137
- production, 127
- real-time monitoring, 134
- recent innovation, 142–145
- regulated additives, 134–136, *135*, *136*

Agrobacterium tumefaciens, 77

analytical techniques, 287, **288**, *see also individual techniques*
- aerogels, 295
- biosensors, 291–292
- C-Dots, 289–290
- laser and laser optics, 292–293
- nanocomposites, 293–295
- practical limitations, 296–297
- rapid microbial techniques, 295–296
- smartphone technology, 290–291

B

Basic Local Alignment Search Tool (BLAST), 227

bio-based polymers, 4

biosensors, 2, 287, 291–292
- advancements, 35
- applications in food, **43**
- aptamer-based, 162
- biological component, 34
- challenges, *49*, 49–50
- colorimetric sensors, 40–41
- components of, 35–36
- detection of heavy metals, 43
- direct recognition sensors, 36
- electrochemical transducers, 39, *39*, **41–42**, 144, 274
- enzymatic biosensors, 37, *37*
- enzyme immobilization, 36
- food component analysis
  - fermentation, 47–48
  - spoilage detection, 48
  - sugars/fats/proteins, 46
  - vitamins/antioxidants, 46–47
- food spoilage, 44
- fruit quality, 43–44
- fruits and vegetables, 42
- gas, 145
- immunosensor technologies, 38, *38*, 44–46, **45**
- indirect recognition sensors, 36
- mass-sensitive, 161
- microarray, 242
- microbial techniques, 159
- nanomaterial-based, 6
- optical biosensor, 39–40, *40*
- piezoelectric, 40
- recent advancements, 50
- solid substrate-based assays, 41
- solution-based assays, 41
- SPR-based, 7, 249
- transducers, 37

built-in sensors, 104

bulk acoustic wave (BAW) resonators, 40, 161

C

carbon dots (C-dots), 1, 14, 145, 289–290

carbon quantum dots (CQDs), 13
- applications
  - food additives and pathogens, **23**, 23–24
  - in food analysis, 22–23
  - heavy metal detection, 24–25, *26*
  - insecticide and antibiotic residues, 25–27
- challenges, 27–28
- characterization, 21–22
- fluorescence quenching/enhancement, 14
- functionalization of
  - covalent bonds, 19–20 (*see also* covalent bonds)

non-covalent bonds, 20–21 (*see also* non-covalent bonds)
mechanism and challenges, *14*
multimodal detection, 15
photocatalytic activity, 15
PL spectroscopy, 22
selective binding and sensing, 14
signal amplification, 15
surface modifications
biological conjugation, 18
click chemistry, 17–18
doping, 18
hybrid nanostructure, 18
in-situ functionalization, 17
layer-by-layer assembly, 18
ligand exchange technique, 16–17
polymer coating, 17
silica coating, 17
cerium and nitrogen co-doped carbon quantum dots (Ce–N–CDs), 27
characterization techniques, 1, 22
charge-coupled device (CCD) detectors, 3, 58
Clostridium botulinum, 162
cloth-based hybridization array system, 222
colorimetric aerogel gas sensor (CAS), 142
colorimetric sensors, 2, 40–41, 106, 119
conventional biosensors, 48
copolymerization reaction, 20
covalent bonds
amide coupling reactions, 19
copolymerization reaction, 20
esterification, 20
Schiff's base reaction, 20, *21*
silylation reaction, 19–20
cross-linking agents, 4, 247
culture-based methods, 6, 7

D

differential gel electrophoresis (DIGE), 215
differential proteomics, 210–211
Digital Image Colorimetry (DIC), 116
direct epifluorescent filter technique (DEFT), 155
direct recognition sensors, 36

E

electrochemical biosensors, 2, 39, 41, **41–42**, 48, 50, 144, 274
electrochemical immunosensors, 38, *38*, 45
electro-chemiluminescence (ECL), 105, 110, 116
electrospray ionization (ESI), 216
electrospray ionization mass spectrometry (ESI-MS), 222–223
enzyme immunosensor, *37*
enzyme-linked immune sorbent assay (ELISA), 6, 164
Escherichia coli, 6, 44, 152, 158–159, 161, 226, 291
esterification, 20
exfoliated nanocomposite, 88

F

fiber optic biosensors, 161–162
flow cytometry, 6, 155–156, 165
fluorescence-based biosensors, 2
fluorescence spectroscopy, 2, 24, 107
fluorescent in situ hybridization (FISH), 155
fluorophores, 21, 107
food analysis, 13
applications, **278–279**
chromatography, 268–269
DNA-based techniques, 266–268, *267*
enzymatic protein hydrolysis, 270–271
future scopes, 280–281
hyphenated techniques, 273–277, *275*, *276*, *277*
imaging technologies, 272–273
mass spectrometry (MS), 269–270
nanocomposites, 91–92
NIR spectroscopy, 271–272 (*see also* near-infrared spectroscopy (NIRS))
parameters of food quality, 266, *266*
proteomics, 268 (*see also* proteomics)
surface-based analytical techniques, 277–278, 279
foodborne illness, 6, 42, 95, 152, 224
foodborne microorganism, 152, *153*, 156, **160**, **162**, **163**
foodomics, 209, 223
food scarcity, 265
Fourier-transform infrared (FTIR) spectroscopy, 7, 8, 22
freeze drying, 4, 129, 134, 145
functional genomics, 210

G

gas chromatography with flame ionization detector (GC-FID), 218
genetically modified organisms (GMOs), 7, 34, 158, 267
genomic technologies, 209–210
Gordonia alkanivorans, 225

H

headspace sampling, 197
high-performance liquid chromatography (HPLC), 34, 213, 265
hyperspectral imaging (HSI), 7
hyphenated techniques, 8, 273–277, *275*, *276*, *277*

I

image processing algorithms, 112
immunochromatography assay (ICA), 245–246, *246*
indirect detection sensors, 36
infrared gas sensors, 2
infrared spectroscopy (IR), 265
intercalated nanocomposite, 88
isotope tagging, 198

L

laser-induced breakdown spectroscopy (LIBS), 3, 287, 292–293
  adulterant detection
    brown rice flour, 71–72
    honey, 71
    in meat, 72
    in milk, 61, *70*
    in tea, 71
  advantages, 58
  applications, **62–69**
  brightness, 57
  cereal quality, 79
  challenges, 80–81
  coherence, 57
  future scope, 81
  genetically modified food, 79–80, *80*
  geographical origin
    honey, 74–75, *76*
    olive oil, 75, 77
  history, 58
  microbial assessment, 77, *78*
  minerals detection
    green leafy vegetables, 72, *73*
    liquid infant formula, 73–74
    poultry meat, 72–73
    wheat flour, 74
  monochromatic light, 57
  parameters influencing signal, 60
  physical states of matter, 60
  principles
    detector, 59
    instrumentation, 59
    laser system, 59
    light collection, 59
    optics, 59
    physical interactions, 60
    sample chamber, 59–60
    spectrometer, 59
  protein evaluation
    in canned tuna fish, 79
    in sheep colostrum, 79
  rice grain, 79
  spectral acquisition, 60–61
  toxins/pesticide residue
    fenthion in soybean oil, 78
    pesticide residue on apple surface, 78
  unidirectional, 58
lateral flow immunoassay (LFIA), 117, 164
liquid-liquid extraction (LLE), 194–195, 198–199
Listeria monocytogenes, 152, 156, 161, 221, 294
localized surface plasmon resonance (LSPR), 240–242, **241**
loop-mediated isothermal amplification (LAMP), 6, 158, 296
luminophores, 110

M

machine learning-based smartphone applications, 5
machine learning models, 112, 134
mass spectrum, 24, 216
matrix-assisted laser desorption/ionization (MALDI), 215
mercaptoundecanoic acid (MUA), 16–17
metal-organic framework (MOF), 36, 46, 142
microfluidic paper-based analytical devices (μPADs), 5
microorganisms, *see* rapid microbiological methods (RMMs)
microwave-assisted synthesis, 1
mid-infrared spectroscopy (MIR), 7
molecular-based techniques, 7, *see also* proteomics
molecularly imprinted polymer (MIP), 22, 44, 247
multiple locus variable number tandem repeat sequence analysis (MLVA), 220–221, *221*
mycotoxins, 8, 102, 152, 159, 226

N

nanocomposites, 3–4
  capacity to revolutionize, 90
  challenges, 97, *98*
  classifications, 88–89, *89*
  components, 88
  consumer acceptance, 99
  environmental impact, 99
  evaluation of toxicity, 98
  food analysis, 13, 91–92
    bio-based nanocomposites, 93
    carbon-based nanocomposites, 93–94
    detecting microorganisms, 95
    detection of contaminants, 95–96
    metal-based nanocomposites, 92
    metal oxide-based nanocomposites, 92–93
    nanosensors, 95
    packaging, 96
    spoilage detection, 96–97

future of, 99–100
regulations, 98–99
synthesis techniques, 90–91
nano-liquid chromatography (nano-LC), 222–223
nanoscience, 87–88
nanosensors, 95, 242
nanotechnology, 9, 87, *see also* nanocomposites
advancements, 99
applications of, *91*
biosensor function, 50
food security and quality, 89–90
nanozinc oxide (ZnO), 4
near-infrared spectroscopy (NIRS), 7, 115
next-generation sequencing (NGS), 6, 218–219, 266
nitrogen-doped fluorescent C-dots (N-FCDs), 289
non-covalent bonds
complexation, 21
electrostatic interactions, 21
π-π interactions, 20–21
nuclear magnetic resonance (NMR) spectroscopy, 175–176, *176*, 218
alcoholic drinks and beverages, 201–202
analysis of mixtures, 176–181
biomarkers, 198
centrifugation
cellular/subcellular analysis, 194
lipidomics, 194
protein structure/dynamics, 194
cinnamon, 203–204
classification of, *182*, *188*
deuterium oxide (D2O) exchange, 201
direct dissolution, 199
edible oil, 199–200
environmental monitoring, 197–198
fat extraction, 201
fish, 204–205
flavor components, 203
fruit extract/pulp/seeds, 198–199
honey, 200
identification of analytes, *179*
isotope tagging, 198
liquid-liquid extraction (LLE), 194–195, 198–199
liquid-state, 189
milk and milk products, 201
multiple spectroscopic techniques, *178*
nectar, 200
nonalcoholic drinks and beverages, 202–203
problems, *180*
processing techniques, **183–186**
protein precipitation, 201
proton count, *187*
quantification of sugars, 203
resolution, 181
saffron, 205
sample filtration, 201
sample preparation methods
decantation/filtration, 193
metabolomics, 194
natural product analysis, 194
sample preparation technique, *177*
sampling device
crew cap *vs.* JY valved tube, 190
molecular weight of analyte, 190
NMR tube, 191
probe field view, 190, *191*
salt-tolerant NMR tubes, 192
screw-thread cap, 192–193
silica tube, 192
sap, 200–201
selectivity, 181
sensitivity, 181
signal-to-noise ratio, *187*
solid phase extraction (SPE), 195–196, 199
solvent-free extraction, 196–198
volatile compounds, 197
nucleic acid sequence-based amplification (NASBA), 158–159

O

one-dimensional gel electrophoresis (1-DE), 215
optical emission spectroscopy (OES), 3

P

pathogenic/spoilage-causing microorganisms, 153, **154**
phase-separated nanocomposite, 89
piezoelectricity, 40
polycarbonate track etching (PCTE) membranes, 246
polymerase chain reaction (PCR), 6, 156
polymer-based nanocomposites, 88, *89*
polyvinyl alcohol (PVA), 4, 93, 116
process analytical technology (PAT), 3
Protected Designation of Origin (PDO), 75
Protected Geographical Indication (PGI), 75
protein identification, 216
proteomics, 210
applications of, **211**
artificial intelligence (AI) technologies, 223
challenges, 228
differential, 210–211
foodborne pathogens, **211**
food spoilage/pathogenic microorganisms, 223–226
functional, 211, 213
future scope, 228–229
machine learning, 223
multi ohmic approaches, 217

omics technologies, **212**
protein identification
bottom-up proteomics, 213, *213*
mass spectrometer, *216*, 216–217
top-down proteomics, *214*, 214–217
recent innovation
analyzing allergens, 226–227
antibiotic use, 227–228
detection of pathogens, 226
food supply chain, 227
target-driven phase
cloth-based hybridization array system, 222
electrospray ionization mass spectrometry (ESI-MS), 222–223
multi locus sequence typing (MLST), 221–222
multiple locus variable number tandem repeat sequence analysis (MLVA), 220–221, *221*
nano-liquid chromatography (nano-LC), 222–223
next-generation sequencing (NGS), 218–219
pulse field gel electrophoresis (PFGE), 219–220, *220*
whole genome sequencing (WGS), 218–219
Pseudomonas fluorescens, 225, 294
pulse field gel electrophoresis (PFGE), 219–220, *220*

Q

quantitative PCR (qPCR), 6, 156–157
quantum dots (QDs), 1, *see also* carbon quantum dots (CQDs)
quartz crystal microbalance (QCM), 40, 161

R

Raman spectroscopy, 7
rapid microbial techniques (RMTs), 287
rapid microbiological methods (RMMs), 5–6
advantages/disadvantages, **166–167**
biosensor-based
aptamer-based biosensors, 162
electronic nasal sensors, 159
fiber optic biosensors, 161–162
foodborne microorganism identification, **162–163**
mass-sensitive biosensors, 161
surface plasmon resonance (SPR), 161
challenges, 169
direct epifluorescent filter technique (DEFT), 155
droplet digital PCR, 157–158
flow cytometry, 155–156
fluorescent in situ hybridization (FISH), 155
food safety assurance, 165
loop-mediated isothermal amplification (LAMP), 158
luminescence-based
bioluminescence/ATP detection, 163
enzyme-linked immune sorbent assay (ELISA), 164
immunological tests, 163–164
lateral flow immunoassay (LFIA), 164
microscopic-based approaches, 153
modern molecular techniques, 168
multiplex PCR, 157
nested PCR, 157
nucleic acid-based methods, 156
nucleic acid sequence-based amplification (NASBA), 158–159, **160**
operational applications, 168
polymerase chain reaction (PCR), 156
real-time/quantitative PCR (qPCR), 156–157
real-time data collection, 5
real-time monitoring systems, 112
relative standard deviations (RSDs), 26
remote monitoring, 5

S

Salmonella enteritidis, 161
Schiff's base reaction, 20, *21*
shotgun proteomics, 213
silver nanoparticles (AgNPs), 3–4
smartphone-based technologies, 5
smartphone-integrated biosensors, 5
smartphone sensors, 104
artificial intelligence, 112–113
bakery, 115–116
challenges, 117–119
colorimetric analysis, 105–107
computer vision, 110–112, *111*
confectionery, 115–116
ECL, 110
electrochemical techniques, 109
fluorescence techniques, 107–108
food and beverages, 113
future of, 119–120
integrated with AI, 106
meat/fish/poultry, 114–115
microscopic imaging technique, 108, **109**
mobile applications, 116–117
nondestructive methods, 114
smartphone spectroscopy, 5
sodium dodecyl sulfate-polyacrylamide gel electrophoresis (SDS-PAGE), 215
solid phase extraction (SPE), 195–196, 199, 200
solid-phase microextraction (SPME), 196
solvent less extraction, 196

solvothermal synthesis, 1
spectral techniques, 6–7
surface acoustic wave (SAW) resonators, 40, 161
surface-based sensing analysis, 7–8, 237–238
  antibiotic residues, 253–254
  challenges, 249–250
  colorimetric probe, 248
  digital imaging technique, 248–249
  food safety/quality analysis, **251**
  immunochromatography assay, 245–246
  LSPR, 240–242, **241**
  metal ions, 253
  microbial pathogens, 250–253
  microfluidic detectors, 242–244, **243**, *244*
  microfluidics system, 246–247
  molecular labeling system, 244–245
  molecularly imprinted polymer (MIP), 247
  pesticides, 253
  SERS technology, 238–239, *239*, 250
  SPR, 239–240, *240*, 253, 254–258, **255,** 256
surface-enhanced Raman spectroscopy (SERS), 7, 104, 237–239, *239*
surface plasmon resonance (SPR), 2, 161, 239–240, *240*, 254, **255**, 256
systematic evolution of exponential enrichment (SELEX) techniques, 244

T

thermal desorption, 196–197
thermoplastic nanocomposites, 89
thermosetting nanocomposites, 89
Trans Proteome Pipeline (TPP), 216
two-dimensional gel electrophoresis (2-DE), 215

U

UV-Vis spectroscopy, 2, 7

V

viable but non-culturable (VBNC) pathogens, 153
volatile organic compounds (VOCs), 5, 96, 142, 197

W

whole genome sequencing (WGS), 7, 218–219

Z

Ziziphus mauritiana (Indian Ber), 77, *78*

For Product Safety Concerns and Information please contact our EU representative GPSR@taylorandfrancis.com Taylor & Francis Verlag GmbH, Kaufingerstraße 24, 80331 München, Germany